Means ADA Compliance Pricing Guide

Cost Data for 75 Essential Projects

*A Collaboration between
Adaptive Environments Center, Inc.
and R.S. Means Engineering Staff*

Means ADA Compliance Pricing Guide

Cost Data for 75 Essential Projects

- 263 ADA Construction Assemblies
- 3100 Unit Costs
- 927 Location Pricing Factors

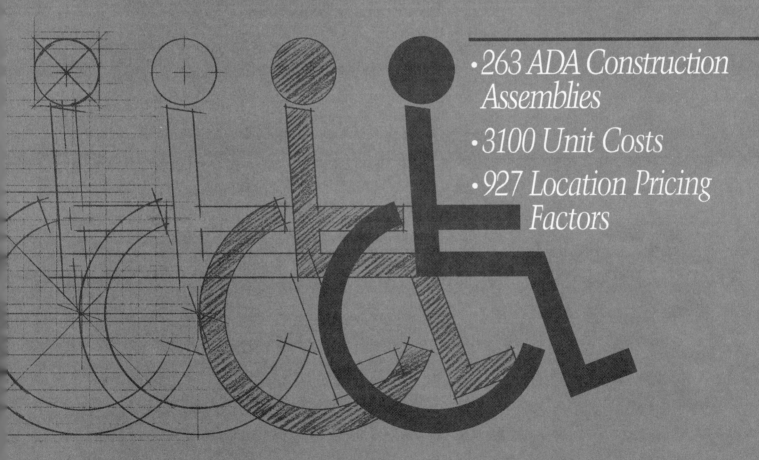

A Collaboration between
Adaptive Environments Center, Inc.
and R.S. Means Engineering Staff

Copyright 1994

R.S. MEANS COMPANY, INC.
CONSTRUCTION PUBLISHERS & CONSULTANTS

100 Construction Plaza
P.O. Box 800
Kingston, MA 02364-0800
(617) 585-7880

The information presented here is intended solely as informational guidance, and is neither a determination of your legal rights or responsibilities under the ADA, nor binding on any agency with enforcement responsibility under the ADA.

In keeping with the general policy of R.S. Means Company, Inc., its authors, editors, and engineers apply diligence and judgment in locating and using reliable sources for the information published. However, no guarantee or warranty can be given, and all responsibility and liability for loss or damage are hereby disclaimed by the authors, editors, engineers and publisher of this publication with respect to the accuracy, correctness, value and sufficiency of the data, methods, and other information contained herein as applied for any particular purpose or use.

The editors for this book were Mary Greene and Robert Mewis. Composition was supervised by Helen Marcella. The book and jacket were designed by Norman R. Forgit.

No part of this publication may be reproduced, stored in a retrieval system, or transmitted in any form or by any means without prior written permission of R. S. Means Company, Inc.

Printed in the United States of America

10 9 8 7 6 5 4 3

Library of Congress Cataloging in Publication Data
ISBN 0-87629-351-8

Table of Contents

Acknowledgments	ix
Introduction	xi
Part One—Getting Started	1
Overview of ADA	3
Using the Projects for Cost Estimating	7
Part Two—Modifications Estimates	9
Parking and Drop-off	
1. Install Accessible Parking Spaces	10
2. Create Accessible Passenger Drop-off	14
Pathways and Curbcuts	
3. Construct New Pathway	16
4. Construct Graded Entrance Pathway	20
5. Modify Existing Pathway	24
6. Install Compliant Gratings	28
7. Install or Modify Curb Cuts	30
Ramps	
8. Construct New Ramp: Straight	34
9. Construct New Ramp: Switch-Back	38
10. Construct New Ramp: Dog-Leg	42
11. Construct New Ramp: Below Grade	46
12. Modify Existing Ramp	50
13. Install or Modify Ramp Railings	54
Lifts	
14. Install Vertical Platform Lift	58
15. Install Stairway Inclined Lift	60
Stairs	
16. Install New Stairs	62
17. Modify Existing Stairs	66
18. Install or Modify Stair Handrails	70
19. Install Under-Stair Barrier	74
Doorways/Entrances	
20. Install New Door: Drywall	76
21. Install New Door: Masonry	78
22. Install New Door: Glass Storefront	84
23. Install Automatic Door Opener	88

24.	Modify Existing Double-Leaf Doors	90
25.	Modify Existing Door	98
26.	Install Sliding Door	102
27.	Enlarge Vestibule	106
28.	Modify Buzzer or Intercom	110

Interior Accessible Route

29.	Remove or Relocate Partitions	114
30.	Install Wing Walls	116
31.	Install Slip-Resistant Flooring Materials	118

Elevators

32.	Install New Elevator: Exterior Shaft	122
33.	Install New Elevator: Interior Shaft	126
34.	Modify Existing Elevator Cab	130
35.	Modify Existing Elevator Hall Signals	134
36.	Elevator Raised and Braille Characters	136

Telephones

37.	Install or Modify Public Telephones	138
38.	Install Public Text Telephones (TTYs/TDDs)	142

Drinking Fountains

39.	Install or Modify Drinking Fountains	144

Alarm Systems & Controls

40.	Install or Modify Controls	146
41.	Install or Modify Outlets	148
42.	Install or Modify Switches	150
43.	Install Audible and Visual Fire Alarms	152

Signage

44.	Install Signage	154
45.	Install Electronic Display Signage	158

Toilet Rooms

46.	Construct Single-User Toilet Rooms	160
47.	Modify Multiple-Stall Toilet Rooms	162
48.	Create Accessible Stall	164
49.	Install New Toilet	166
50.	Modify Existing Toilet	168
51.	Install New Sink	170
52.	Modify Existing Sink	172
53.	Install or Modify Grab Bars	176
54.	Install or Modify Toilet Room Dispensers	178
55.	Children's Accessible Bathroom Fixtures	180

Bathing Facilities

56.	Install Roll-In Shower	184
57.	Modify Existing Shower	186
58.	Replace Tub with Roll-In Shower	190
59.	Modify Existing Tub	192
60.	Create Accessible Gang Showers	196

Retail/Service

61.	Create Accessible Counter	198
62.	Create Accessible Aisles	200

63.	Modify Dining Area	202
64.	Create Accessible Dressing Rooms	206

Assembly Areas
65.	Create Accessible Theater Seating	208
66.	Install Assistive Listening Systems	210

Safety
67.	Install Detectable Warnings	214
68.	Install Emergency Communication Device	218

Kitchens
69.	Modify Kitchenette	220

Storage
70.	Modify Closet	224

Recreation
71.	Install Accessible Play Area Pathways	226
72.	Create Beach Access	230
73.	Install Swimming Pool Access	234
74.	Create Accessible Trails	236

Lodging
75.	Create Accessible Sleeping Rooms	238

Part Three—Case Studies — 243
Existing Bathroom Rehab — 244
Historic Entry Modifications — 246

Part Four—Unit Costs — 249

Part Five—Location Adjustment Factors — 321

Part Six—Appendix, Glossary and Index — 329

Appendix
- Resources — 331
- Cross Reference: Pricing Guide/ADAAG — 337
- Cross Reference: ADAAG/Pricing Guide — 339
- Abbreviations — 341

Glossary — 345

Index — 347

Acknowledgments

This book was co-authored by Adaptive Environments Center, Inc. and R.S. Means Company, Inc.

The Project Editor for Adaptive Environments was Elaine Ostroff. The text author was Joshua Barnett, R.A. Editors and reviewers affiliated with Adaptive Environments included: James Gary Cronburg, R.A., Soni Gupta, Ruth Lehrer, and Ed Neubauer.

R.S. Means provided all of the cost estimates in Part II, the Unit Costs (Part IV), the Location Adjustment Factors (Part V), and the cost- and estimating-related text. Robert Mewis was the cost estimator and technical editor for this project. Mary Greene was Means Project Editor. Kevin Foley, Gary Hoitt, and Kathy Rodriguez provided technical reviews. Some of the illustrations were adapted by Lois Sansone, under the direction of R.S. Means.

Barrier Free Environments also contributed to this publication, with a review and editorial comments from Jim Bostrom, and illustrations by Rex Pace.

Portions of this book were reviewed by Ellen Harland of the Public Access Office, Civil Rights Division, Department of Justice, Harry Beyer of the Pike Institute, and the following people who provided review at various stages: Richard Duncan, Center for Accessible Housing; Gayle Epp, Epp Associates; Paul Grayson, Environments for Living; Rich Lemoin, University of Massachusetts; Malcolm Meltzer; and Chris Palames, Independent Living Resources.

We also wish to thank the directors and staff of the ADA Technical Assistance Centers for their review and suggestions.

Portions of this publication appear in a book produced for the National Institute on Disability Rehabilitation and Research, under subcontracts to Barrier Free Environments and the United Cerebral Palsy Associations. Portions of Part II were based on the Modifications catalog developed by Gayle Epp of Welch & Epp Associates for the Division of Capital Planning and Operations of the Commonwealth of Massachusetts, August 1986.

Introduction

The Purpose of This Publication

Two of the most frequently asked questions about the Americans with Disabilities Act are: "What do I have to do?" and "How much will it cost me?" Not knowing the actual cost of building modifications for accessibility compliance greatly impedes the implementation of ADA, since people often assume they can't afford it. To meet the need for detailed information on access construction and costs, this publication was written to help facility managers, owners, contractors, designers, and users budget renovations for accessibility.

How the Book Is Organized

This book is a guide to the costs of typical ADA compliance modifications. Like all renovations, many access modifications require on-site construction estimating services or design services in addition to the actual construction. The book should help identify these kinds of situations.

The first step in budgeting for ADA compliance projects is to survey the facility to identify barriers to access. Once a survey has been conducted, this book can be used to estimate the cost of barrier removal and to help prioritize which barriers can be removed on a certain schedule.

Use of the book involves three basic steps: 1. Surveying the facility to identify barriers to accessibility; 2. Looking up the specific items in the book; 3. Using the projects to identify necessary modifications and determine cost estimates. To make it easier to view the modifications in terms of use, and to coordinate them with other possible renovations, the projects listed in this book follow a person's typical route of travel up to and through a facility. To help correlate projects with the ADA Accessibility Guidelines (ADAAG), a Cross Reference is included at the back of the book to list where construction information on an ADAAG item can be found in this book. The main body of the book is divided into projects, which are single access modifications that might be constructed in a facility. These fall into three categories:

- *General Access Modifications:* Typical ADA compliance access modifications.
- *Specialized Access Modifications:* Access elements included in specific types of facilities.
- *Case Studies:* Examples of the barrier removal and access construction process. These illustrate how a range of solutions to barrier removal might be devised.

Individual Projects

Each individual project is divided into three parts: illustration, text, and cost estimates—to convey the use of the design element, regulations, construction, and cost information. The illustration shows a typical access element with relevant information, such as installation, dimensions, or use. The text gives an overview of the requirements, design considerations, and level of difficulty involved for any project. The estimate(s) list the materials and labor involved for construction of the project in a typical situation, often with several variations in design or materials used, based on varying site conditions or other options. All projects contain the following information:

- ADA requirements: Where a particular element might be required by ADA, and the specific ADA technical standards for compliance;
- Design suggestions, considerations, or discussion of the particular requirement's application or limits;
- Key items: A summary of materials and/or products necessary to execute a project;
- Level of difficulty (for performing the work required to construct the modification): A very rough approximation of the amount of expense and labor involved in executing a particular access modification in a typical situation. *Low* = minimal cost, usually doesn't require a building permit, and may be performed by building maintenance staff; *Moderate* = more difficult, could involve skilled tradespeople and obtaining a building permit; *High* = expensive, involves structural and/or utility work, specialized tradespeople, and building permits.
- Cost estimates for each accessibility project.

Additional References

Following the case study projects are additional references. Following the Unit Cost section are "Resources," which provides government agency addresses and phone numbers for obtaining technical assistance on ADA issues. It also lists nongovernment-produced checklists and publications that are useful in performing ADA surveys and providing additional technical assistance.

A glossary of terms defines ADA and access terminology, and an abbreviations list clarifies design and construction terms used in the estimates.

The first, the Unit Cost section, contains individual construction line items that allow you to make adjustments to the project estimates in the main part of the book. For example, in project No. 20, "Install New Door," it is assumed that some demolition is required of an exterior wall constructed of horizontal wood siding, with metal studs and gypsum wall board on the interior. If, in fact, your wall is constructed of different materials, such as brick or concrete, you can substitute the appropriate demolition line item from the Unit Cost section for the demolition line item in the project. At the beginning of the Unit Cost section is a full explanation of how to find and apply the information you need.

To account for regional variations in construction costs, we have included location factors for 900 cities throughout the U.S. The "Location Factors" section provides percentage multipliers to be applied to the total cost (including contractor's overhead and profit) for each project. For example, if your total project cost is $1,000, and you are located in Atlanta, Georgia, your location factor is .86; therefore, your project cost is $860.

Reminders

Each situation is different. While the most likely projects, with a variety of likely site conditions and building materials are presented, it is not possible to cover every circumstance requiring access modifications. The projects included represent the *most common* building element modifications needed to bring a facility into compliance with a particular requirement of the ADA Standards (ADAAG). Additional modifications not included here might be necessary to bring a facility into compliance with ADA, or to make it more accessible.

Most states already had access regulations in force when ADA was passed, and many municipalities have additional accessibility requirements. Some local regulations may be stricter than ADA (for instance, requiring ramps to be 48" wide instead of 36"). Some states require a variance for a platform lift; most prohibit them on designated fire egress stairs. Plumbing codes in some localities prohibit a unisex accessible restroom. In case of conflict between regulations, the stricter requirements govern the design. State and local codes must **always** be investigated prior to executing any construction.

The estimates are intended as a guide to budgeting access modifications. They are not intended to be absolutely comprehensive or definitive. While many alternatives in materials and site conditions have been presented, there are still other materials that can be used if they comply with ADA. More expensive materials, such as those often used in historic situations, will, of course, affect the total installation price.

Prices also vary between cities and regions; therefore, a national cost index based on over 900 Zip Code locations has been included at the back of the book for estimate adjustments.

Material costs vary according to supply and demand, usually up (but occasionally down). Labor costs vary widely, depending on the difficulty of a particular renovation, prevailing local wage laws, and the availability of labor.

The size of the job or amount of overall work to be done also affects the cost. Construction unit costs are always less when renovation is more extensive and more modifications are done at once, due to factors such as the contractor's travel and setup time. This should be considered when deciding whether or not to proceed with more than one modification for access. In the long run, hiring a contractor to build a ramp and widen three doorways at the same time is usually cheaper than doing them a year apart. In the project estimates, where a total unit cost such as *per square foot* or *per linear foot* is given, this price assumes a quantity of work sufficient to cover contractor's costs and a fair profit. For projects of minimal size, the contractor's markup might be significantly higher. As with all renovations, a range of bids can help determine the best price. Again, the pricing in the book cannot be absolute; a contractor's estimate can vary considerably from a price obtained from this (or any other) book due to a number of nonquantifiable factors.

It must be remembered that the main goal is to eliminate discrimination based on accessibility and to create a facility usable by a larger group of people with a wider range of abilities. Even though the ADA Standards are technical standards and not a building code, the same consideration exists when applying the ADA Standards as with a building code: these are only *minimum* requirements. It is often possible, and desirable, to exceed the minimum standards given. Also, the ADA Standards are not a design guide, only the minimum technical requirements necessary to ensure a minimum level of accessibility compliance. Actual design will be determined by the individual situation.

Above all, the usefulness of input from people with disabilities is key to the successful design of access modifications. It is possible (and all too common) to have an element that is compliant but may not work for an individual. Whenever possible, a design should be reviewed by potential users with disabilities to determine if it is, in fact, the optimal solution for removal of a particular barrier.

Summary

ADA addresses a long-standing need. There is a very long history of inaccessible architecture to overcome. What we have tried to stress in the introduction to this book, and what each project emphasizes, is that barrier removal is not complicated and that removing a barrier for a particular group of people removes it for everyone. Modifications such as curb cuts, ramps, lever door handles, paddle sink faucets, audible elevator signals, and visual fire alarms are just some of the accessibility projects that make a facility easier and safer to use for all people. With this overview of ADA, and a survey of your facility, the access renovation projects included here will help make your facility more compliant with ADA, accessible by people with disabilities, and more open to everyone.

Part One
Getting Started

Part I introduces two important concepts that will apply to all of the projects that follow. First, there is an overview of the Americans With Disabilities Act. This text outlines the four Titles of the ADA, and the roles of the various federal agencies who issued and are enforcing its standards. The basic requirements of the ADA are spelled out for new construction, existing facilities, and renovations. This section also describes the types of facilities covered by ADA and how the law applies to each. Some guidance is also given on the facility survey that precedes the modifications.

The second component in Part I is guidance for estimating using the projects presented in this book. Here, the elements of each project are defined, with guidance to help readers apply the given information to arrive at an accurate cost estimate for their own particular material requirements and site conditions. The 75 Modification Project Estimates follow in Part II.

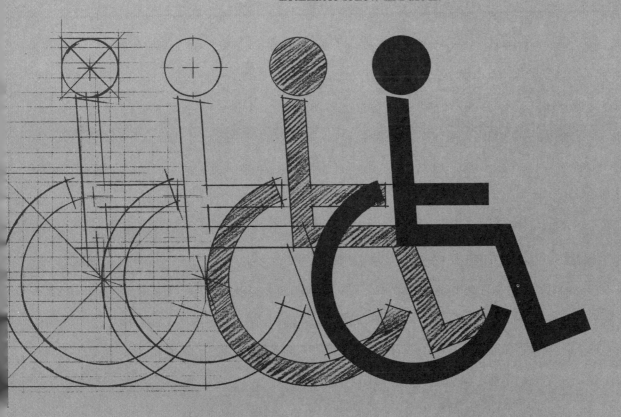

Overview of ADA

What is ADA?

The Americans with Disabilities Act (ADA) is a federal civil rights act enacted in 1990 prohibiting discrimination against people with disabilities. There are five sections, or "titles," which cover different aspects of discrimination: Title I, Employment; Title II, State and Local Government; Title III, Public Accommodations and Commercial Facilities; Title IV, Telecommunications; and Title V, which covers miscellaneous provisions of the law. Titles I, II, and III all have sections that deal with construction: Titles II and III proactively, Title I in response to specific requests.

What is the Process for Complying with ADA?

There is a basic process for complying with ADA:
1. Learning about the requirements of ADA and how they apply to a facility or program;
2. Conducting a survey to identify barriers;
3. Establishing a list of potential modifications for barrier removal, including changes to policies, facilities, and cost estimates;
4. Removing existing barriers.

Although the ADA Standards (ADAAG) are written like a building code, unlike other codes there is no formal sign-off for ADA compliance. This means that facility owners need to know how their facilities are covered under ADA and which modifications might be required. The following is a brief overview of ADA requirements for different facility types to assist in the costing process.

What are ADA's Standards for Accessible Design?

The ADA Standards for Accessible Design are the enforceable standards issued by the Department of Justice as part of the Final Rule for Title III. The Architectural and Transportation Barriers Compliance Board (ATBCB) developed the ADA Accessibility Guidelines, or ADAAG, to serve as minimum guidelines for the Department of Justice's Standards for Accessible Design. The Standards are often referred to as ADAAG, and in each of the projects in this book, we provide the ADAAG reference citation for ease of use. When altering any building or space, it is important to use the DOJ Final Rule where you will find not only the Standards, but all of the requirements for barrier removal and alterations.

The ADA Standards for Accessible Design established minimum technical requirements for the design and construction of buildings and facilities. They were written with a clear intent: to increase the level of accessibility in the built environment in new construction, alterations, and existing facilities.

The Department of Justice is the agency that enforces ADA Title II and Title III, and ultimately enacts all changes and additions to all aspects of the regulations for both, including the ADA Standards for Accessible Design. Additional sections are being developed and will gradually become incorporated into the ADA Standards. The next sections to be added to the Guidelines in 1994 include requirements and scoping for additional facility types in State and Local Government; Judicial, Legislative and Regulatory Facilities; Detention and Correctional Facilities; Accessible Residential Housing (housing authorities); and Public Rights of Way. Also under development are scoping and technical requirements for Children's Environments, and Recreation and Parks. For clarification on Title II and Title III Regulations, including the ADA Standards, contact the Department of Justice. Also, the ATBCB has a number of design-related technical assistance documents (see *Resources* section).

It is important to remember that since the ADA is an anti-discrimination civil rights act, and not a building code, there is more to compliance than the minimum technical requirements. For full background on the ADA, contact the ADA Technical Assistance Centers (see *Resources*). Although the law is complex, the basic provisions for compliance in construction are relatively simple:

- **For new construction and additions:** Both publicly-used and employee-only spaces have to comply with the ADA Standards (as defined in the Americans with Disabilities Act Accessibility Guidelines, or ADAAG). Only "non-occupiable" employee spaces, such as elevator pits, are exempt.

- **For renovations:** Any alterations to existing facilities, both public and employee-only common spaces, have to comply with the ADA Standards (ADAAG) unless it is "technically infeasible" to do so. Alterations to an "area containing a primary function" trigger accessibility requirements for modifying the path of travel to the area.
- **For existing facilities:**
  1. *Employee-only facilities or spaces:* Under Title I, employers with 25 or more employees (15 or more after July, 1994) must make "reasonable accommodations" for employees. Reasonable accommodations are made according to individual need and may include architectural modifications, but no modifications are required if not requested.
  2. *Government facilities:* Under Title II, state and local governments must have all their *programs* accessible. Modifications to existing buildings are required when administrative changes to programs are not sufficient to create access.
  3. *Public accommodations (Privately owned businesses serving the public):* Under Title III, businesses must address use of their services through removing existing barriers where it is readily achievable to do so, that is, "easily accomplished with little expense." Readily achievable modifications are required even when no other renovations are planned.

With such broad requirements, complications arise in deciding exactly what construction complies "to the maximum extent feasible," what are "reasonable accommodations," what is "an area of primary function" and what barriers are "readily achievable" to remove, since each varies greatly depending on particular circumstances. (Refer to the Glossary at the back of the book for the definitions given in ADA.) As written, the regulations allow facility owners or managers some leeway in deciding how to comply with the different provisions that apply to their facilities, but it also places responsibility for compliance on them. For complete information refer to the ADA regulations and technical assistance manuals identified in the resource section at the back of the book, which also gives information on access resources and accessibility checklists to help identify existing barriers.

While the ADA Standards are the primary means for meeting the requirements of ADA, ADAAG Section 2.2, Equivalent Facilitation, states "Departures from particular technical and scoping requirements of this guideline by the use of other designs and technologies are permitted where the alternative designs and technologies used will provide substantially equivalent or greater access to and usability of the facility." This is most useful in considering accessibility in relation to projects like children's toilet room fixtures, which is not covered in ADAAG. As stated, the ADA Standards as defined in the ADA Accessibility Guidelines are the minimum requirements that are used to establish accessibility.

Facilities Covered By ADA

All types of facilities are covered under ADA, with the exception of privately-owned permanent housing and religious facilities and private clubs, which are exempt from Title III. Most states already have accessibility regulations in place that are enforced as part of the local building codes. The ADA Accessibility Guidelines are based on the American National Standards Institute's *Accessible and Usable Buildings and Facilities (CABO/ANSI 117.1–1992)*, which in its current form is incorporated into all three national model building codes. There are differences, so it shouldn't be assumed that compliance with ANSI A117.1 or a model code is the same as compliance with ADA. ADA does not require any scoping or technical standards to exceed those required by the the ADA Standards (ADAAG), but if sections of the local access codes are stricter than ADA, the stricter regulation always governs.

Title I, Employment

New Construction, Additions, and Alterations: Title I does not specifically address architectural accessibility. Design of employee areas for new construction and alterations is covered in the Title III regulations and ADAAG [also Uniform Federal Accessibility Standards (UFAS) for Title II facilities].

Existing Facilities: Employees with disabilities in companies with more than 25 employees (15 employees after July 1994) may request that the employer provide "reasonable accommodation" to allow them to carry out their essential job functions. A "reasonable accommodation" may or may not be architectural in nature, depending on the individual's needs. What determines a "reasonable accommodation" is worked out between employer and employee; criteria have been set by the federal Equal Employment Opportunity Commission (EEOC). Reasonable accommodations need not be expensive; examples are raising a desk to allow knee space for a wheelchair user, providing lever door handles, increasing lighting levels, or installing an accessible parking space.

Title II, State and Local Government Services

New Construction and Additions: Design of new state or local government buildings is covered by ADA, and all new construction and additions must be fully compliant. Title II regulations allow the UFAS to be used as an alternative to the ADA Accessibility Guidelines, providing that only one standard is used in any facility. ADAAG is similar to UFAS in its technical requirements, but has more substantial scoping differences.

Alterations: All alterations to spaces used by the public and employee-only areas must comply with the ADA Standards (ADAAG) or UFAS unless it is "technically infeasible" to do so (UFAS has a different formula from ADAAG). If technically infeasible, the alteration must comply "to the maximum extent feasible." "Technically infeasible" is defined as having little likelihood of being done because a major structural member would have to be moved or because an existing physical or site constraint prohibits compliance.

Existing Facilities: Title II requires state and local governments to make their programs accessible. This might not require alterations to an existing facility. For instance, if a town hall is inaccessible, it might be possible to hold meetings in an accessible high school. If, however, a facility is unique in its program (such as one library that serves a given area) the building has to be made accessible in order to achieve program access. As with new construction and additions, either ADAAG or UFAS is acceptable as a design standard for alterations to existing facilities.

Title III, Public Accommodations and Commercial Facilities Operated by Private Entities

Title III distinguishes between privately-owned businesses that invite the public in to purchase goods and services (*public accommodations*) and those that don't (*commercial facilities*). Public accommodations are required to remove barriers in existing facilities where it is readily achievable to do so. Title III regulations list twelve categories of public accommodations:

1. Places of lodging (an inn, hotel, motel, or other place of lodging);
2. Establishments serving food or drink (a restaurant, bar, or other establishment serving food or drink);
3. Places of exhibition or entertainment (a motion picture house, theater, concert hall, stadium, or other place of exhibition or entertainment);
4. Places of public gathering (an auditorium, convention center, lecture hall, or other place of public gathering);
5. Sales or rental establishments (a bakery, grocery store, clothing store, hardware store, shopping center, or other sales or rental establishment);
6. Service establishments (a laundromat, dry cleaner, bank, barber shop, beauty salon, travel service, shoe repair service, funeral parlor, gas station, office of an accountant or lawyer, pharmacy, insurance office, professional office of a health care provider, hospital, or other service establishment);
7. Stations used for public transportation (a terminal, depot, or other station used for specified public transportation);
8. Places for public display or collection (a museum, library, gallery, or other place of public display or collection);
9. Places of recreation (a park, zoo, amusement park, or other place of recreation);
10. Places of education (a nursery, elementary, secondary, undergraduate, or postgraduate private school or other place of education);
11. Social service establishments (a day care center, senior citizen center, homeless shelter, food bank, adoption agency, or other social service center establishment);
12. Places of exercise and recreation (a gymnasium, health spa, bowling alley, golf course, or other place of exercise or recreation).

All such establishments have to comply with the requirements for public accommodations; only private clubs and religious establishments are exempt (but any public accommodations leasing spaces from them have to comply).

Business establishments which do not fall into any of the above categories are classified by Title III as "commercial facilities." If part of the facility serves as a public accommodation (such as a tour of a factory), that portion of the facility must be accessible. Commercial facilities must meet the requirements for new construction, additions, and alterations even if a new building or addition will not be open to the public. In new construction, employee common spaces must be fully accessible. Work areas (such as a lab or office floor) must be on an accessible route, and an employee must be able to approach, enter, and exit the area. Individual work spaces are modified on an as-needed basis under Title I, Employment.

New Construction and Additions: All new public accommodations and commercial facilities must be accessible and comply with the ADA Standards (ADAAG). Individual employee-only work spaces (such as a lab station) do not have to comply, but work areas (such as a lab) have to be on an accessible route of travel, and any employee has to be able to approach, enter, and exit the space. Only non-occupiable spaces, such as catwalks and elevator pits, are exempted.

Alterations: All alterations to public and employee-only areas must comply with the ADA Standards (ADAAG) unless it is "technically infeasible" to do so. If so, the alteration has to comply "to the maximum extent feasible."

Alterations to an area containing a primary function trigger additional accessibility requirements in both public accommodations and commercial facilities. A primary function is defined as "a major activity for which the facility is intended," such as a bank's customer service area, or the dining area of a cafeteria. Spaces such as mechanical rooms, entrances, and rest rooms are not areas of primary function, meaning that alterations to these kinds of spaces do not trigger additional accessibility requirements. If an area of primary function is being altered, the path of travel to the area of primary function must be brought into compliance with the ADA Standards (ADAAG) at a cost up to 20% of the total cost of the area's renovation.

For leased places of public accommodation, both the landlord and tenant are responsible for compliance. Allocation of responsibility may be determined by the lease or other contract. ADA does not state who is directly responsible for removing barriers either in existing facilities or during renovations, but places the responsibility for ADA compliance on the contractual agreement between landlord and tenant.

Existing Facilities: For public accommodations where no renovations are planned, all barriers must be removed if it is readily achievable to do so. "Readily achievable" is defined as "easily accomplished with little expense," and this varies depending on the situation. What is readily achievable for a national chain store might not be readily achievable for a single-owner grocery store.

Determining what barrier removal measures would be considered readily achievable must be on a case-by-case basis. Factors to consider include the nature and cost of the remedial action; the organization's financial resources; the size of the organization (number of facilities and employees), and type of operation. The Department of Justice regulations to help determine what modifications might be undertaken recommend the following order of priorities in removing barriers:

1. Access into the facility
2. Access to where goods and services are made available to the public
3. Access to rest rooms if provided for public use
4. Other amenities

These recommendations, given in section 36.304(c) of the Title III regulations, can vary in different facilities depending on what service is being provided. Examples of readily achievable modifications are installing grab bars in a rest room, adding a lever to a door knob, or putting up Braille signage. Modifications which might be readily achievable depending on the individual circumstances include installing a ramp up a small flight of stairs, installing a platform lift, or replacing a sink. Installing a new elevator probably would not be considered a readily achievable modification. Barrier removal that is not readily achievable can be addressed at a later date during building modernization, as part of a facility's ongoing obligation to remove barriers.

Surveying the Facility

Surveying the facility is the first step in costing accessibility modifications. ADA regulations do not specify one particular method of identifying barriers, but surveying with an accessibility checklist can be very useful (several are listed in the *Resources* section at the back of the book). When surveying a facility for accessibility it is vital to remember that wheelchair access is only one factor to consider. To comply with ADA, people with other mobility impairments such as balance or stamina problems, and people with visual, hearing, or cognitive impairments must also be accommodated: The ADA Standards (ADAAG) include requirements which improve access for people who are blind (raised character and Braille signage), people with hearing impairments (visual fire alarms), people with low fine motor control (lever handles or paddle faucets), control heights for people with limited reach, and many others.

Surveying a facility for accessibility involves identifying spaces, routes of travel, and individual items which are not usable by persons with disabilities and may not be compliant with the ADA Standards (ADAAG). It is important to know the entire list of access issues within a facility to be able to determine which barriers can be removed. Knowing the areas of the facility that need to be surveyed is key: Public areas in public accommodations need to be surveyed to determine barrier removal needs. The existing barriers then need to be prioritized as to severity in impeding access and ease of removal. The Department of Justice recommendations for barrier removal are just that. If an entrance is not fully compliant, but is still usable as an accessible entrance, and there is no accessible rest room in the facility, the rest room would be the higher priority in terms of barrier removal, even though it is listed as a lower priority. The ultimate decision on priorities rests with management.

Involving people with disabilities is critical in setting priorities for barrier identification and removal. To quote directly from the Department of Justice preamble to the ADA Title III Final Rule, discussion of *Section 36.304 Removal of Barriers:* "The Department recommends that this process include appropriate consultation with individuals with disabilities or organizations representing them. A serious effort at self-assessment and consultation can diminish the threat of litigation and save resources . . . The Department recommends . . . the development of an implementation plan designed to achieve compliance with the ADA's barrier removal requirements before they become effective on January 26, 1992. Such a plan, if appropriately designed and diligently executed, could serve as evidence of a good faith effort to comply."

As stated, this publication is not a survey tool. In some cases, such as an inaccessible entrance that can be ramped, a survey might not be necessary; but even in these cases, other issues could exist (in fact probably do) that should be identified. Again, the most valuable information comes from a user.

Using the Projects for Cost Estimating

Project Budgeting

Once a barrier has been identified, various design solutions are needed to determine how much each will cost. If a flight of stairs presents a barrier, several critical questions need to be asked: Can a ramp be built? Is there enough room to accommodate a compliant ramp? Would the ramp block the stairs? Could a lift be installed on the stairs? Is the flight of stairs wide enough? Would a lift impede required fire egress? Once a possible design solution (or list of solutions) is determined, the particulars of the modifications need to be added to the appropriate project.

As with any renovation, there are two main influences on the cost: requirements and design. (There is always more than one way of doing what's needed.) While this publication prices different materials, actual design decisions are made by the facility owner or operator. A ramp is required to have a maximum slope of 1 in 12, but whether the ramp is made of wood, concrete, or granite is a design decision based on existing conditions and the project budget. Budget is a major factor in determining design, whether or not the renovations are legally mandated, and accessibility is no exception. With this in mind, the following is a description of the accessibility modifications cost estimation process.

Applying the Projects

The information contained in each project is intended to identify where a barrier might need to be removed, and the standards to which it needs to be designed, to comply with the requirements listed in the ADA Standards (ADAAG). Each project includes the following sections:

1. The **illustration** shows how a modification would be used in a typical application.
2. **Where Applicable** lists those facilities and situations which are covered by the ADA. For instance, the "Construct New Pathways" project lists all new public and common-use pathways as needed to comply with the listed sections of ADAAG.
3. The **ADAAG Reference** lists the regulation sections which apply to this project
4. The **Design Requirements** summarizes the minimum technical standards to which this project must conform.
5. The **Design Suggestions** are recommendations that could make the item easier to use. The recommendations may exceed the minimum ADA requirements, and should be consulted prior to designing to incorporate the best use criteria from the outset. For instance, one design suggestion for a ramp is to have a slope shallower than 1:12 (the maximum slope allowed by ADAAG), therefore making it easier to use by a wider range of people. This modification increases the ramp's length, and therefore its cost.

    The list of design suggestions for any project is by no means comprehensive; potential users are the best source for design suggestions. They may know of similar designs that have been built, and owners or users of that facility might be able to say what has (or hasn't) worked about the design. The constraints imposed by both existing conditions and the budget will help determine if it is, in fact, possible to apply any of the recommendations.
6. The **Key Items** section lists components needed to execute the design. As with the other sections, this list can grow depending on the particulars of the situation. For example, a ramp always needs a slip-resistant surface, structure, and rails, but it may also include a set of steps (this is recommended, since some people find stairs easier to use than a ramp), an overhang either over the entrance or the entire ramp, or a heated surface to melt snow. These features should be evaluated and a full list of chosen elements should be completed prior to costing.
7. **Level of Difficulty** is included to help determine whether it will be necessary to hire an outside contractor and what kind of professional services might be needed. It is important to know just what trades, if any, will need to be called in for planning and scheduling various barrier removal projects. In most cases, a building permit will be necessary prior to making any modifications. In *all* cases, modifications will need to comply with all local building and fire safety codes, regardless of the level of difficulty or the involvement of a designer on the project.
8. **Estimates:** After your modification design is set, the appropriate project

estimate should be selected from the projects section of this book. A cost is given for each project installation as defined. In many cases (such as in the estimates for various ramp configurations) a cost per unit is also given, so that you can easily arrive at a total cost that addresses your own design requirements.

For each ramp estimate, a total project installation cost is given. Underneath that cost is a "per linear foot" cost that can be multiplied by the number of feet your project length requires. For instance, if a ramp will be 24' long and constructed of wood, the total cost given "per linear foot" of this type of ramp would be multiplied by 24.

Other units might also be used, depending on the project. Widening an existing concrete pathway is estimated per square foot of added area. If an existing 60' long pathway is to be widened by 1', multiply the cost per square foot for that modification by 60.

Each project should give an idea of how a modification would be used, where it might be needed, the minimum design standards, necessary construction, and how much it will cost.

It must be reiterated that the costs given are for typical accessibility modifications and common existing conditions details. Each situation is different; for instance, in the following example, the cost of modifying a concrete walk will depend on the thickness of the concrete, whether it is reinforced or has an integral footing, and whether it is edged with a stone curb. While many of the given projects include a number of alternate estimates for varying materials or designs, it is impossible to include a cost for modifying every possible detail. In our example, we assume an unreinforced 4" concrete slab laid over a tamped gravel substrate with regular control joints, which is a common method of installing a concrete walkway. Costs for modifications will vary based on actual conditions, which should be accounted for to ensure that the estimate is as accurate as possible.

Example

The following example shows how project estimates may be combined to produce a total modification cost.

One item included in an accessibility survey of an existing public facility was a 60' long concrete pathway from an accessible doorway to a street crossing. The pathway was at a running slope of less than 1:20, had a cross-slope of less than 1:50, and was in good condition, but was surprisingly narrow—30"—and needed to be widened. It was decided that it should be widened to 48", even though 36" would have been compliant. Also there was no curb cut or curb ramp at the curb. To bring the pathway into compliance and to bring it up to the facility's standards, it was to be widened by 18" for its entire length and a curb cut needed to be installed at the end of the pathway. To obtain a cost estimate for these modifications, the relevant projects from this publication would be:

5a. Widen existing concrete pathway

7a. Install new curb cut (concrete sidewalk)

Widen 60 feet of concrete
pathway by 18" at $6/S.F.   = $540

Install 1 curb cut
at $1,033/each   = $1,033

TOTAL RENOVATION   = $1,573

Adjustments to modification totals can be expected, based on a specific site and the amount of work being performed at one time.

Part Two
Modifications Estimates

Part II contains the actual ADA Modification Project Estimates. There are 75 basic projects, many of which include several estimates of alternative materials or site conditions. Altogether, there are 260 separate estimates. Refer to the preceding pages, "Using the Projects for Cost Estimating," for guidance on applying the projects to your own facility. The Location Factors (Part V) will help you to adjust the total project cost to your particular location. If your project calls for special materials or extra work that are not described in one of the standard 75 estimates or the accompanying alternates, refer to the Unit Costs in Part IV. There you will find over 3,000 construction line items that can be used to adapt an existing project to your special requirements.

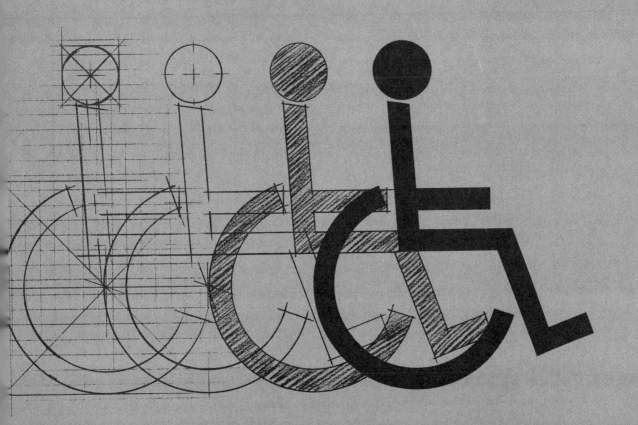

1
Install Accessible Parking Spaces

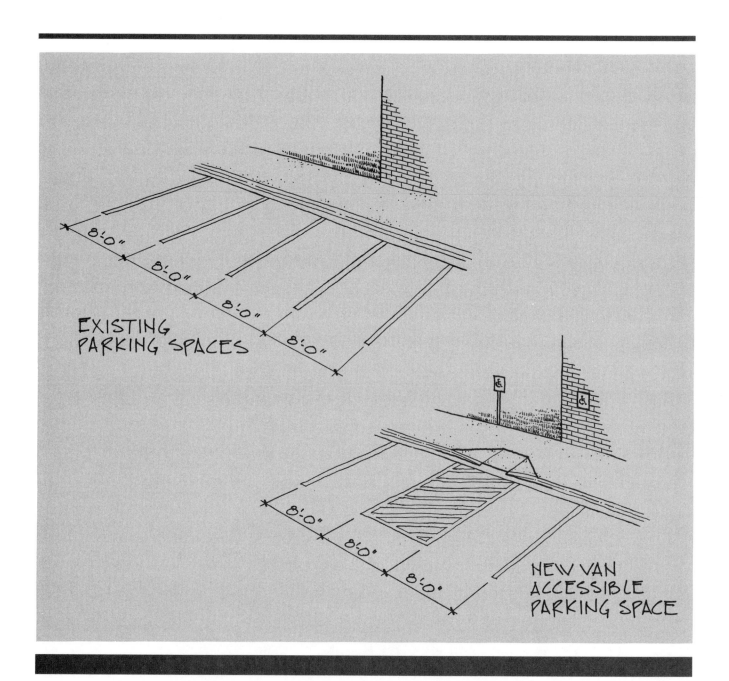

Accessible parking spaces are often the first part of an accessible route of travel. Proper design and location can create the difference between an accessible and inaccessible facility, as well as ensuring the safety of people using them.

ADAAG References
4.1.2 (5) (a) and (b), Accessible Sites and Exterior Facilities, New Construction
4.6 Parking and Passenger Loading Zones

Where Applicable
All parking lots for visitors or employees.

Design Requirements
- Number of accessible spaces as designated by ADAAG 4.1.2 (5) (a) and (b). Accessible spaces can be dispersed among multiple lots with accessible entrances.
- Spaces on an accessible route with curb cut, if necessary.
- Spaces located closest to accessible entrance to the building, or to the lot entrance if no particular building is served.
- Spaces 8' wide, with 5' access aisle (two spaces can share one aisle).
- One designated van-accessible space for every eight accessible spaces (never less than one), an 8' space with 8' access aisle.
- 98" minimum height clearance at the van-accessible space.
- Surfaces to be stable, slip-resistant, maximum cross slope 2% (1:50).
- Signage with accessibility symbol.

Design Suggestions
ADAAG shows perpendicular spaces in its illustrations, but it is possible to use angled or parallel spaces if access aisle requirements can be met. Look at the route of travel first: Spaces should be located so that people with disabilities are not forced into vehicular traffic. Vans equipped with wheelchair lifts usually exit to the passenger side (some exit at the rear), so if only one van space is required, the wide access aisle should be on the passenger side of the space. Locate accessibility signs to always be visible to the car's driver and not blocked by cars, poles, etc.

Key Items
Paving material (asphalt or concrete), striping paint, signs, curb cut material if necessary.

Level of Difficulty
Low to moderate. Re-striping, ground markings, or installing signage can be done by facility staff or contractors. Installing a curb cut is usually done by a mason or concrete contractor.

Estimates

Add new accessible spaces

Description	Quantity	Unit	Work Hours	Material
Saw cutting, conc., per inch of depth (20 L.F., 4" deep)	80.000	L.F.	1.360	20.00
Site demolition, remove granite curb	14.000	L.F.	0.952	0.00
Site demolition, remove concrete sidewalk	5.340	S.Y.	0.838	0.00
Site demolition, remove asphalt paving	1.670	S.Y.	0.124	0.00
Painted line removal	100.000	L.F.	2.200	35.00
Install granite curbing, 6" x 18"	14.000	L.F.	1.736	235.20
Gravel base, 4" deep	48.000	S.F.	0.480	5.76
Install 4" concrete sidewalk, broom finish	48.000	S.F.	1.920	43.20
Miscellaneous asphalt patching	1.670	S.Y.	0.890	2.45
Line painting, latex, yellow, 4" wide	20.000	L.F.	0.040	1.00
Line painting, gore lines	100.000	S.F.	1.600	79.00
Install signs	2.000	Ea.	0.914	17.50
Install sign post	1.000	Ea.	0.160	27.00
Totals			13.214	466.11

Total for two accessible spaces including general contractor's overhead and profit: $1,611

Copyright R. S. Means Company, Inc., 1994

1. Install Accessible Parking Spaces (continued)

Re-stripe existing parking lot

Description	Quantity	Unit	Work Hours	Material
Painted line removal	263.000	L.F.	5.786	92.05
Line painting, latex, yellow, 4" wide	20.000	L.F.	0.040	1.00
Line painting, gore lines	500.000	S.F.	8.000	395.00
Totals			13.826	488.05

Total per space including general contractor's overhead and profit	**$17**
Total per one hundred spaces including general contractor's overhead and profit	**$1,697**

Copyright R. S. Means Company, Inc, 1994

Notes

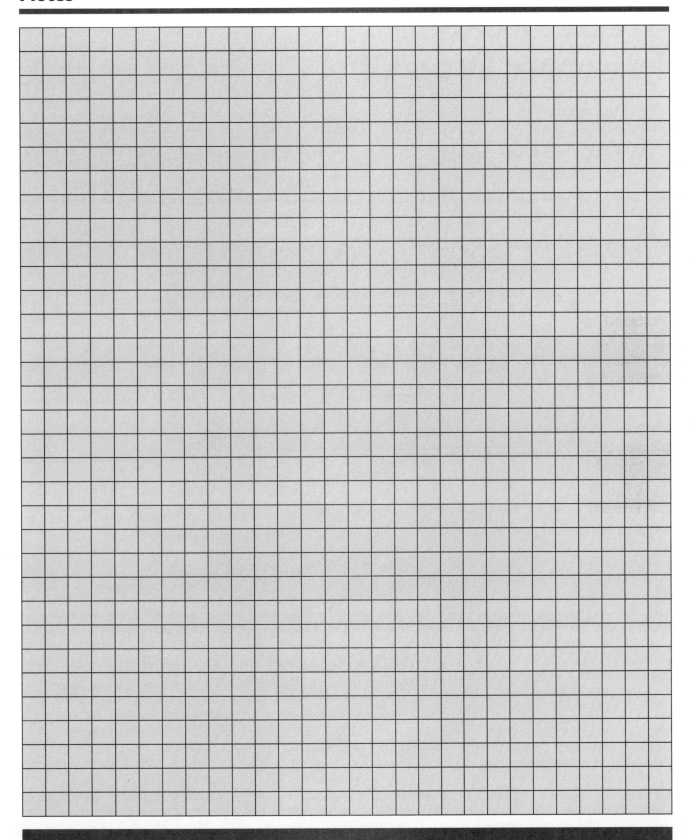

2 Create Accessible Passenger Drop-Off

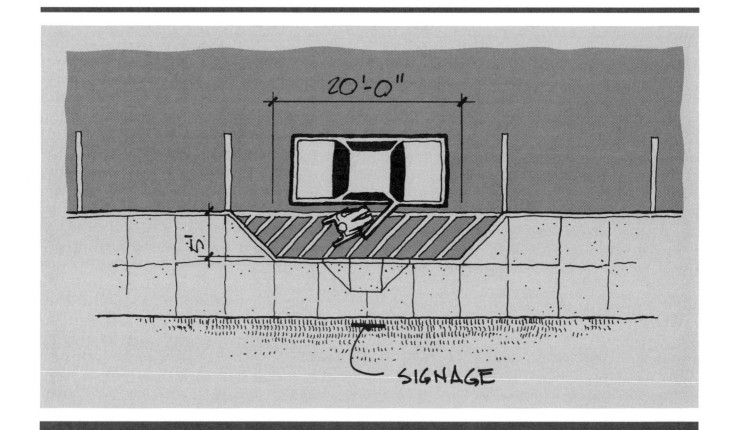

A facility which requires vehicular access but doesn't have an accessible passenger drop-off can force a user with a disability to either get out in traffic or not use the facility at all. A safe, accessible drop-off creates an area where all users can get out of or into their cars or vans, and get directly onto an accessible route.

ADAAG References
4.6 Parking and Passenger Loading Zones
4.6.6 Passenger Loading Zones

Where Applicable
Loading zones not required; if provided, at least one must comply with the requirements listed below.

Design Requirements
- 5' × 20' access aisle adjacent to vehicle space.
- Curb cut if necessary to provide access from vehicle space to sidewalk.
- Level surface, 2% (1:50) slope maximum.
- 114" height clearance minimum at space and along vehicular route to and from the space.

- Surfaces to be stable, slip-resistant, with a maximum cross slope of 2% (1:50).
- Signage with accessibility symbol.

Design Suggestions

Visibility is important at a drop-off; both the vehicle space and the access aisle should be clearly striped. Locate drop-offs as near as possible to accessible entrances, especially where accessible parking is not available.

Key Items

Paving material(s), striping, signage.

Level of Difficulty

Moderate to high, depending on existing site conditions. Requires planning, and possibly extensive paving work. Can require design drawings, approved by local building department.

Estimates

Stripe vehicle space and access aisle, install pole-mounted signage, install concrete curb cut

Description	Quantity	Unit	Work Hours	Material
Saw cutting, conc., per inch of depth (20 L.F., 4" deep)	80.000	L.F.	1.360	20.00
Site demolition, remove granite curb	14.000	L.F.	0.952	0.00
Site demolition, remove concrete sidewalk	5.340	S.Y.	0.838	0.00
Site demolition, remove asphalt paving	1.670	S.Y.	0.124	0.00
Install granite curbing, 6" x 18"	14.000	L.F.	1.736	235.20
Gravel base, 4" thick	48.000	S.F.	0.480	5.76
Install 4" concrete sidewalk, broom finish	48.000	S.F.	1.920	43.20
Miscellaneous asphalt patching	1.670	S.Y.	0.890	2.45
Line painting, gore lines	100.000	S.F.	1.600	79.00
Install sign	1.000	Ea.	0.457	8.75
Install sign post	1.000	Ea.	0.160	27.00
Totals			10.517	421.36

Total per each including general contractor's overhead and profit **$1,379**

Stripe vehicle space and access aisle, install pole-mounted signage, install asphalt curb cut

Description	Quantity	Unit	Work Hours	Material
Saw cutting, asphalt, to 3" deep	20.000	L.F.	0.420	4.20
Site demolition, remove granite curb	14.000	L.F.	0.952	0.00
Site demolition, remove asphalt paving	7.000	S.Y.	0.413	0.00
Install granite curbing, 6" x 18"	14.000	L.F.	1.736	235.20
Gravel base, 4" thick	48.000	S.F.	0.480	5.76
Asphalt sidewalk, 2-1/2" thick	5.340	S.Y.	0.390	19.17
Miscellaneous asphalt patching	1.670	S.Y.	0.890	2.45
Line painting, gore lines	100.000	S.F.	1.600	79.00
Install sign	1.000	Ea.	0.457	8.75
Install sign post	1.000	Ea.	0.160	27.00
Totals			7.498	381.53

Total per each including general contractor's overhead and profit **$1,122**

Copyright R. S. Means Company, Inc, 1994

3
Construct New Pathway

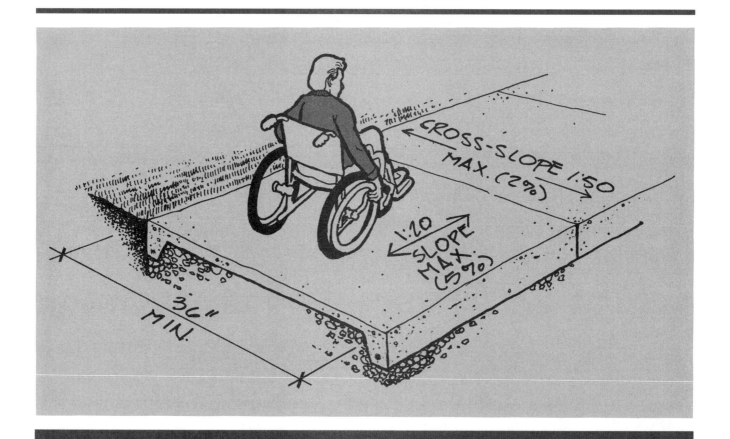

New pathways are needed to create an accessible route where there is an inaccessible route of travel, such as a stepped or a steep walkway. New accessible pathways are also useful where there is *no* designated route of travel, such as in parks or recreation areas with soft surfaces. Accessible pathways provide access not only for people who use wheelchairs, but also for those with bicycles and hand trucks.

ADAAG References
4.3 Accessible Route
4.4 Protruding Objects
4.5 Ground and Floor Surfaces

Where Applicable
Wherever a new exterior accessible route of travel is installed.

Design Requirements
- 36" minimum width.
- Stable, firm, slip-resistant surface.
- Differences in level between 1/4" and 1/2" beveled at maximum slope of 1:2.
- No abrupt changes in level greater than 1/2".
- No gratings with openings greater than 1/2" across in the lesser dimension.

- Maximum 5% running slope (1:20).
- Maximum 2% cross-slope (1:50).
- 80" minimum height clearance.
- No objects 27" a.f.f. or higher protruding more than 4" into pathway without warnings below.
- 60" × 60" passing space every 200' if pathway is less than 60" wide.

Design Suggestions

36" width is a minimum. 48" is recommended, and 60" is convenient for passing. By the same token, 5% slope is a maximum for a pathway (unless a ramp with handrails is provided). Shallower slopes are easier to use. Avoid unit pavers (e.g., bricks) if possible. If used, install unit pavers on a firm base; a firm substrate concrete is preferred, although it is more expensive than tamped sand or stone dust. Avoid placing utilities (e.g., drainage gratings) in the path of travel, even if the dimensions comply. Similarly, locate amenities such as phones adjacent to rather than in the path of travel, but within reach from the accessible path.

Key Items

Paving materials, substrate.

Level of Difficulty

Moderate to high, depending on material used. Pavers or asphalt easier to install than concrete. This project usually involves some landscaping of areas adjacent to the path. Trades involved: paving contractors, masons, landscapers.

Estimates

Install three-foot-wide concrete pathway

Description	Quantity	Unit	Work Hours	Material
Excavating	0.110	C.Y.	0.020	0.00
Gravel base, 4" thick	3.000	S.F.	0.030	0.36
Install 4" concrete sidewalk, broom finish	3.000	S.F.	0.120	2.70
Install topsoil, 4" deep	0.230	S.Y.	0.002	0.43
Install sod	0.002	M.S.F.	0.010	0.31
Totals			0.182	3.80

Total per linear foot including general contractor's overhead and profit — **$16**

Install three-foot-wide asphalt pathway

Description	Quantity	Unit	Work Hours	Material
Excavating	0.100	C.Y.	0.018	0.00
Gravel base, 4" thick	3.000	S.F.	0.030	0.36
Install 2-1/2" thick asphalt sidewalk	0.340	S.Y.	0.025	1.22
Install topsoil, 4" deep	0.230	S.Y.	0.002	0.43
Install sod	0.002	M.S.F.	0.010	0.31
Totals			0.085	2.32

Total per linear foot including general contractor's overhead and profit — **$9**

Copyright R. S. Means Company, Inc., 1994

3. Construct New Pathway (continued)

Install three-foot-wide brick paver pathway over tamped earth base

Description	Quantity	Unit	Work Hours	Material
Excavating	0.083	C.Y.	0.015	0.00
Gravel base, 4" thick	3.000	S.F.	0.030	0.36
Install 4" × 8" × 2-1/4" brick pavers	3.000	S.F.	0.435	7.35
Install topsoil, 4" deep	0.230	S.Y.	0.002	0.43
Install sod	0.002	M.S.F.	0.010	0.31
Totals			0.492	8.45

Total per linear foot including general contractor's overhead and profit — **$39**

Install three-foot-wide brick paver pathway over stone dust base

Description	Quantity	Unit	Work Hours	Material
Excavating	0.083	C.Y.	0.015	0.00
Install stone dust base	0.340	S.Y.	0.003	0.71
Install 4" × 8" × 2-1/4" brick pavers	3.000	S.F.	0.435	7.35
Install topsoil, 4" deep	0.230	S.Y.	0.002	0.43
Install sod	0.002	M.S.F.	0.010	0.31
Totals			0.465	8.80

Total per linear foot including general contractor's overhead and profit — **$39**

Install three-foot-wide brick paver pathway over concrete base

Description	Quantity	Unit	Work Hours	Material
Excavating	0.140	C.Y.	0.025	0.00
Install 4" concrete mud slab	3.000	S.F.	0.120	2.70
Gravel base, 4" thick	3.000	S.F.	0.030	0.36
Install 4" × 8" × 2-1/4" brick pavers	3.000	S.F.	0.435	7.35
Install topsoil, 4" deep	0.230	S.Y.	0.002	0.43
Install sod	0.002	M.S.F.	0.010	0.31
Totals			0.622	11.15

Total per linear foot including general contractor's overhead and profit — **$51**

Copyright R. S. Means Company, Inc, 1994

Notes

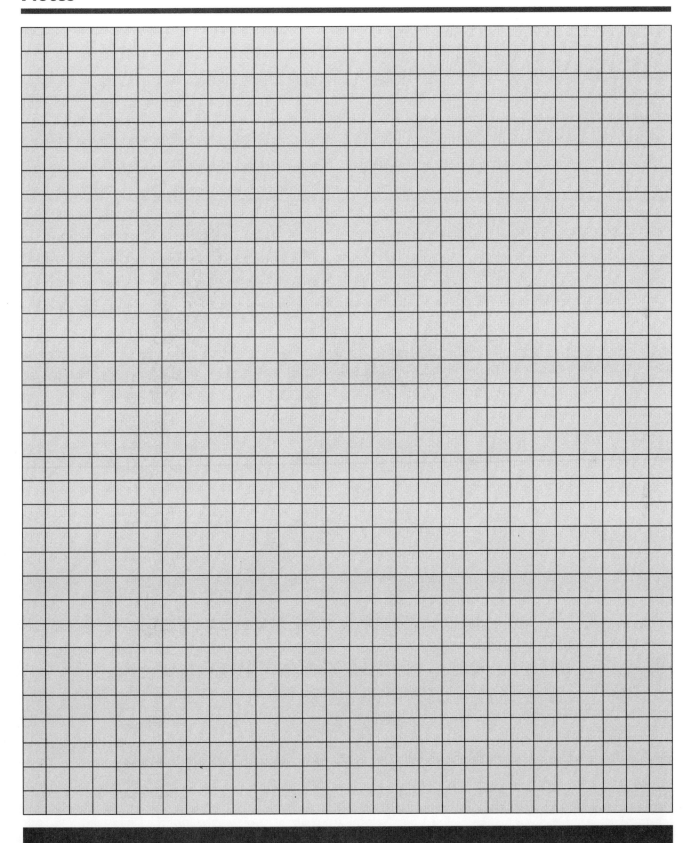

4 Construct Graded Entrance Pathway

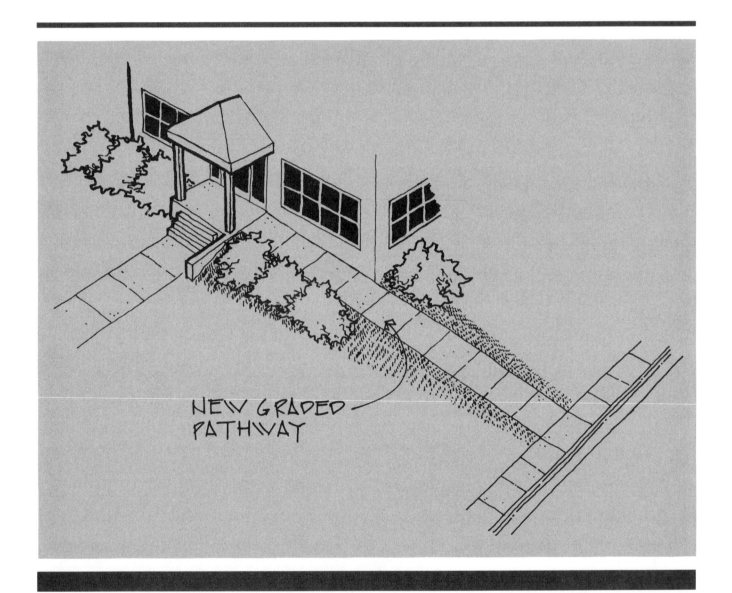

Where an entrance is not on the same grade as an accessible route of travel, constructing a graded pathway may be an alternative to a ramp. Since graded pathways have a slope no greater than 1 in 20, they have the added advantage of being usable by people with a wider range of ability. Graded pathways can be created with landscaping which can be useful in situations where a ramp might be visually disruptive (as in some historic facilities).

ADAAG References
4.3 Accessible Route
4.4 Protruding Objects
4.5 Ground and Floor Surfaces

Where Applicable
Wherever a new exterior accessible route of travel is constructed at a slope of no more than 1:20.

Design Requirements
- 36" wide minimum.
- Surface stable, firm, and slip-resistant.
- Differences in level between 1/4" and 1/2" beveled at maximum slope of 1:2.
- No differences in level greater than 1/2".
- No gratings with holes greater than 1/2" across perpendicular to the direction of travel.
- Maximum 5% slope (1:20).
- Maximum 2% cross-slope (1:50) for drainage.
- 80" minimum height clearance.
- No objects 27" a.f.f. or higher protruding more than 4" into pathway without warnings below.
- 60" × 60" passing space every 200' if pathway is less than 60" wide.

Design Suggestions
Pathways graded at a slope less than 1:20 can avoid the use of ramps. In new construction or additions, the first floor elevation can often be set to accommodate a pathway. In renovations, a graded pathway can be advantageous where a ramp would be difficult to incorporate into the existing building style, as in historic situations, or where a ramp would be prohibitively expensive. (In cases of historic buildings, the pathway should be pulled away from the building to avoid damage to existing features). Generally, pathways are usable by people with a wider range of mobility impairments than ramps and are therefore a recommended design option for creating access. However, some people with stamina difficulties find ramps easier to use. Without an entrance that is level with the existing grade, no solution can accommodate everyone. It is important to consider the height difference between one end of the route of travel and the entrance: Excessively long pathways can create very circuitous accessible routes of travel, and in such situations, ramps might be preferable.

Use the 36" width cited in ADAAG as a minimum. 48" is recommended, and 60" is convenient for passing. By the same token, the 5% slope is a maximum. Shallower slopes are easier to use. Avoid unit pavers (e.g., bricks) if possible. If used, install unit pavers on a firm base; stone dust is recommended, and concrete preferred (the firm substrate is more expensive). Avoid placing utilities (e.g., drainage gratings) in the path of travel, even if the dimensions comply. Similarly, locate amenities, such as phones, adjacent to, rather than in, the path of travel but within reach from the accessible path. Ensure that drainage is sufficient to prevent puddling on the pathway.

Key Items
Paving materials, substrate.

Level of Difficulty
Moderate to high, depending on material. Pavers or bituminous easier to install than concrete. Usually involves landscape planning. Trades involved: paving contractors, masons, landscapers.

Estimates

Install three-foot-wide concrete pathway on graded ramp (total rise, 1 foot)

Description	Quantity	Unit	Work Hours	Material
Gravel fill	0.300	Ton	0.000	0.88
Gravel delivery charge	0.200	C.Y.	0.019	0.00
Hand spread gravel	0.300	Ton	0.164	0.00
Compaction (4 passes)	0.200	C.Y.	0.011	0.00
Gravel base, 4" thick	3.000	S.F.	0.030	0.36
Install 4" concrete sidewalk, broom finish	3.000	S.F.	0.120	2.70
Install topsoil, 4" deep	0.500	S.Y.	0.005	0.94
Install sod	0.005	M.S.F.	0.025	0.78
Totals			0.374	5.66

Total per linear foot including general contractor's overhead and profit	$29

4. Construct Graded Entrance Pathway (continued)

Install three-foot-wide asphalt pathway on graded ramp (total rise, 1 foot)

Description	Quantity	Unit	Work Hours	Material
Gravel fill	0.300	Ton	0.000	0.88
Gravel delivery charge	0.200	C.Y.	0.019	0.00
Hand spread gravel	0.300	Ton	0.164	0.00
Compaction (4 passes)	0.200	C.Y.	0.011	0.00
Gravel base, 4" thick	3.000	S.F.	0.030	0.36
Install 2-1/2" thick asphalt sidewalk	0.340	S.Y.	0.025	1.22
Install topsoil, 4" deep	0.500	S.Y.	0.005	0.94
Install sod	0.005	M.S.F.	0.025	0.78
Totals			0.279	4.18

Total per linear foot including general contractor's overhead and profit: $22

Install three-foot-wide asphalt paver pathway over tamped earth base on graded ramp (total rise, 1 foot)

Description	Quantity	Unit	Work Hours	Material
Gravel fill	0.300	Ton	0.000	0.88
Gravel delivery charge	0.200	C.Y.	0.019	0.00
Hand spread gravel	0.300	Ton	0.164	0.00
Compaction (4 passes)	0.200	C.Y.	0.011	0.00
Gravel base, 4" thick	3.000	S.F.	0.030	0.36
Install 8" x 8" x 2" asphalt block pavers	3.000	S.F.	0.369	10.50
Install topsoil, 4" deep	0.500	S.Y.	0.005	0.94
Install sod	0.005	M.S.F.	0.025	0.78
Totals			0.623	13.46

Total per linear foot including general contractor's overhead and profit: $55

Install three-foot-wide brick paver pathway over tamped earth base on graded ramp (total rise, 1 foot)

Description	Quantity	Unit	Work Hours	Material
Gravel fill	0.300	Ton	0.000	0.88
Gravel delivery charge	0.200	C.Y.	0.019	0.00
Hand spread gravel	0.300	Ton	0.164	0.00
Compaction (4 passes)	0.200	C.Y.	0.011	0.00
Gravel base, 4" thick	3.000	S.F.	0.030	0.36
Install 4" x 8" x 2-1/4" brick pavers	3.000	S.F.	0.435	7.35
Install topsoil, 4" deep	0.500	S.Y.	0.005	0.94
Install sod	0.005	M.S.F.	0.025	0.78
Totals			0.689	10.31

Total per linear foot including general contractor's overhead and profit: $53

Copyright R. S. Means Company, Inc., 1994

Install three-foot-wide brick paver pathway over stone dust base on graded ramp (total rise, 1 foot)

Description	Quantity	Unit	Work Hours	Material
Gravel fill	0.300	Ton	0.000	0.88
Gravel delivery charge	0.200	C.Y.	0.019	0.00
Hand spread gravel	0.300	Ton	0.164	0.00
Compaction (4 passes)	0.200	C.Y.	0.011	0.00
Install stone dust base	0.340	S.Y.	0.003	0.71
Install 4" × 8" × 2-1/4" brick pavers	3.000	S.F.	0.435	7.35
Install topsoil, 4" deep	0.500	S.Y.	0.005	0.94
Install sod	0.005	M.S.F.	0.025	0.78
Totals			0.662	10.66

Total per linear foot including general contractor's overhead and profit: $52

Install three-foot-wide brick paver pathway over concrete base on graded ramp (total rise, 1 foot)

Description	Quantity	Unit	Work Hours	Material
Gravel fill	0.300	Ton	0.000	0.88
Gravel delivery charge	0.200	C.Y.	0.019	0.00
Hand spread gravel	0.300	Ton	0.164	0.00
Compaction (4 passes)	0.200	C.Y.	0.011	0.00
Install 4" concrete mud slab	3.000	S.F.	0.120	2.70
Gravel base, 4" thick	3.000	S.F.	0.030	0.36
Install 4" × 8" × 2-1/4" brick pavers	3.000	S.F.	0.435	7.35
Install topsoil, 4" deep	0.500	S.Y.	0.005	0.94
Install sod	0.005	M.S.F.	0.025	0.78
Totals			0.809	13.01

Total per linear foot including general contractor's overhead and profit: $63

Copyright R. S. Means Company, Inc, 1994

5
Modify Existing Pathway

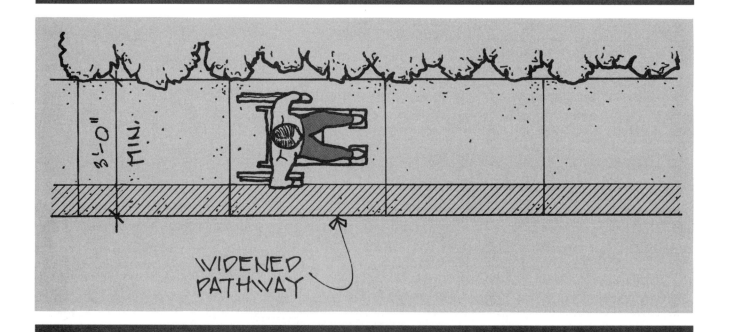

Narrow, uneven, or steeply pitched pathways can be difficult for anyone to use and impassible for people with mobility or visual impairments or who can't maneuver around or see a large crack in the walkway. Fixing an existing path of travel to make it level and smooth is often one of the easiest and most useful modifications that can be made to create an accessible facility.

ADAAG References
4.3 Accessible Route
4.4 Protruding Objects
4.5 Ground and Floor Surfaces

Where Applicable
Wherever an exterior pathway needs to be widened, smoothed, regraded or resurfaced.

Design Requirements
- 36" minimum width.
- Stable, firm, slip-resistant surface.
- Differences in level between 1/4" and 1/2" beveled at maximum slope of 1:2.
- No abrupt changes in level greater than 1/2".
- No grates with openings greater than 1/2" across in the lesser dimension.
- Maximum 5% running slope (1:20).
- Maximum 2% cross-slope (1:50).
- 80" minimum height clearance.
- No objects 27" a.f.f. or higher protruding more than 4" into pathway without warnings below.
- 60" × 60" passing space every 200' if pathway is less than 60" wide.

Design Suggestions
Use 36" width as a minimum. 48" is recommended, and 60" is convenient for passing. By the same token, 5% slope is a maximum unless a ramp is provided. Shallower slopes are easier to use. Avoid unit pavers (e.g., bricks) if possible. If used, install unit pavers on a firm base; such as a firm substrate concrete is preferred even though it is

more expensive. Avoid placing utilities (e.g., drainage gratings) in the path of travel, even if the dimensions comply. Similarly, locate amenities such as phones adjacent to rather than in the path of travel, but within reach from the accessible path.

Key Items
Paving materials, substrate.

Level of Difficulty
Moderate, depending on the material used. Trades involved: paving contractors, masons, landscapers.

Estimates

Widen existing concrete pathway

Description	Quantity	Unit	Work Hours	Material
Excavating	0.037	C.Y.	0.007	0.00
Gravel base, 4" thick	1.000	S.F.	0.010	0.12
Install 4" concrete sidewalk, broom finish	1.000	S.F.	0.040	0.90
Install topsoil, 4" deep	0.110	S.Y.	0.001	0.21
Install sod	0.001	M.S.F.	0.005	0.16
Totals			0.063	1.39

Total per square foot including general contractor's overhead and profit: $6

Widen existing asphalt pathway

Description	Quantity	Unit	Work Hours	Material
Excavating	0.034	C.Y.	0.006	0.00
Gravel base, 4" thick	1.000	S.F.	0.010	0.12
Install 2-1/2" thick asphalt sidewalk	0.110	S.Y.	0.008	0.39
Install topsoil, 4" deep	0.110	S.Y.	0.001	0.21
Install sod	0.001	M.S.F.	0.005	0.16
Totals			0.030	0.88

Total per square foot including general contractor's overhead and profit: $3

Widen existing asphalt block paver pathway

Description	Quantity	Unit	Work Hours	Material
Excavating	0.028	C.Y.	0.005	0.00
Gravel base, 4" thick	1.000	S.F.	0.010	0.12
Install 8" x 8" x 2" asphalt block pavers	1.000	S.F.	0.123	3.50
Install topsoil, 4" deep	0.110	S.Y.	0.001	0.21
Install sod	0.001	M.S.F.	0.005	0.16
Totals			0.144	3.99

Total per square foot including general contractor's overhead and profit: $14

Copyright R. S. Means Company, Inc., 1994

5. Modify Existing Pathway *(continued)*

Widen existing brick paver pathway

Description	Quantity	Unit	Work Hours	Material
Excavating	0.028	C.Y.	0.005	0.00
Gravel base, 4" thick	1.000	S.F.	0.010	0.12
Install 4" × 8" × 2-1/4" brick pavers	1.000	S.F.	0.145	2.45
Install topsoil, 4" deep	0.110	S.Y.	0.001	0.21
Install sod	0.001	M.S.F.	0.005	0.16
Totals			0.166	2.94

Total per square foot including general contractor's overhead and profit — **$13**

Add concrete to correct pathway cross slope

Description	Quantity	Unit	Work Hours	Material
Install concrete walk, up to 4" thick, broom finish	1.000	S.F.	0.040	0.90
Totals			0.040	0.90

Total per square foot including general contractor's overhead and profit — **$3**

Remove concrete pathway, add gravel to correct pathway cross slope, install new concrete pathway

Description	Quantity	Unit	Work Hours	Material
Concrete sidewalk demolition	0.110	S.Y.	0.018	0.00
Gravel base, 4" thick	1.000	S.F.	0.010	0.12
Install concrete walk, up to 4" thick, broom finish	1.000	S.F.	0.040	0.90
Totals			0.068	1.02

Total per square foot including general contractor's overhead and profit — **$5**

Patch existing concrete pathway (4" thick)

Description	Quantity	Unit	Work Hours	Material
Saw cutting, per inch of depth (6 linear feet, 4" deep)	24.000	L.F.	0.408	6.00
Concrete demolition	1.000	S.Y.	0.160	0.00
Install concrete pathway, up to 4" thick, broom finish	9.000	S.F.	0.360	8.10
Totals			0.928	14.10

Total per square foot including general contractor's overhead and profit — **$9**

Total per 3' × 3' section including general contractor's overhead and profit — **$80**

Copyright R. S. Means Company, Inc., 1994

Patch existing asphalt pathway (2-1/2" thick)

Description	Quantity	Unit	Work Hours	Material
Saw cutting (3 L.F. per S.F., 1" deep)	7.500	L.F.	0.158	1.58
Asphalt demolition	0.110	S.Y.	0.008	0.00
Miscellaneous asphalt patching	0.110	S.Y.	0.059	0.16
Totals			0.225	1.74

Total per square yard including general contractor's overhead and profit — **$18**

Patch existing asphalt paver pathway

Description	Quantity	Unit	Work Hours	Material
Remove damaged asphalt block pavers	0.110	S.Y.	0.010	0.00
Install new asphalt block pavers	1.000	S.F.	0.123	3.50
Totals			0.133	3.50

Total per square foot including general contractor's overhead and profit — **$13**

Patch existing brick paver pathway

Description	Quantity	Unit	Work Hours	Material
Remove brick pavers	0.110	S.Y.	0.010	0.00
Install 4" × 8" × 2-1/4" brick pavers	1.000	S.F.	0.145	2.45
Totals			0.155	2.45

Total per square foot including general contractor's overhead and profit — **$12**

Relocate objects in path (e.g., bench, bolted to surface)

Description	Quantity	Unit	Work Hours	Material
Remove bench	1.000	Ea.	1.600	0.00
Re-install bench	1.000	Ea.	3.200	0.00
Totals			4.800	0.00

Total per each including general contractor's overhead and profit — **$208**

Install detectable warning in concrete pathway at traffic crossing or drop off

Description	Quantity	Unit	Work Hours	Material
Saw cutting (3 L.F. per S.F., 1" deep)	12.000	L.F.	0.204	3.00
Concrete demolition	0.110	S.Y.	0.018	0.00
Install 4" concrete sidewalk	1.000	S.F.	0.040	0.90
Install patterned surface finish	1.000	S.F.	0.020	0.00
Totals			0.282	3.90

Total per square foot including general contractor's overhead and profit — **$25**

Copyright R. S. Means Company, Inc, 1994

6 Install Compliant Gratings

A feature as seemingly innocuous as a grating can pose a severe inconvenience, if not an outright safety hazard, to someone with a mobility or visual impairment. Grates with large openings in a path of travel act as a barrier to people who use wheelchairs, canes, walkers, or crutches, or who have balance impairments (as well as high heels). Attention to details like grating location and design helps to create a truly accessible facility.

ADAAG Reference
4.5.4 Gratings

Where Applicable
Any grating located within a common pedestrian route of travel.

Design Requirements
- No openings greater than 1/2" across in the lesser dimension.
- Grates with elongated holes placed such that the long dimension is perpendicular to the direction of travel.
- Top of grate flush with surrounding paving.

Design Suggestions
Wherever possible, locate grates out of the most commonly used routes of travel. If holes are larger than allowed by ADA, an inexpensive, though not ideal solution would be to fill holes in tree grates flush with the surrounding surface, either with gravel or wood chips, to provide a relatively smooth surface (this is useful no matter how big the grate openings are). Storm grates and ventilation grilles have very specific requirements for the amount of open area, which need to be considered and perhaps re-sized prior to replacement with an accessible grating.

Key Items
Metal gratings.

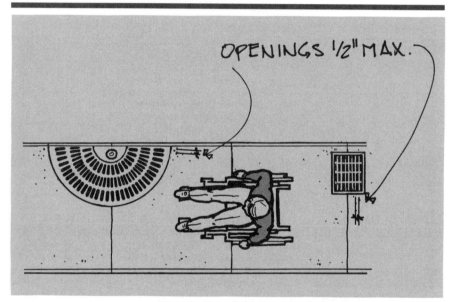

Level of Difficulty

Low to moderate. Existing gratings can sometimes be replaced by facility maintenance crews. A manufacturer's representative, civil or structural engineer may want to verify opening dimension(s). New gratings are installed by a contractor.

Estimates

Replace trench drain

Description	Quantity	Unit	Work Hours	Material
Remove trench drains	1.000	Ea.	1.000	0.00
Install new trench drains, modular, 12" x 12"	1.000	Ea.	2.000	150.00
Totals			3.000	150.00

Total per linear foot of 1 foot wide trench drain including general contractor's overhead and profit: $412

Replace 5' diameter cast iron tree grate

Description	Quantity	Unit	Work Hours	Material
Remove tree grate	1.000	Ea.	0.480	0.00
Install cast iron tree grate, 5' diameter	1.000	Ea.	0.960	285.00
Totals			1.440	285.00

Total per each including general contractor's overhead and profit: $575

Replace catch basin frame and cover

Description	Quantity	Unit	Work Hours	Material
Remove catch basin frame and cover	1.000	Ea.	1.846	0.00
Install new catch basin frame and cover	1.000	Ea.	3.077	176.00
Totals			4.923	176.00

Total per each including general contractor's overhead and profit: $588

Copyright R. S. Means Company, Inc, 1994

7
Install or Modify Curb Cuts

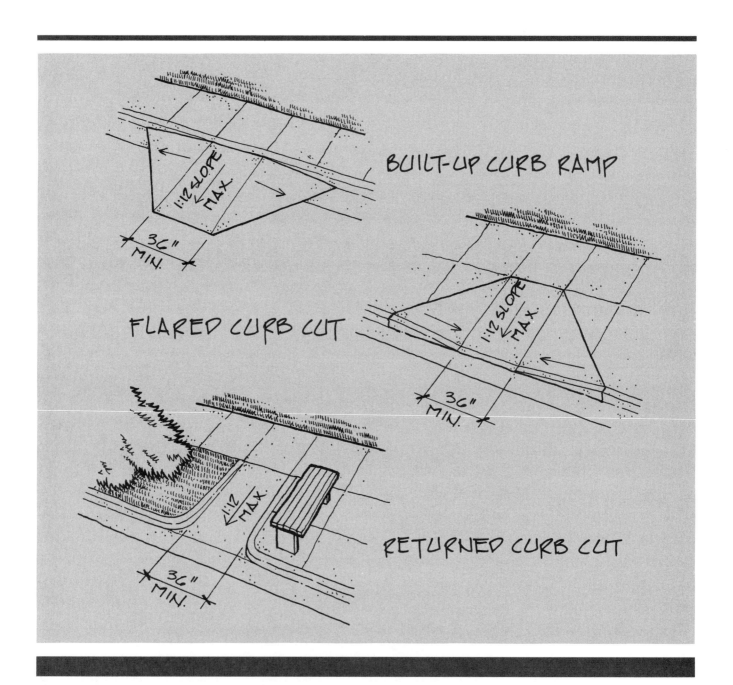

There is probably no other example of an access modification that is so vital both to people with mobility impairments and all other users. To someone using a wheelchair, a curb without a curb cut is as difficult to use as a curb four feet high would be to a walking person. Curb cuts also remove a barrier for people with baby strollers or shopping carts and for those on bicycles. Proper design and location of curb cuts is essential in creating an exterior-accessible route of travel.

ADAAG Reference

4.7 Curb Ramps

(Note: ADAAG uses the term "curb ramp" to describe any sloped surface used to create access at a curb. Common terminology is to call a slope cut into a pathway a curb cut, and a slope added to a pathway a curb ramp. We have used the accepted construction term here, so ADAAG 4.7 Curb Ramps refers to the curb cuts described below.)

Where Applicable

Wherever an accessible pedestrian route crosses a curb with a vertical drop.

Design Requirements

- 36" minimum width.
- Slip-resistant, stable, firm surface.
- Slope no greater than 1:12, with a 1:10 slope maximum for flared sides.
- Returned sides allowed only where pedestrians would not normally walk across the curb cut.
- Lip flush with street paving.
- Location within crosswalk (if crosswalk provided), not obstructed by vehicles.
- Diagonal curb cuts at corners: 48" clear space at base, 24" curb length on each side within crosswalk.
- No obstructions, such as utility poles, within curb cut blocking accessible approach.

Design Suggestions

When installing cuts on a street corner, two per corner are recommended. This allows direct crossing of the street without entering traffic, which can be a danger with a single curb cut located diagonally on the corner. It is also important to ensure adequate drainage, since the base of a curb cut is a natural low point. Only slip-resistant materials should be used, such as thermal finished stone, broom-finished concrete, or unglazed pavers. Concrete or asphalt are the recommended paving materials. It helps if the curb cut paving material is different from the sidewalk; the concrete can be stained (rather than painted) to provide visual contrast. If pavers are used for the curb cut, they should be installed over a concrete base to minimize future heaving. The paving material can be scored perpendicular to the route of travel to create additional slip-resistance. This is difficult with asphalt or stone, but easier with concrete. (Concrete must, however, be maintained to prevent cracking).

Key Items

Paving materials.

Level of Difficulty

Moderate to high. May require heavy demolition. Must ensure no disruption of under-street utilities. Trades involved: paving contractor.

Estimates

Install new curb cut (concrete sidewalk)

Description	Quantity	Unit	Work Hours	Material
Saw cutting, conc. per inch of depth (20 L.F., 4" deep)	80.000	L.F.	1.360	20.00
Site demolition, remove stone curb	14.000	L.F.	0.952	0.00
Site demolition, remove concrete paving	5.340	S.Y.	0.838	0.00
Site demolition, remove asphalt paving	1.670	S.Y.	0.124	0.00
Install granite curbing, 6" x 18"	14.000	L.F.	1.736	235.20
Install gravel base, 4" thick	48.000	S.F.	0.480	5.76
Install concrete walk, 4" thick, broom finish	48.000	S.F.	1.920	43.20
Miscellaneous asphalt patching	1.670	S.Y.	0.890	2.45
Totals			8.300	306.61

Total per each including general contractor's overhead and profit: $1,033

Copyright R. S. Means Company, Inc., 1994

7. Install or Modify Curb Cuts (continued)

Install new curb cut (asphalt sidewalk)

Description	Quantity	Unit	Work Hours	Material
Saw cutting, asphalt	80.000	L.F.	1.680	16.80
Site demolition, remove stone curb	14.000	L.F.	0.952	0.00
Site demolition, remove asphalt paving	7.000	S.Y.	0.518	0.00
Install granite curbing, 6" x 18"	14.000	L.F.	1.736	235.20
Install gravel base, 4" thick	48.000	S.F.	0.480	5.76
Install 2-1/2" asphalt sidewalk	5.340	S.Y.	0.390	19.17
Miscellaneous asphalt patching	1.670	S.Y.	0.890	2.45
Totals			6.646	279.38

Total per each including general contractor's overhead and profit — **$879**

Install new curb cut (brick paver sidewalk on concrete base)

Description	Quantity	Unit	Work Hours	Material
Remove brick pavers	9.000	S.Y.	0.801	0.00
Site demolition, remove concrete base	6.000	S.Y.	0.960	0.00
Site demolition, remove stone curb	14.000	L.F.	0.952	0.00
Site demolition, remove asphalt paving	1.670	S.Y.	0.124	0.00
Install granite curbing, 6" x 18"	14.000	L.F.	1.736	235.20
Install 4" concrete mud slab	81.000	S.F.	3.240	72.90
Install 4" x 8" x 2-1/4" brick pavers	81.000	S.F.	11.745	198.45
Miscellaneous asphalt patching	1.670	S.Y.	0.890	2.45
Totals			20.448	509.00

Total per each including general contractor's overhead and profit — **$1,921**

Install new curb cut (brick paver sidewalk on stone dust base)

Description	Quantity	Unit	Work Hours	Material
Remove brick pavers	9.000	S.Y.	0.801	0.00
Hand excavation of paver base	0.450	C.Y.	0.900	0.00
Site demolition, remove stone curb	14.000	L.F.	0.952	0.00
Site demolition, remove asphalt	1.670	S.Y.	0.124	0.00
Install granite curbing, 6" x 18"	14.000	L.F.	1.736	235.20
Install stone dust base	6.000	S.Y.	0.060	12.48
Install 4" x 8" x 2-1/4" brick pavers	81.000	S.F.	11.745	198.45
Miscellaneous asphalt patching	1.670	S.Y.	0.890	2.45
Totals			17.208	448.58

Total per each including general contractor's overhead and profit — **$1,649**

Copyright R. S. Means Company, Inc., 1994

Patch lip at base of curb cut (asphalt paving)

Description	Quantity	Unit	Work Hours	Material
Site demolition, remove asphalt	1.670	S.Y.	0.124	0.00
Miscellaneous asphalt patching	1.670	S.Y.	0.890	2.45
Totals			1.014	2.45

Total per each including general contractor's overhead and profit: $55

Install flared sides, existing curb cut (concrete sidewalk)

Description	Quantity	Unit	Work Hours	Material
Saw cutting, conc. per inch of depth (16 L.F., 4" deep)	64.000	L.F.	1.088	16.00
Site demolition, remove stone curb	25.000	L.F.	1.700	0.00
Site demolition, remove concrete paving	3.340	S.Y.	0.524	0.00
Site demolition, remove asphalt paving	1.670	S.Y.	0.124	0.00
Miscellaneous fill at curb removal	12.500	S.F.	0.100	4.38
Install granite curbing, 6" x 18"	14.000	L.F.	1.736	235.20
Install gravel base, 4" thick	30.000	S.F.	0.300	3.60
Install concrete walk, 4" thick, broom finish	30.000	S.F.	1.200	27.00
Miscellaneous asphalt patching	1.670	S.Y.	0.890	2.45
Totals			7.662	288.63

Total per each including general contractor's overhead and profit: $958

Install flared sides, existing curb cut (asphalt sidewalk)

Description	Quantity	Unit	Work Hours	Material
Saw cutting, asphalt	64.000	L.F.	1.344	13.44
Site demolition, remove stone curb	25.000	L.F.	1.700	0.00
Site demolition, remove asphalt paving	5.000	S.Y.	0.370	0.00
Miscellaneous fill at curb removal	12.500	S.F.	0.100	4.38
Install granite curbing, 6" x 18"	14.000	L.F.	1.736	235.20
Install gravel base, 4" thick	30.000	S.F.	0.300	3.60
Install 2-1/2" asphalt sidewalk	3.340	S.Y.	0.244	11.99
Miscellaneous asphalt patching	1.670	S.Y.	0.890	2.45
Totals			6.684	271.06

Total per each including general contractor's overhead and profit: $865

Install new curb ramp

Description	Quantity	Unit	Work Hours	Material
Install conc. curb ramp, to 6" thick, broom finish	54.000	S.F.	2.538	67.50
Totals			2.538	67.50

Total per each including general contractor's overhead and profit: $247

Copyright R. S. Means Company, Inc, 1994

8 Construct New Ramp: Straight

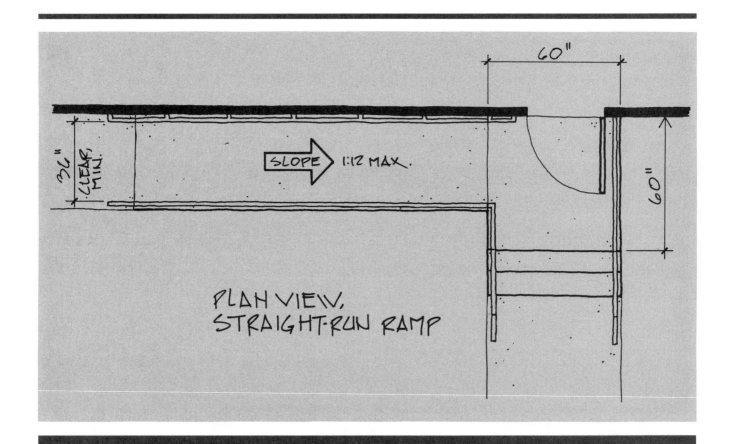

Ramps are the most common method of creating an accessible route of travel when bridging a height difference up to several feet. A straight ramp is the easiest to use since it requires no turns by the user. Straight ramps, like all ramps, are accessible not only to people who use wheelchairs, but also to people with baby strollers, shopping carts, bicycles, and almost any other vehicle with wheels that doesn't use gas.

ADAAG Reference
4.8 Ramps

Where Applicable
Public entrances and routes of travel where steps or different levels exist which are being made accessible and a ramp is installed with no turns.

Design Requirements
- 36" minimum clear width between handrails or curbs.
- Maximum slope 1:12, or one inch of rise for every 12 inches of run. (Any slope between 1:20 and 1:12 is a ramp).
- Base flush with adjacent paving.
- A level landing at least every 30" of rise.

- Slip-resistant surface.
- 60" level area at base, all landings, and top, as wide as the ramp; 60" × 60" minimum landing wherever ramp changes direction.
- Handrails on both sides, 1-1/4" to 1-1/2" diameter, continuous, round or oval, 12" level extensions top and bottom, 34"–38" above ramp surface, 1-1/2" from the wall, ends returned to a post or wall.
- Curbs or rails at any drop-offs on both sides of the ramp.
- No pooling of water on ramp surface.

Design Suggestions

Straight-run ramps: A straight-run ramp is efficient for use since it requires no turns. A very long straight-run ramp can be a strong visual element, however, and if this is a consideration, an alternate configuration can help minimize the impact.

General Ramp Design: When making an entrance accessible, it should be determined if ramps are necessary. In new construction or additions, floor heights can be established so that neither steps nor a ramp are necessary at entrances. Graded pathways at a slope of 1:20 or less might be possible. If a ramp is necessary, choose the location carefully. Always look to first making the main or public entrance accessible. If that is not possible, another entrance may be modified and proper signage installed to direct people to the accessible entrance. The 1:12 slope cited in ADAAG is a *maximum*, so even if a graded pathway is not possible, a ramp with a shallower slope might be. A ramped entrance or route should also have stairs adjacent to it. This integrates the ramp into the common route of travel, and stairs are easier for some people to use.

It helps acceptance and use of the ramp if the ramp is integrated into the existing building aesthetic, rather than constructed of completely different materials and details. Plantings or other landscaping can help integrate the ramp into the surrounding environment. All materials must be slip-resistant, a critical issue for wood ramps, which can be slippery when wet. Even though it may be expensive, consider the possibility of covering the ramp for weather protection. This helps keep the ramp free of water, snow, ice, etc., and makes maintenance easier. Detail railings for continuity (Under-rail brackets work best; be sure that there is 1-1/2" clearance between the rail and the wall or other surface). Above all, remember that ramps get used by people with baby strollers, shopping carts, hand trucks, etc., and are a helpful addition to an existing facility.

Key Items

- Ramp material: painted wood, treated lumber, concrete.
- Rails: pipe rails, metal rails, wood banisters or rails, uprights, attachments.
- Possibly requires slip-resistant applied materials: sand paint, sandpaper strips.

Level of Difficulty

Moderate to high. A wood ramp requires skilled carpentry. Concrete requires foundation work. Can require design drawings and building permit.

8. Construct New Ramp: Straight (continued)

Estimates

Install new painted wood straight ramp

Description	Quantity	Unit	Work Hours	Material
Hand excavating for post footings	3.000	C.Y.	6.000	0.00
Conc. forms, 12" dia. tubes (16 forms, 4' deep)	64.000	L.F.	13.632	201.60
Hand backfilling around post footings	1.000	C.Y.	0.727	0.00
Concrete, material only	2.000	C.Y.	0.000	110.00
Place concrete for footings, direct chute	2.000	C.Y.	1.746	0.00
Install two-piece galvanized steel post foot	16.000	Ea.	0.992	52.80
4" × 4" post framing	0.172	M.B.F.	5.292	176.30
2" × 8" joist framing	0.384	M.B.F.	4.208	220.80
1" × 8" board decking	272.000	SF Flr.	4.352	239.36
2" × 6" railing enclosure	544.000	L.F.	7.072	299.20
Handrail stock	136.000	L.F.	13.600	122.40
Drilling bolt holes	408.000	Inch	7.344	0.00
Bolts, nuts & washers	102.000	Ea.	5.814	88.74
Handrail bracket	34.000	Ea.	8.500	544.00
Anchor layout and drilling, per inch of depth	8.000	Ea.	1.280	0.96
1/2" anchors	4.000	Ea.	0.400	7.12
Painting (2 coats)	2160.000	L.F.	23.760	345.60
Totals			104.719	2408.88

Total per linear foot including general contractor's overhead and profit — **$159**

Total per each 60-foot ramp including general contractor's overhead and profit — **$9,515**

Copyright R. S. Means Company, Inc., 1994

Install new pressure-treated wood straight ramp

Description	Quantity	Unit	Work Hours	Material
Hand excavating for post footings	3.000	C.Y.	6.000	0.00
Conc. forms, 12" diameter tubes (16 forms, 4' deep)	64.000	L.F.	13.632	201.60
Hand backfilling around post footings	1.000	C.Y.	0.727	0.00
Place concrete for footings, direct chute	2.000	C.Y.	1.746	0.00
Concrete, material only	2.000	C.Y.	0.000	110.00
Install two-piece galvanized steel post foot	16.000	Ea.	0.992	52.80
4" × 4" post framing, w/ pressure-treated lumber	0.172	M.B.F.	5.292	245.10
2" × 8" joist framing, w/ pressure-treated lumber	0.384	M.B.F.	4.208	309.12
1" × 8" board decking	272.000	SF Flr.	4.352	320.96
2" × 6" railing enclosure	544.000	L.F.	7.072	402.56
Handrail stock	136.000	L.F.	13.600	122.40
Drilling bolt holes	408.000	Inch	7.344	0.00
Bolts, nuts & washers	102.000	Ea.	5.814	88.74
Handrail bracket	34.000	Ea.	8.500	544.00
Anchor layout and drilling, per inch of depth	8.000	Ea.	1.280	0.96
1/2" anchors	4.000	Ea.	0.400	7.12
Totals			80.959	2405.36

Total per linear foot including general contractor's overhead and profit: $140

Total per each 60-foot ramp including general contractor's overhead and profit: $8,383

Install new concrete straight ramp

Description	Quantity	Unit	Work Hours	Material
Excavation	90.000	C.Y.	13.320	0.00
Concrete forms, footings	272.000	SFCA	17.952	103.36
Concrete forms, walls	1824.000	SFCA	89.376	547.20
Reinforcing (@ 50 lbs./C.Y.)	0.825	Ton	8.800	400.13
Place concrete, direct chute	33.000	C.Y.	17.589	0.00
Concrete, material only	33.000	C.Y.	0.000	1815.00
Backfilling	67.000	C.Y.	9.916	0.00
Gravel under slab	5.000	C.Y.	0.235	17.50
Slab for ramp	5.000	C.Y.	4.380	287.50
Aluminum pipe railing (2 rail)	136.000	L.F.	27.200	1645.60
Totals			188.768	4816.29

Total per linear foot including general contractor's overhead and profit: $329

Total per each 60-foot ramp including general contractor's overhead and profit: $19,766

Copyright R. S. Means Company, Inc, 1994

9 Construct New Ramp: Switch-Back

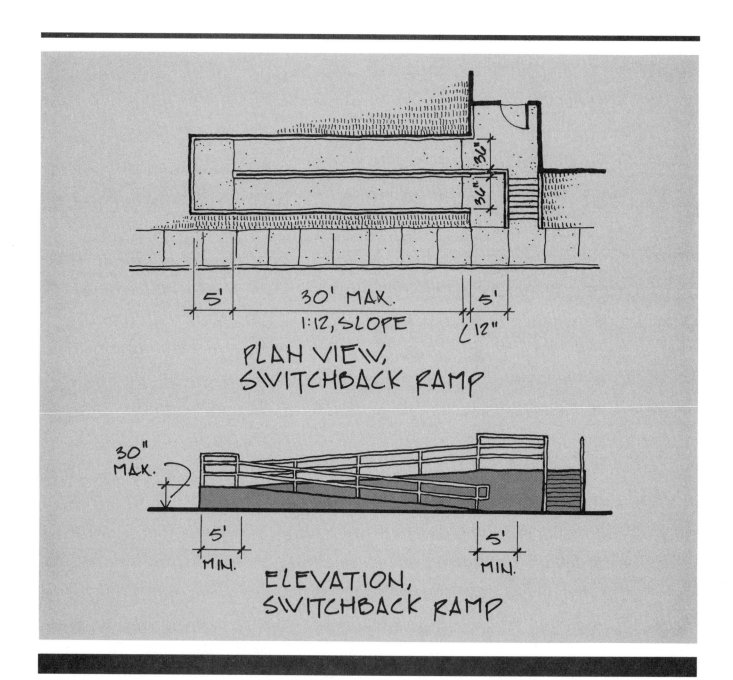

By adding additional ramp length, a switch-back ramp increases the potential height of a ramp. It also enables a ramp to fit in a tighter area than a straight-run or dog-leg ramp, and can be useful in situations where there are obstructions or a lack of open land.

ADAAG Reference

4.8 Ramps

Where Applicable

Public entrances and routes of travel where steps or different levels exist which are being made accessible and a ramp is installed with one leg parallel to the other (or where more than two legs of a ramp are parallel to each other).

Design Requirements

- 36" minimum clear width between handrails or curbs.
- Maximum slope 1:12, or one inch of rise for every 12 inches of run. (Any slope between 1:20 and 1:12 is a ramp).
- Base flush with adjacent paving.
- A level landing at least every 30" of rise.
- Slip-resistant surface.
- 60" level area at base, all landings, and top, as wide as the ramp; 60" × 60" minimum landing wherever ramp changes direction.
- Handrails on both sides, 1-1/4" to 1-1/2" diameter, continuous, round or oval, 12" level extensions top and bottom, 34"–38" above ramp surface, 1-1/2" from the wall, ends returned to a post or wall.
- Curbs or rails at any drop-offs on both sides of ramp.
- No pooling of water on ramp surface.

Design Suggestions

Switch-back ramps: Switch-back ramps can be installed in areas where a straight-run or dog-leg ramp won't fit. As with dog-leg ramps, switch-back ramps require at least one turn, so if possible the ramp can be widened to the width of the landing to make it easier to use. If a switch-back ramp has multiple turns, care should be taken to ensure that the ramp entrance is on an accessible route (preferably the main route of travel) and is clearly visible.

General Ramp Design: When making an entrance accessible, it should be determined if ramps are necessary. In new construction or additions, floor heights can be established so that neither steps nor a ramp are necessary at entrances. Graded pathways at a slope of 1:20 or less might be possible. If a ramp is necessary, choose the location carefully. Always look to first making the main or public entrance accessible. If that is not possible, another entrance may be modified and proper signage installed to direct people to the accessible entrance. The 1:12 slope cited in ADAAG is a *maximum*, so even if a graded pathway is not possible, a ramp with a shallower slope might be. A ramped entrance or route should also have stairs adjacent to it. This integrates the ramp into the common route of travel, and stairs are easier for some people to use.

It helps acceptance and use of the ramp if the ramp is integrated into the existing building aesthetic, rather than constructed of completely different materials and details. Plantings or other landscaping can help integrate the ramp into the surrounding environment. All materials must be slip-resistant, a critical issue for wood ramps, which can be slippery when wet. Even though it may be expensive, consider the possibility of covering the ramp for weather protection. This helps keep the ramp free of water, snow, ice, etc., and makes maintenance easier. Detail railings for continuity (Under-rail brackets work best; be sure that there is 1-1/2" clearance between the rail and the wall or other surface). Above all, remember that ramps get used by people with baby strollers, shopping carts, hand trucks, etc., and are a helpful addition to an existing facility.

Key Items

- Ramp material: painted wood, treated lumber, concrete.
- Rails: pipe rails, metal rails, wood banisters or rails, uprights, attachments.
- Possibly requires slip-resistant surfaces applied materials: sand paint, sandpaper strips.

Level of Difficulty

Moderate to high. A wood ramp requires skilled carpentry. Concrete requires foundation work. Can require design drawings and building permit.

9. Construct New Ramp: Switch-Back (continued)

Estimates

Install new painted wood switch-back ramp

Description	Quantity	Unit	Work Hours	Material
Hand excavating for post footings	3.000	C.Y.	6.000	0.00
Conc. forms, 12" diameter tubes (16 forms, 4' deep)	52.000	L.F.	11.076	163.80
Hand backfilling around post footings	1.000	C.Y.	0.727	0.00
Place concrete for footings, direct chute	2.000	C.Y.	1.746	0.00
Concrete, material only	2.000	C.Y.	0.000	110.00
Install two-piece galvanized steel post foot	16.000	Ea.	0.992	52.80
4" × 4" post framing	0.140	M.B.F.	4.308	143.50
2" × 8" joist framing	0.486	M.B.F.	5.326	279.45
1" × 8" board decking	312.000	SF Flr.	4.992	274.56
Handrail stock	144.000	L.F.	14.400	129.60
Drilling bolt holes	432.000	Inch	7.776	0.00
Bolts, nuts & washers	108.000	Ea.	6.156	93.96
Handrail bracket	36.000	Ea.	9.000	576.00
2" × 6" railing enclosure	564.000	L.F.	7.332	310.20
Anchor layout and drilling, per inch of depth	16.000	Ea.	2.560	1.92
1/2" anchors	8.000	Ea.	0.800	14.24
Painting (2 coats)	2325.000	L.F.	25.575	372.00
Totals			108.766	2522.03

Total per linear foot including general contractor's overhead and profit	**$165**
Total per each 60-foot ramp including general contractor's overhead and profit	**$9,921**

Copyright R. S. Means Company, Inc., 1994

Install new pressure-treated wood switch-back ramp

Description	Quantity	Unit	Work Hours	Material
Hand excavating for post footings	3.000	C.Y.	6.000	0.00
Conc. forms, 12" diameter tubes (13 forms, 4' deep)	52.000	L.F.	11.076	163.80
Hand backfilling around post footings	1.000	C.Y.	0.727	0.00
Place concrete for footings, direct chute	2.000	C.Y.	1.746	0.00
Concrete, material only	2.000	C.Y.	0.000	110.00
Install two-piece galvanized steel post foot	13.000	Ea.	0.806	42.90
4" x 4" post framing, w/pressure-treated lumber	0.140	M.B.F.	4.308	199.50
2" x 8" joist framing, w/ pressure-treated lumber	0.486	M.B.F.	5.326	391.23
1" x 8" board decking	312.000	SF Flr.	4.992	368.16
Handrail stock	144.000	L.F.	14.400	129.60
Drilling bolt holes	432.000	Inch	7.776	0.00
Bolts, nuts & washers	108.000	Ea.	6.156	93.96
Handrail bracket	36.000	Ea.	9.000	576.00
2" x 6" railing enclosure	564.000	L.F.	7.332	417.36
Anchor layout and drilling, per inch of depth	16.000	Ea.	2.560	1.92
1/2" anchors	8.000	Ea.	0.800	14.24
Totals			83.005	2508.67

Total per linear foot including general contractor's overhead and profit	**$145**
Total per each 60-foot ramp including general contractor's overhead and profit	**$8,671**

Install new concrete switch-back ramp

Description	Quantity	Unit	Work Hours	Material
Excavation	60.000	C.Y.	8.880	0.00
Concrete forms, footings	248.000	SFCA	16.368	94.24
Concrete forms, walls	1752.000	SFCA	85.848	525.60
Reinforcing (@ 50 lbs./C.Y.)	0.800	Ton	8.534	388.00
Place concrete, direct chute	32.000	C.Y.	17.056	0.00
Concrete, material only	32.000	C.Y.	0.000	1760.00
Backfilling	38.000	C.Y.	5.624	0.00
Gravel under slab	6.000	C.Y.	0.282	21.00
Slab for ramp, 4" thick	6.000	C.Y.	5.256	345.00
Aluminum pipe rail (2 rail)	150.000	L.F.	30.000	1815.00
Totals			177.848	4948.84

Total per linear foot including general contractor's overhead and profit	**$317**
Total per each 60-foot ramp including general contractor's overhead and profit	**$19,047**

Copyright R. S. Means Company, Inc, 1994

10 Construct New Ramp: Dog-Leg

In addition to all the standard advantages of a ramp, a dog-leg ramp can double the height of a straight-run ramp, and can be a useful configuration for circumventing an obstruction (like a tree). They may also be installed at the corner of a building.

ADAAG Reference
4.8 Ramps

Where Applicable
Public entrances and routes of travel where steps or different levels exist which are being made accessible and a ramp is installed with one leg perpendicular to the other (or where more than two legs of a ramp are perpendicular to each other).

Design Requirements
- 36" minimum clear width between handrails or curbs.
- Maximum slope 1:12, or one inch of rise for every 12 inches of run. (Any slope between 1:20 and 1:12 is a ramp).
- Base flush with adjacent paving.
- A level landing for at least every 30" of rise.
- Slip-resistant surface.
- 60" level area at base, all landings, and top, as wide as the ramp; 60" × 60" minimum landing wherever ramp changes direction.
- Handrails on both sides, 1-1/4" to 1-1/2" diameter, continuous, round or oval, 12" level extensions top and bottom, 34"–38" above ramp surface, 1-1/2" from the wall, ends returned to a post or wall.
- Curbs or rails at any drop-offs on both sides of ramp.
- No pooling of water on ramp surface.

Design Suggestions
Dog-leg ramps: Dog-leg ramps require at least one turn, which requires a landing larger than the ramp to comply with ADA. This can be awkward visually, so if possible the ramp can be widened to the width of the landing. This will also make it easier to use. Dog-leg ramps are often installed

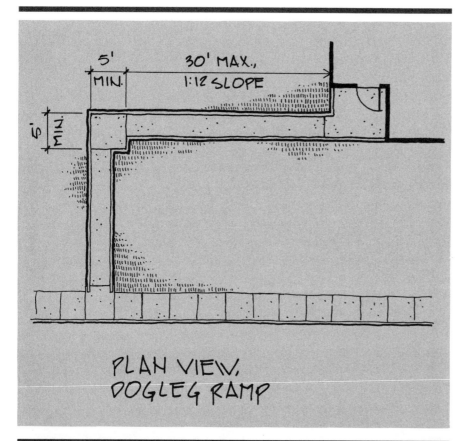

PLAN VIEW, DOGLEG RAMP

around the corner of a building, but if this locates the base of the ramp away from the main approach (not a desirable solution), clear signage is needed to direct people to the ramp entrance. It is always desirable to locate the ramp entry along the typical route of travel to the facility.

General Ramp Design: When making an entrance accessible, it should be determined if ramps are necessary. In new construction or additions, floor heights can be established so that neither steps nor a ramp are necessary at entrances. Graded pathways at a slope of 1:20 or less might be possible. If a ramp is necessary, choose the location carefully. Always look to first making the main or public entrance accessible. If that is not possible, another entrance may be modified and proper signage installed to direct people to the accessible entrance. The 1:12 slope cited in ADAAG is a *maximum*, so even if a graded pathway is not possible, a ramp with a shallower slope might be. A ramped entrance or route should also have stairs. This integrates the ramp into the common route of travel, and stairs are easier for some people to use.

It helps acceptance and use of the ramp if the ramp is integrated into the existing building aesthetic, rather than constructed of completely different materials and details. Plantings or other landscaping can help integrate the ramp into the surrounding environment. All materials must be slip-resistant, a critical issue for wood ramps, which can be slippery when wet. Even though it may be expensive, consider the possibility of covering the ramp for weather protection. This helps keep the ramp free of water, snow, ice, etc., and makes maintenance easier. Detail railings for continuity (Under-rail brackets work best; be sure that there is 1-1/2" clearance between the rail and the wall or other surface). Above all, remember that ramps get used by people with baby strollers, shopping carts, hand trucks, etc., and are a helpful addition to an existing facility.

Key Items

- Ramp material: painted wood, treated lumber, concrete.
- Rails: pipe rails, metal rails, wood banisters or rails, uprights, attachments.
- Possibly requires slip-resistant surfaces applied materials: sand paint, sandpaper strips.

Level of Difficulty

Moderate to high. A wood ramp requires skilled carpentry. Concrete requires foundation work. Can require design drawings and building permit.

10. Construct New Ramp: Dog-Leg (continued)

Estimates

Install new painted wood dog-leg ramp

Description	Quantity	Unit	Work Hours	Material
Hand excavating for post footings	3.000	C.Y.	6.000	0.00
Conc. forms, 12" diameter tubes (16 forms, 4' deep)	64.000	L.F.	13.632	201.60
Hand backfilling around post footings	1.000	C.Y.	0.727	0.00
Place concrete for footings, direct chute	2.000	C.Y.	1.746	0.00
Concrete, material only	2.000	C.Y.	0.000	110.00
Install two-piece galvanized steel post foot	16.000	Ea.	0.992	52.80
4" × 4" post framing	0.172	M.B.F.	5.292	176.30
2" × 8" joist framing	0.384	M.B.F.	4.208	220.80
1" × 8" board decking	272.000	SF Flr.	4.352	239.36
Handrail stock	136.000	L.F.	13.600	122.40
Drilling bolt holes	408.000	Inch	7.344	0.00
Bolts, nuts & washers	102.000	Ea.	5.814	88.74
Handrail bracket	34.000	Ea.	8.500	544.00
2" × 6" railing enclosure	544.000	L.F.	7.072	299.20
Anchor layout and drilling, per inch of depth	8.000	Ea.	1.280	0.96
1/2" anchors	4.000	Ea.	0.400	7.12
Painting (2 coats)	2160.000	L.F.	23.760	345.60
Totals			104.719	2408.88

Total per linear foot including general contractor's overhead and profit	**$159**
Total per each 60-foot ramp including general contractor's overhead and profit	**$9,515**

Copyright R. S. Means Company, Inc., 1994

Install new pressure-treated wood dog-leg ramp

Description	Quantity	Unit	Work Hours	Material
Hand excavating for post footings	3.000	C.Y.	6.000	0.00
Conc. forms, 12" diameter tubes (16 forms, 4' deep)	64.000	L.F.	13.632	201.60
Hand backfilling around post footings	1.000	C.Y.	0.727	0.00
Place concrete for footings, direct chute	2.000	C.Y.	1.746	0.00
Concrete, material only	2.000	C.Y.	0.000	110.00
Install two-piece galvanized steel post foot	16.000	Ea.	0.992	52.80
4" × 4" post framing, w/ pressure-treated lumber	0.172	M.B.F.	5.292	245.10
2" × 8" joist framing, w/ pressure-treated lumber	0.384	M.B.F.	4.208	309.12
1" × 8" board decking	272.000	SF Flr.	4.352	320.96
Handrail stock	136.000	L.F.	13.600	122.40
Drilling bolt holes	108.000	Inch	1.944	0.00
Bolts, nuts & washers	102.000	Ea.	5.814	88.74
Handrail bracket	34.000	Ea.	8.500	544.00
2" × 6" railing enclosure	544.000	L.F.	7.072	402.56
Anchor layout and drilling, per inch of depth	4.000	Ea.	0.640	0.48
1/2" anchors	4.000	Ea.	0.400	7.12
Totals			74.919	2404.88

Total per linear foot including general contractor's overhead and profit **$134**

Total per each 60-foot ramp including general contractor's overhead and profit **$8,062**

Install new concrete dog-leg ramp

Description	Quantity	Unit	Work Hours	Material
Excavation	90.000	C.Y.	13.320	0.00
Concrete forms, footings	272.000	SFCA	17.952	103.36
Concrete forms, walls	1824.000	SFCA	89.376	547.20
Reinforcing (@ 50 lbs./C.Y.)	0.825	Ton	8.800	400.13
Place concrete, direct chute	33.000	C.Y.	17.589	0.00
Concrete, material only	33.000	C.Y.	0.000	1815.00
Backfilling	67.000	C.Y.	9.916	0.00
Gravel under slab	5.000	C.Y.	0.235	17.50
Slab for ramp, 4" thick	5.000	C.Y.	4.380	287.50
Aluminum pipe rail (2 rail)	136.000	L.F.	27.200	1645.60
Totals			188.768	4816.29

Total per linear foot including general contractor's overhead and profit **$329**

Total per each 60-foot ramp including general contractor's overhead and profit **$19,766**

Copyright R. S. Means Company, Inc, 1994

11
Construct New Ramp: Below Grade

Many existing buildings were constructed with the first floor level above grade level to allow for drainage, snow buildup, or because it was always done that way. Some buildings, however, have an entrance below grade. Those (with proper drainage) can often be made accessible with a ramp that connects to the accessible route from the street or parking lot.

ADAAG Reference

4.8 Ramps

Where Applicable

Public entrances and routes of travel where steps or different levels exist which are being made accessible and a ramp is installed below existing ground level.

Design Requirements

- 36" minimum clear width between handrails or curbs.
- Maximum slope 1:12, or one inch of rise for every 12 inches of run. (Any slope between 1:20 and 1:12 is a ramp).
- Base flush with adjacent paving.
- A level landing for at least every 30" of rise.
- Slip-resistant surface.
- 60" level area at base, all landings, and top, as wide as the ramp; 60" × 60" minimum landing wherever ramp changes direction.
- Handrails on both sides, 1-1/4" to 1-1/2" diameter, continuous, round or oval, 12" level extensions top and bottom, 34"–38" above ramp surface, 1-1/2" from the wall, ends returned to a post or wall.
- Curbs or rails at any drop-offs on both sides of ramp.
- No pooling of water on ramp surface.

Design Suggestions

Below-grade ramps: Below-grade ramps need to comply with all the design requirements and suggestions as other ramps. If the existing entrance area is not large enough to accommodate a ramp, excavation will be needed, along with retaining walls and drainage; otherwise, water will drain to the base of the ramp. Even if excavation is needed, a ramp should be considered rather than a lift which would require maintenance (excavation will also require some landscaping). Buildings often have entrances on different levels; if a building's main entrance is below grade, but a secondary entrance is on grade, the secondary entrance can be used as a compliant entrance if it is not a service entrance (such as a loading dock), is along an accessible route, leads to an accessible route, and is clearly located with signage at all entrances.

General Ramp Design: When making an entrance accessible, it should be determined if ramps are necessary. In new construction or additions, floor heights can be established so that neither steps nor a ramp are necessary at entrances. Graded pathways at a slope of 1:20 or less might be possible. If a ramp is necessary, choose the location carefully. Always look to first making the main or public entrance accessible. If that is not possible, another entrance may be modified and proper signage installed to direct people to the accessible entrance. The 1:12 slope cited in ADAAG is a *maximum*, so even if a graded pathway is not possible, a ramp with a shallower slope might be. A ramped entrance or route should also have stairs. This integrates the ramp into the common route of travel, and stairs are easier for some people to use.

It helps acceptance and use of the ramp if the ramp is integrated into the existing building aesthetic, rather than constructed of completely different materials and details. Plantings or other landscaping can help integrate the ramp into the surrounding environment. All materials must be slip-resistant, a critical issue for wood ramps, which can be slippery when wet. Even though it may be expensive, consider the possibility of covering the ramp for weather protection. This helps keep the ramp free of water, snow, ice, etc., and makes maintenance easier. Detail railings for continuity (Under-rail brackets work best; be sure that there is 1-1/2" clearance between the rail and the wall or other surface). Above all, remember that ramps get used by people with baby strollers, shopping carts, hand trucks, etc., and are a helpful addition to an existing facility.

Key Items

- Ramp material: painted wood, treated lumber, concrete.
- Rails: pipe rails, metal rails, wood banisters or rails, uprights, attachments.
- Possibly requires slip-resistant surfaces applied materials: sand paint, sandpaper strips.

Level of Difficulty

Moderate to high. A wood ramp requires skilled carpentry. Concrete requires foundation work. Possible excavation work involved. Can require design drawings and building permit.

11. Construct New Ramp: Below Grade (continued)

Estimate

Install new concrete below-grade switch-back ramp

Description	Quantity	Unit	Work Hours	Material
Excavation	115.000	C.Y.	17.020	0.00
Remove concrete wall at existing stairs	64.000	S.F.	21.312	0.00
Concrete forms, footings	240.000	SFCA	15.840	91.20
Concrete forms, walls	1440.000	SFCA	70.560	432.00
Reinforcing (@ 50 lbs./C.Y.)	0.600	Ton	6.400	291.00
Place concrete, direct chute	24.000	C.Y.	12.792	0.00
Concrete, material only	24.000	C.Y.	0.000	1320.00
2 coat bituminous dampproofing	360.000	S.F.	5.760	32.40
Backfilling	31.000	C.Y.	4.588	0.00
Gravel under slab	6.000	C.Y.	0.282	21.00
Slab for ramp, 4" thick	6.000	C.Y.	5.256	345.00
Aluminum pipe rail (2 rail)	90.000	L.F.	18.000	1089.00
Aluminum wall pipe railing	150.000	L.F.	22.500	945.00
Haul excess material	60.000	C.Y.	5.640	0.00
Totals			205.950	4566.60

Total per linear foot including general contractor's overhead and profit	**$347**
Total per each 60-foot ramp including general contractor's overhead and profit	**$20,815**

Copyright R. S. Means Company, Inc, 1994

Notes

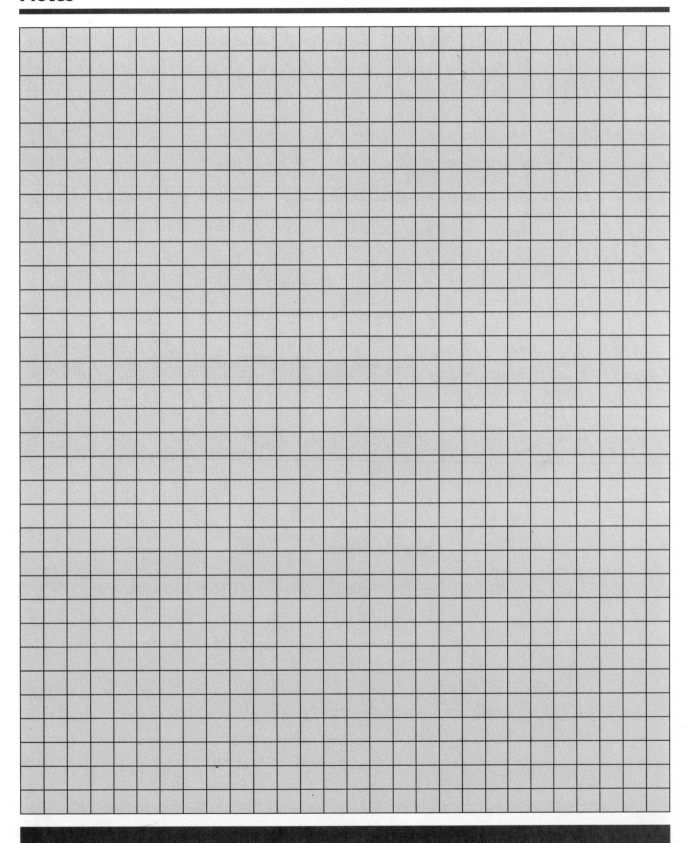

12
Modify Existing Ramp

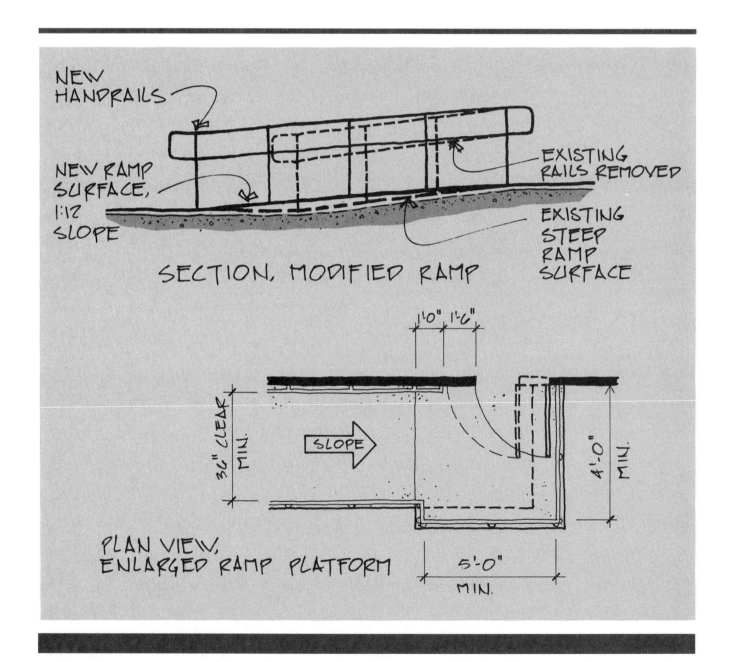

Some of the modifications to a ramp can be expensive, and still not create a fully accessible ramp. Before assuming that the existing ramp should be altered, assess the cost of modifications versus replacing the existing ramp with a new one (or determine if another accessible means of vertical circulation is necessary or possible).

ADAAG Reference
4.8 Ramps

Where Applicable
All existing ramps at public entrances or routes of travel.

Design Requirements
- 36" minimum clear width between handrails or curbs.
- Maximum slope 1:12, or one inch of rise for every 12 inches of run. (Any slope between 1:20 and 1:12 is a ramp).
- Base flush with adjacent paving.
- A level landing for at least every 30" of rise.
- Slip-resistant surface.
- 60" level area at base, all landings, and top, as wide as the ramp; 60" × 60" minimum landing wherever ramp changes direction.
- Handrails on both sides, 1-1/4" to 1-1/2" diameter, continuous, round or oval, minimum 12" level extensions top and bottom, 34"–38" above ramp surface, 1-1/2" from the wall, ends returned to a post or wall.
- Handrails.
- Curbs or rails at any drop-offs.
- No allowance for pooling of water on ramp surface.

Design Suggestions
To increase usability and minimize maintenance, consider the possibility of covering the ramp. As with new construction, select a railing attachment that allows continuous grasping: Under-rail brackets work best.

Key Items
- For applied materials including: special coatings, such as sand paint or sandpaper strips.
- For modifying the slope, adding surface materials: concrete, asphalt, wood to match existing.
- For changing rails: pipe rails, metal rails, wood banisters, uprights, attachments.
- For increasing the size of the landing: ramp materials, and new rails to match.

Level of Difficulty
Varies. Low for creating a slip-resistant surface. Moderate for changing handrails. Moderate to high for widening the platform. High for decreasing the slope.

Estimates

Repave concrete ramp to shallower slope, including 1-1/2" pipe handrails

Description	Quantity	Unit	Work Hours	Material
Torch cut handrail supports	8.000	Ea.	0.304	0.00
4" sidewalk, broom finish	36.000	S.F.	1.440	32.40
Aluminum pipe rail (2 rail)	24.000	L.F.	4.800	290.40
Totals			6.544	322.80

Total per linear foot including general contractor's overhead and profit	$82
Total per 12 linear feet including general contractor's overhead and profit	$984

Copyright R. S. Means Company, Inc., 1994

12. Modify Existing Ramp (continued)

Repave asphalt ramp to shallower slope, including 1-1/2" pipe handrails

Description	Quantity	Unit	Work Hours	Material
Torch cut handrail supports	8.000	Ea.	0.304	0.00
Install asphalt sidewalk, 2-1/2" thick	5.340	S.Y.	0.390	19.17
Aluminum pipe rail (2 rail)	24.000	L.F.	4.800	290.40
Totals			5.494	309.57

Total per linear foot including general contractor's overhead and profit: $76

Total per 12 linear feet including general contractor's overhead and profit: $910

Widen switch-back concrete ramp (including handrails)

Description	Quantity	Unit	Work Hours	Material
Excavation	35.000	C.Y.	5.180	0.00
Torch cut handrail supports	20.000	Ea.	0.760	0.00
Concrete forms, footings	78.000	SFCA	5.148	29.64
Concrete forms, walls	546.000	SFCA	26.754	163.80
Reinforcing (@ 50 lbs./C.Y.)	0.300	Ton	3.200	145.50
Place concrete, direct chute	12.000	C.Y.	6.396	0.00
Concrete, material only	12.000	C.Y.	0.000	660.00
Backfilling	23.000	C.Y.	3.404	0.00
Aluminum pipe rail (2 rail)	78.000	L.F.	15.600	943.80
Totals			66.442	1942.74

Total per linear foot including general contractor's overhead and profit: $125

Total per each 60-foot ramp including general contractor's overhead and profit: $7,496

Resurface existing ramp (sand paint)

Description	Quantity	Unit	Work Hours	Material
Prepare surface	1.000	S.Y.	0.003	0.00
Nonskid pavement renewal	1.000	S.Y.	0.023	0.55
Totals			0.026	0.55

Total per square yard including general contractor's overhead and profit: $2

Resurface existing ramp (sandpaper strips)

Description	Quantity	Unit	Work Hours	Material
Anti-skid sandpaper strips (6" x 24"), one per L.F.	1.000	L.F.	0.067	3.15
Totals			0.067	3.15

Total per linear foot including general contractor's overhead and profit: $9

Copyright R. S. Means Company, Inc., 1994

Enlarge pressure-treated wood ramp platform

Description	Quantity	Unit	Work Hours	Material
Remove board decking	16.000	S.F.	0.576	0.00
Remove handrails	32.000	L.F.	0.320	0.00
Remove framing	28.000	L.F.	0.476	0.00
Remove 4" × 4" posts	16.000	L.F.	0.320	0.00
Hand excavating for post footings	0.500	C.Y.	1.000	0.00
Conc. forms, 12" diameter tubes (2 forms, 4' deep)	8.000	L.F.	1.704	25.20
Hand backfilling around post footings	0.250	C.Y.	0.182	0.00
Place concrete for footings, direct chute	0.250	C.Y.	0.218	0.00
Concrete, material only	0.250	C.Y.	0.000	13.75
Install two-piece galvanized steel post foot	2.000	Ea.	0.124	6.60
4" × 4" post framing, w/ pressure-treated lumber	0.022	M.B.F.	0.677	31.35
2" × 8" joist framing, w/ pressure-treated lumber	0.014	M.B.F.	0.149	10.95
1" × 8" board decking	25.000	SF Flr.	0.400	29.50
2" × 6" railing enclosure	44.000	L.F.	0.572	32.56
Handrail stock	44.000	L.F.	4.400	39.60
Drilling bolt holes	66.000	Inch	1.188	0.00
Bolts, nuts & washers	33.000	Ea.	1.881	28.71
Handrail bracket	11.000	Ea.	2.750	176.00
Anchor layout and drilling, per inch of depth	8.000	Ea.	1.280	0.96
1/2" anchors	4.000	Ea.	0.400	7.12
Totals			**18.617**	**402.30**

Total per square foot including general contractor's overhead and profit **$66**

Total per each 25-square-foot landing including general contractor's overhead and profit **$1,660**

Widen concrete ramp platform (including handrails)

Description	Quantity	Unit	Work Hours	Material
Excavation	4.000	C.Y.	0.592	0.00
Torch cut handrail supports	3.000	Ea.	0.114	0.00
Concrete forms, footings	12.000	SFCA	0.792	4.56
Concrete forms, walls	36.000	SFCA	1.764	10.80
Reinforcing (@ 50 lbs./C.Y.)	0.050	Ton	0.533	24.25
Place concrete, direct chute	2.000	C.Y.	1.066	0.00
Concrete, material only	2.000	C.Y.	0.000	110.00
Backfilling	2.000	C.Y.	0.296	0.00
Aluminum pipe rail (2 rail)	12.000	L.F.	2.400	145.20
Totals			**7.557**	**294.81**

Total per square foot including general contractor's overhead and profit **$39**

Total per each 25-square-foot platform including general contractor's overhead and profit **$980**

Copyright R. S. Means Company, Inc, 1994

13
Install or Modify Ramp Railings

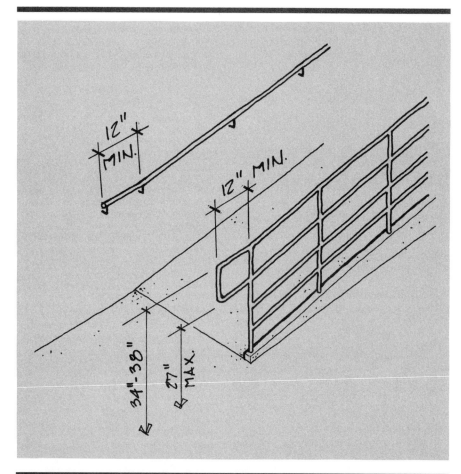

The design of a handrail is just as vital to a ramp's usability as the slope and surface. Rails provide an essential gripping surface that enables people with a range of abilities to use a ramp. Without proper rails, ramps, especially in exterior situations, are just pitched walkways that can be very difficult to use. Adding or replacing railings on a ramp is an essential modification that increases compliance and accessibility.

ADAAG References
4.8 Ramps
4.26 Handrails, Grab Bars, and Tub and Shower Seats

Where Applicable
Both sides of all ramps.

Design Requirements
- 36" minimum width between the rails.
- Rails 1-1/4" to 1-1/2" diameter, continuous, round or oval, 12" level extensions top and bottom, 34"–38" above surface, 1-1/2" from the wall, ends returned to post or wall.
- Rails solid and fixed, not to rotate in their fittings.

Design Suggestions
Railings should be attached in such a way that a person's grasp is uninterrupted. Brackets under the rails work best, with the rails separated from the uprights instead of resting on them. This allows someone to use the rail without letting go. Also, especially on older buildings, the design should incorporate the ramp into the existing aesthetic, using color, ornament, etc., as necessary. When specifying rails, check nominal dimensions against actual requirements. Pipe rails specified with a 1-1/2" diameter can have an actual diameter of 1-7/8" or more (although the Federal Access Board accepts this different size as it is an industry standard). Metal rails are very durable, but exterior rails can become quite hot in summer and dangerously cold in winter. Some pipe rails have an

insulated coating or sleeve to alleviate these problems, and these should be installed where wood rails are inappropriate.

Key Items
Pipe rails, metal rails, wood banisters or rails, uprights, attachments.

Level of Difficulty
Low to moderate. Attaching rails to masonry can require a contractor to set the bolts. Welding extensions to existing pipe requires a skilled welder.

Estimates

Bolt pipe rail to wood surface

Description	Quantity	Unit	Work Hours	Material
Aluminum pipe rail (2 rail)	1.000	L.F.	0.200	12.10
Strap brackets (2 per upright)	0.500	Ea.	0.061	0.80
Drilling bolt holes	1.500	Inch	0.027	0.00
Bolts, nuts & washers	1.000	Ea.	0.057	0.87
Totals			0.345	13.77

Total per linear foot including general contractor's overhead and profit — **$46**

Set pipe rail in concrete

Description	Quantity	Unit	Work Hours	Material
Pipe rail set in concrete (2 rail)	1.000	L.F.	0.200	12.10
Totals			0.200	12.10

Total per linear foot including general contractor's overhead and profit — **$35**

Set hand-forged wrought iron railing in concrete

Description	Quantity	Unit	Work Hours	Material
Wrought iron railing	1.000	L.F.	1.000	340.00
Totals			1.000	340.00

Total per linear foot including general contractor's overhead and profit — **$653**

Mount wood dowel railing on brass brackets (wood uprights existing)

Description	Quantity	Unit	Work Hours	Material
Handrail stock	1.000	L.F.	0.100	0.90
Drilling bolt holes	3.000	Inch	0.054	0.00
Bolts, nuts & washers	0.750	Ea.	0.043	0.65
Handrail bracket	0.250	Ea.	0.063	4.00
Totals			0.260	5.55

Total per linear foot including general contractor's overhead and profit — **$23**

Copyright R. S. Means Company, Inc., 1994

13. Install or Modify Ramp Railings *(continued)*

Mount wood dowel railing on brass brackets (pipe rail uprights)

Description	Quantity	Unit	Work Hours	Material
Handrail stock	1.000	L.F.	0.100	0.90
Drilling bolt holes	3.000	Inch	0.054	0.00
Bolts, nuts & washers	0.750	Ea.	0.043	0.65
Handrail bracket	0.250	Ea.	0.063	4.00
Totals			0.260	5.55

Total per linear foot including general contractor's overhead and profit — **$23**

Attach pipe railing to brick wall

Description	Quantity	Unit	Work Hours	Material
Pipe rail attached to brick wall	1.000	L.F.	0.150	6.30
Totals			0.150	6.30

Total per linear foot including general contractor's overhead and profit — **$22**

Attach wood dowel railing to brick wall

Description	Quantity	Unit	Work Hours	Material
Handrail stock	1.000	L.F.	0.100	0.90
Drilling bolt holes, per inch of depth	2.250	Ea.	0.360	0.27
Expansion bolts & shields	0.750	Ea.	0.075	1.34
Handrail bracket	0.250	Ea.	0.063	4.00
Totals			0.598	6.51

Total per linear foot including general contractor's overhead and profit — **$43**

Weld handrail extensions to existing pipe

Description	Quantity	Unit	Work Hours	Material
Fabricate 1-1/2" diameter pipe extensions	1.000	Ea.	3.080	25.20
Install pipe extensions on site	1.000	Ea.	2.000	0.00
Totals			5.080	25.20

Total per each including general contractor's overhead and profit — **$366**

Copyright R. S. Means Company, Inc., 1994

Attach handrail extensions to end posts (wood)

Description	Quantity	Unit	Work Hours	Material
Handrail stock	2.000	L.F.	0.200	1.80
Drilling bolt holes	24.000	Inch	0.432	0.00
Bolts, nuts & washers	6.000	Ea.	0.342	5.22
Handrail bracket	2.000	Ea.	0.500	32.00
Additional wood posts	0.011	M.B.F.	0.338	11.28
Totals			1.812	50.30

Total per each including general contractor's overhead and profit $183

Copyright R. S. Means Company, Inc, 1994

14
Install Vertical Platform Lift

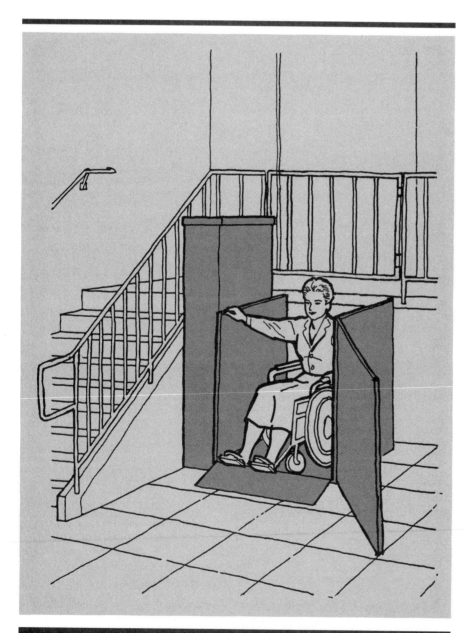

Vertical platform lifts are useful for creating access where a ramp would not fit or would be too long, an elevator would be too expensive, and a stair lift would block an egress stair. Platform lifts can be installed inside or outside and can be open or enclosed, and are a standard item that can create accessible vertical circulation in many retrofit situations.

ADAAG References
4.1.3 (5) Accessible Buildings, New Construction
4.11 Platform Lifts (Wheelchair Lifts)

Where Applicable
Allowed in existing buildings to create accessible vertical circulation between levels where installation of an elevator or ramp is not possible.

Design Requirements
- Accessible route to and off lift, top and bottom.
- Lift large enough to accommodate wheelchair (30" × 48").
- Clear space at top and bottom of lift (30" × 48"), plus any maneuvering space required at swinging gate or door.
- Stable, firm, slip-resistant surface at approaches, top and bottom.
- Controls within reach range (48" a.f.f. maximum for forward reach, 54" a.f.f. for side reach), and operable with a closed fist (lift should be independently operable).
- In compliance with American Society of Mechanical Engineers (ASME) A17.1, Part 20.

Design Suggestions/Considerations
Review the possibility of using elevators or ramps at locations where lifts might be installed. Lifts can be used as a means of creating vertical access where an elevator is too expensive or a ramp takes up too much space, but they are usually limited to wheelchair users by local safety codes (but not by ADA). Also, the on/off direction of travel must

be a straight line, since a lift platform is usually not large enough to allow a right angle turn. The maximum height of a vertical lift is usually 8', but 12' heights are available. ASME A17.1, which is referenced in ADAAG, sets strict technical requirements as to length of travel and methods of operation; local codes may have additional requirements, and should be consulted prior to any installation. Any lift should be usable independently (unassisted), but where this is not possible, such as in schools or certain businesses, convenient assistance must be available. A call button within reach should be installed to signal assistance. Lifts will also need servicing, periodic testing, staff training, and protection against vandalism.

Key Items
- Accessible route to lift.
- Lift equipment.
- Concrete pad, for exterior applications.
- Enclosure if desired, prefab or custom. Important in older building applications.
- Call button, if necessary

Level of Difficulty
Moderate to high. Installation usually done by lift manufacturer's representatives. Requires preparatory electrical wiring. Possible concrete work for pad, carpentry if custom enclosure.

Estimates

Install exterior unenclosed lift with concrete pad

Description	Quantity	Unit	Work Hours	Material
Excavation	0.600	C.Y.	0.089	0.00
Concrete formwork	32.000	SFCA	4.384	16.00
15" thick concrete pad	16.000	S.F.	0.464	38.88
Unenclosed lift	1.000	Ea.	16.000	4250.00
Totals			20.937	4304.88

Total per each including general contractor's overhead and profit: $8,571

Install exterior enclosed lift with concrete pad

Description	Quantity	Unit	Work Hours	Material
Excavation	0.600	C.Y.	0.089	0.00
Concrete formwork	32.000	SFCA	4.384	16.00
15" thick concrete pad	16.000	S.F.	0.464	38.88
Enclosed lift	1.000	Ea.	32.000	10,100.00
Totals			36.937	10,154.88

Total per each including general contractor's overhead and profit: $19,568

Install interior unenclosed lift

Description	Quantity	Unit	Work Hours	Material
Interior unenclosed lift	1.000	Ea.	16.000	4250.00
Totals			16.000	4250.00

Total per each including general contractor's overhead and profit: $8,210

Copyright R. S. Means Company, Inc, 1994

15
Install Stairway Inclined Lift

It is common for older buildings to have no elevator, floor-to-floor distances too great to ramp, and no room for a vertical lift. Even where the vertical distance between levels is not excessive, there is often no room to install a ramp. In such situations, installing a lift on a stair can create access between levels with a minimum of modification to the building.

ADAAG References
4.1.3 (5) Accessible Buildings, New Construction
4.11 Platform Lifts (Wheelchair Lifts)

Where Applicable
Allowed in existing buildings to create accessible vertical circulation between levels where installation of an elevator or ramp is not possible.

Design Requirements
- Platform on an accessible route at top and bottom.
- Lift large enough to accommodate wheelchair (30" × 48").
- Clear space at top and bottom at lift approach (30" × 48").
- Stable, firm, slip-resistant surface at approaches, top and bottom.
- Controls within reach range, and operable with a closed fist (lift should be independently operable).
- Stair wide enough to accommodate lift in down position without intruding into required clear fire egress widths.

Design Suggestions/Considerations
Make sure that other options (such as elevators or ramps) aren't feasible. As with vertical platform lifts, inclined lifts are usually restricted by local code to use by people in wheelchairs. Some lifts are capable of traveling up to six flights of stairs continuously, but such installations are expensive. Fire egress widths are calculated with the lift in the operating (down) position (about 40"; folded up, the lifts are about 13" wide). It is assumed by fire marshalls that the

lift will be in use during an emergency. This prevents the use of inclined lifts on fire egress stairs where the lift in the down position would intrude on the necessary stair width determined by the local fire safety codes. As with vertical lifts, independent (unassisted) usage of a lift is often not possible, such as in schools or some businesses. A call button within reach should be installed to signal for assistance. In such cases, convenient assistance must be readily available. Lifts will also need servicing, periodic testing, staff training, and protection against vandalism.

Key Items

Lift, operating mechanism, call button, if necessary.

Level of Difficulty

Moderate to high. Usually installed by manufacturer or manufacturer's representative, with preliminary electrical work required.

Estimates

Install stairway lift, straight run

Description	Quantity	Unit	Work Hours	Material
Straight run stair lift	1.000	Ea.	16.000	8400.00
Totals			16.000	8400.00

Total per each including general contractor's overhead and profit $15,304

Install stairway lift, one turn

Description	Quantity	Unit	Work Hours	Material
One turn stair lift	1.000	Ea.	80.000	13,300.00
Totals			80.000	13,300.00

Total per each including general contractor's overhead and profit $27,481

Copyright R. S. Means Company, Inc, 1994

16 Install New Stairs

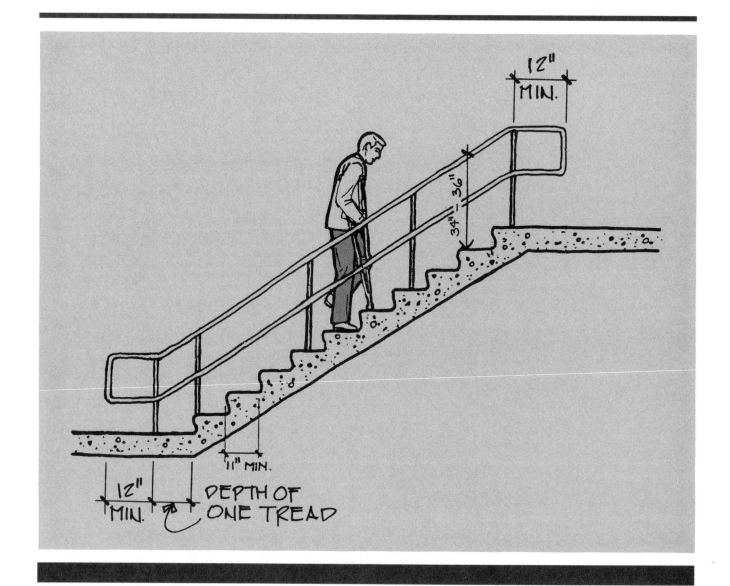

Stairs are sometimes missed in an access survey since they are not wheelchair accessible, but stairs are such a heavily used means of vertical circulation that attention to design is critical in creating an accessible facility.

An inaccessible stair which would be difficult to modify may, in some cases, be better replaced with an accessible stair that can greatly increase a facility's usability and safety.

ADAAG References
4.1.3 (4), Accessible Buildings, New Construction
4.9 Stairs
4.26 Handrails, Grab Bars, and Tub and Shower Seats

Where Applicable
All stairs connecting levels that are not accessible by elevators, ramps, or lifts.

Design Requirements
- Minimum treads width of 11".
- Slip-resistant surface.
- No protruding nosings greater than 1-1/2".
- No open risers.
- Rails on both sides, 34"–38" above nosings, 1-1/4"–1-1/2" diameter, round or oval, continuous inside rail, 1-1/2" from wall.
- If rails are not continuous, 12" extensions at the top parallel to the floor, and 12" plus width of one tread extensions at the bottom, sloped for the width of one tread and parallel to the floor for the remaining distance.
- No pooling water on exterior stairs.

Design Suggestions
Although ADAAG covers only stairs that connect levels not connected by an accessible means of vertical access, it is strongly recommended that all stairs conform to ADA regulations, since stairs are a heavily used element by people with a wide range of abilities.

There are some reasons to consider complete replacement of a stair: if it is too steep, has open risers and has specially slippery tread surfaces.

Key Items
Foundation/footings, stairs (wood, concrete, metal pan), rails (wood, pipe, metal).

Level of Difficulty
Moderate to high, depending on the materials used. Installation of wood stairs requires skilled carpenters; concrete stairs require foundation and formwork; metal stairs require welding. This project may require design drawings, and will require building permit.

Estimates

Install painted wood stairs

Description	Quantity	Unit	Work Hours	Material
Stair stringers, 2" x 12"	0.144	M.B.F.	8.861	100.80
Pine risers, 3/4" x 7-1/2"	64.000	L.F.	7.744	86.40
Aluminum pipe rail (2 rail)	18.000	L.F.	3.600	217.80
Strap brackets (2 per upright)	10.000	Ea.	1.220	16.00
Drilling bolt holes	30.000	Inch	0.540	0.00
Bolts, nuts & washers	20.000	Ea.	1.140	17.40
Pipe rail attached to brick wall	18.000	L.F.	2.700	113.40
Painting, treads, risers, and stringers	160.000	L.F.	6.720	12.80
Totals			32.525	564.60

Total per riser including general contractor's overhead and profit	**$176**
Total per 16 risers including general contractor's overhead and profit	**$2,816**

Copyright R. S. Means Company, Inc., 1994

16. Install New Stairs (continued)

Install pressure-treated wood stairs

Description	Quantity	Unit	Work Hours	Material
Stair stringers, 2" × 12"	0.144	M.B.F.	8.861	141.12
Pine risers, 3/4" × 7-1/2"	64.000	L.F.	7.744	86.40
2" × 6" stair treads	0.120	M.B.F.	2.560	91.80
Aluminum pipe rail (2 rail)	18.000	L.F.	3.600	217.80
Strap brackets (2 per upright)	10.000	Ea.	1.220	16.00
Drilling bolt holes	30.000	Inch	0.540	0.00
Bolts, nuts & washers	20.000	Ea.	1.140	17.40
Pipe rail attached to brick wall	18.000	L.F.	2.700	113.40
Painting, risers	60.000	L.F.	2.520	4.80
Totals			30.885	688.72

Total per riser including general contractor's overhead and profit	**$186**
Total per 16 risers including general contractor's overhead and profit	**$2,968**

Install concrete stairs

Description	Quantity	Unit	Work Hours	Material
Freestanding concrete stairs	64.000	LF Nose	38.400	316.80
Safety treads	64.000	L.F.	4.480	512.00
Aluminum pipe rail (2 rail)	18.000	L.F.	3.600	217.80
Aluminum wall railing	18.000	L.F.	2.700	113.40
Totals			49.180	1160.00

Total per riser including general contractor's overhead and profit	**$292**
Total per 16 risers including general contractor's overhead and profit	**$4,671**

Install metal pan stairs, concrete fill

Description	Quantity	Unit	Work Hours	Material
Metal pan stairs with railings	16.000	Riser	17.072	1040.00
Non-slip concrete fill for treads	64.000	S.F.	5.120	172.80
Totals			22.192	1212.80

Total per riser including general contractor's overhead and profit	**$224**
Total per 16 risers including general contractor's overhead and profit	**$3,584**

Copyright R. S. Means Company, Inc, 1994

Notes

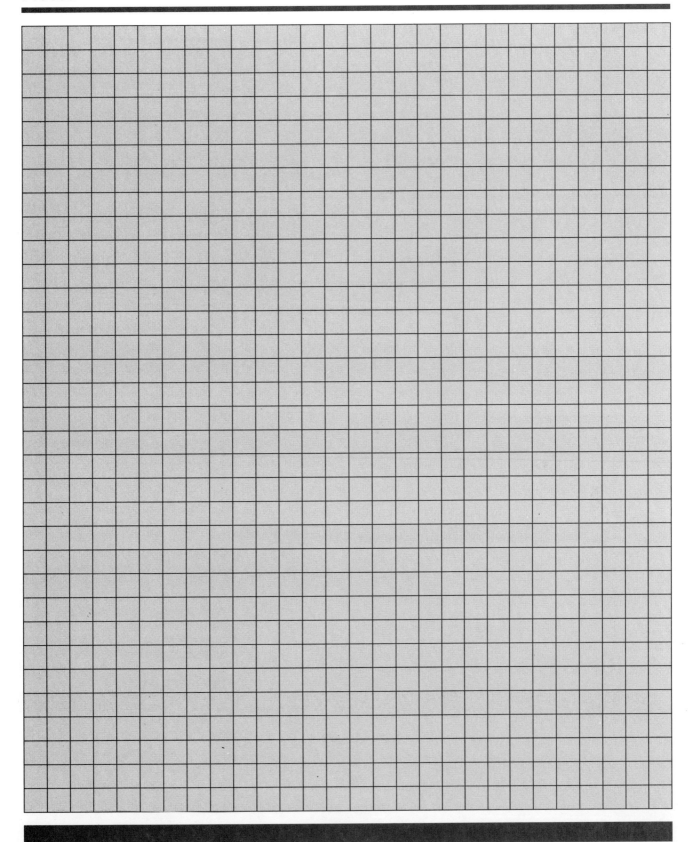

17 Modify Existing Stairs

Inaccessible stairs can make travel between levels extremely difficult for people with balance, grasping, or visual impairments. This situation can be dangerous, especially when stairs are used for emergency evacuation. Handrails that are hard to grasp or missing, protruding nosings, and slippery surfaces can, in fact, make the stair awkward and unsafe for all users. Accessible modifications to a flight of stairs result in a much more accessible and safe building element for all users.

ADAAG References
4.9 Stairs

Where Applicable
All stairs connecting levels not accessible by elevators, ramps, or lifts.

Design Requirements
- Treads 11" deep minimum.
- Slip-resistant surface.
- No protruding nosings greater than 1-1/2".
- No open risers.

Design Suggestions
It is strongly recommended that *all* stairs conform to ADA regulations. For adding slip-resistant surfaces, pick a material that will stand up to serious wear, such as carpet for interior installations or rubber treads for exterior applications. Applied stair treatments must be carefully installed to resist detaching. It is possible to meet the slip-resistance criteria in ADAAG with coatings such as sand paint, but these will require frequent maintenance and are not as durable as applied tread materials.

Key Items
Wood risers or bevels for nosings; wood or metal risers; slip-resistant surfaces (sandpaper strips, carpet).

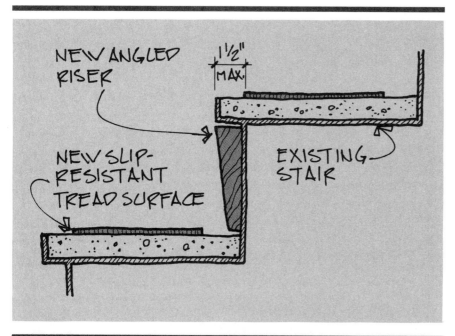

Level of Difficulty

Low to moderate. Modifying metal stairs requires welding or bolting; modifications to wood stair requires carpentry; applying sandpaper strips or carpet usually doesn't require contracted labor.

Estimates

Fill in open stair riser (wood stairs)

Description	Quantity	Unit	Work Hours	Material
3/4" × 7-1/2" stair risers	1.000	L.F.	0.121	1.35
Painting	1.000	S.F.	0.013	0.03
Totals			0.134	1.38

Total per linear foot of riser including general contractor's overhead and profit: $9

Fill in open stair riser (metal pan stairs)

Description	Quantity	Unit	Work Hours	Material
Plate steel for risers (8" × 4', 6.81 lbs./L.F.)	0.272	C.W.T.	0.000	8.84
Welding	0.200	Hr.	0.200	0.55
Painting, both sides	8.000	L.F.	0.152	0.32
Totals			0.352	9.71

Total per linear foot of riser including general contractor's overhead and profit: $42

Bevel stair nosing (wood)

Description	Quantity	Unit	Work Hours	Material
3/4" × 7-1/2" stair risers	1.000	L.F.	0.121	1.35
Painting	1.000	L.F.	0.013	0.03
Totals			0.134	1.38

Total per linear foot of riser including general contractor's overhead and profit: $9

Copyright R. S. Means Company, Inc., 1994

17. Modify Existing Stairs (continued)

Bevel stair nosing (concrete)

Description	Quantity	Unit	Work Hours	Material
Remove concrete nosing	0.120	C.F.	0.074	0.00
Drill 1/2" holes for reinf. pockets (per inch of depth)	8.000	Ea.	1.280	0.96
Wedge anchors for reinforcing support	2.000	Ea.	0.124	3.16
Deformed reinforcing dowels	2.000	Ea.	0.256	1.98
Steel angle edging for stair nosing	4.000	L.F.	0.388	13.20
Form board (4 L.F.)	4.000	S.F.	1.012	4.08
2" nonshrink grout	4.000	S.F.	1.280	23.40
Totals			4.414	46.78

Total per linear foot of riser including general contractor's overhead and profit — **$81**

Total per 4-foot-wide riser including general contractor's overhead and profit — **$324**

Bevel stair nosing (metal pan)

Description	Quantity	Unit	Work Hours	Material
Plate steel for risers (8" × 4', 6.81 lbs./L.F.)	0.272	C.W.T.	0.000	8.84
Welding	0.200	Hr.	0.200	0.55
Painting, both sides	8.000	L.F.	0.152	0.32
Totals			0.352	9.71

Total per linear foot of riser including general contractor's overhead and profit — **$11**

Total per 4-foot-wide riser including general contractor's overhead and profit — **$42**

Add non-slip surface to stair tread (sandpaper strips)

Description	Quantity	Unit	Work Hours	Material
Anti-skid sandpaper strips (6" × 24"), one per riser	1.000	Riser	0.067	3.15
Totals			0.067	3.15

Total per riser including general contractor's overhead and profit — **$9**

Copyright R. S. Means Company, Inc, 1994

Notes

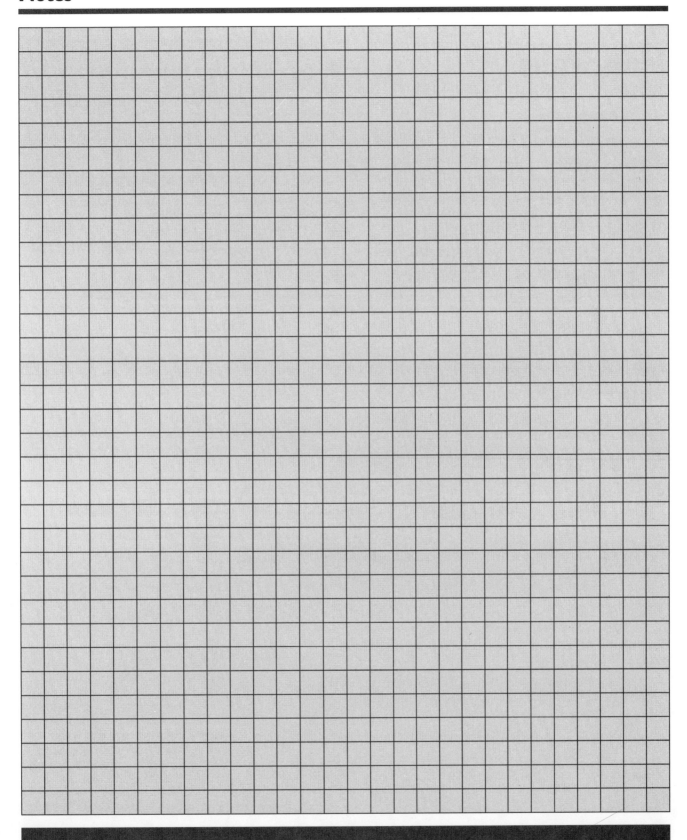

18
Install or Modify Stair Handrails

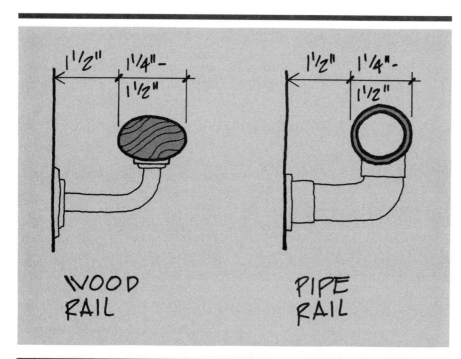

Handrails on stairs are often relatively easy to replace or add. It is strongly recommended that *all* stairs have compliant handrails, whether or not they are required by ADAAG.

ADAAG References
4.9 Stairs
4.26 Handrails, Grab Bars, and Tub and Shower Seats

Where Applicable
All stairs connecting levels not connected by elevators.

Design Requirements
- Rails on both sides.
- 34"-38" above nosings.
- 1-1/4"–1-1/2" diameter, round or oval.
- Continuous inside rail.
- 1-1/2" from wall.
- If rails are not continuous, 12" extensions at the top parallel to the floor, and 12" plus width of one tread extensions at the bottom, sloped for the width of one tread and parallel to the floor for the remaining distance.
- Ends returned to wall, floor, rail or post.
- Rails fixed in their supports.

Design Suggestions
On dog-leg or switchback stairs, both inside and outside rails should be continuous at a landing. Under-rail brackets work best, with the rails separated from the uprights instead of resting on them. This allows someone to use the rail without letting go. Check nominal dimensions against actual requirements when specifying rails. Metal rails specified at 1-1/2" (nominal dimension) will have an actual diameter of 1-7/8" or greater. Rails are a highly visible element. Especially on older buildings, it is important to choose design options that will incorporate a ramp into the existing aesthetic, such as posts with the same design as stair newel posts.

Key Items
Rails (wood, pipe, metal), uprights, anchors.

Level of Difficulty
Moderate to high, depending on the material used. Installing wood rails requires carpentry skill; pipe or metal rail installation requires welding or bolting, possibly to a masonry surface.

Estimates

Add new wall-mounted pipe rail

Description	Quantity	Unit	Work Hours	Material
1-1/2" wall-mounted aluminum pipe rail	1.000	L.F.	0.150	6.30
Totals			0.150	6.30

Total per linear foot including general contractor's overhead and profit: $22

Add new freestanding pipe rail

Description	Quantity	Unit	Work Hours	Material
1-1/2" freestanding aluminum pipe rail (2 rail)	1.000	L.F.	0.200	12.10
Totals			0.200	12.10

Total per linear foot including general contractor's overhead and profit: $35

Add new wall-mounted wood dowel rail

Description	Quantity	Unit	Work Hours	Material
Handrail stock	1.000	L.F.	0.100	0.90
Drilling bolt holes	1.500	Inch	0.027	0.00
Bolts, nuts & washers	0.750	Ea.	0.043	0.65
Handrail bracket	0.250	Ea.	0.063	4.00
Totals			0.233	5.55

Total per linear foot including general contractor's overhead and profit: $22

Add new wood dowel handrail including pipe rail uprights bolted to wood

Description	Quantity	Unit	Work Hours	Material
Aluminum pipe railing	0.750	L.F.	0.150	9.08
Handrail stock	1.000	L.F.	0.100	0.90
Handrail bracket	0.250	Ea.	0.063	4.00
Drilling bolt holes for bracket	1.500	Inch	0.027	0.00
Bolts, nuts & washers	0.750	Ea.	0.043	0.65
Strap brackets (2 per upright)	0.500	Ea.	0.061	0.80
Drilling bolt holes	1.500	Inch	0.027	0.00
Bolts, nuts & washers	1.000	Ea.	0.057	0.87
Totals			0.528	16.30

Total per linear foot including general contractor's overhead and profit: $59

Copyright R. S. Means Company, Inc., 1994

18. Install or Modify Stair Handrails (continued)

Add new wall-mounted metal rail

Description	Quantity	Unit	Work Hours	Material
1-1/2" wall-mounted steel railing	1.000	L.F.	0.150	4.60
Totals			0.150	4.60

Total per linear foot including general contractor's overhead and profit — **$19**

Add new freestanding metal railing

Description	Quantity	Unit	Work Hours	Material
1-1/2" freestanding pipe rail (2 rail)	1.000	L.F.	0.200	8.25
Totals			0.200	8.25

Total per linear foot including general contractor's overhead and profit — **$29**

Add 12" extension to existing wall-mounted pipe railing

Description	Quantity	Unit	Work Hours	Material
Fabricate 1-1/2" diameter pipe extensions	1.000	Ea.	3.080	12.60
Install pipe extensions on site	1.000	Ea.	2.000	0.00
Drilling bolt holes, per inch of depth	6.000	Inch	0.960	0.72
Expansion bolts & shields	3.000	Ea.	0.300	5.34
Totals			6.340	18.66

Total per each including general contractor's overhead and profit — **$425**

Add 12" extension to existing freestanding pipe railing

Description	Quantity	Unit	Work Hours	Material
Fabricate 1-1/2" diameter pipe extensions	1.000	Ea.	3.080	25.20
Install pipe extensions on site	1.000	Ea.	2.000	0.00
Totals			5.080	25.20

Total per each including general contractor's overhead and profit — **$366**

Add 12" extension to existing wall-mounted wood railing

Description	Quantity	Unit	Work Hours	Material
Handrail stock	1.000	L.F.	0.100	0.90
Drilling bolt holes	12.000	Inch	0.216	0.00
Bolts, nuts & washers	6.000	Ea.	0.342	5.22
Handrail bracket	2.000	Ea.	0.500	32.00
Totals			1.158	38.12

Total per each including general contractor's overhead and profit — **$127**

Copyright R. S. Means Company, Inc., 1994

Add 12" extension to existing freestanding wood railing

Description	Quantity	Unit	Work Hours	Material
Handrail stock	2.000	L.F.	0.200	1.80
Drilling bolt holes	24.000	Inch	0.432	0.00
Bolts, nuts & washers.	6.000	Ea.	0.342	5.22
Handrail bracket	2.000	Ea.	0.500	32.00
Additional wood posts	0.011	M.B.F.	0.338	11.28
Totals			1.812	50.30

Total per each including general contractor's overhead and profit — **$183**

Add 12" extension to existing wall-mounted metal railing

Description	Quantity	Unit	Work Hours	Material
Fabricate 1-1/2" diameter pipe extensions	1.000	Ea.	3.080	12.60
Install pipe extensions on site	1.000	Ea.	2.000	0.00
Drilling bolt holes, per inch of depth	6.000	Inch	0.960	0.72
Expansion bolts & shields	3.000	Ea.	0.300	5.34
Totals			6.340	18.66

Total per each including general contractor's overhead and profit — **$425**

Add 12" extension to existing freestanding metal railing

Description	Quantity	Unit	Work Hours	Material
Fabricate 1-1/2" diameter pipe extensions	1.000	Ea.	3.080	25.20
Install pipe extensions on site	1.000	Ea.	2.000	0.00
Totals			5.080	25.20

Total per each including general contractor's overhead and profit — **$366**

Copyright R. S. Means Company, Inc, 1994

19 Install Under-Stair Barrier

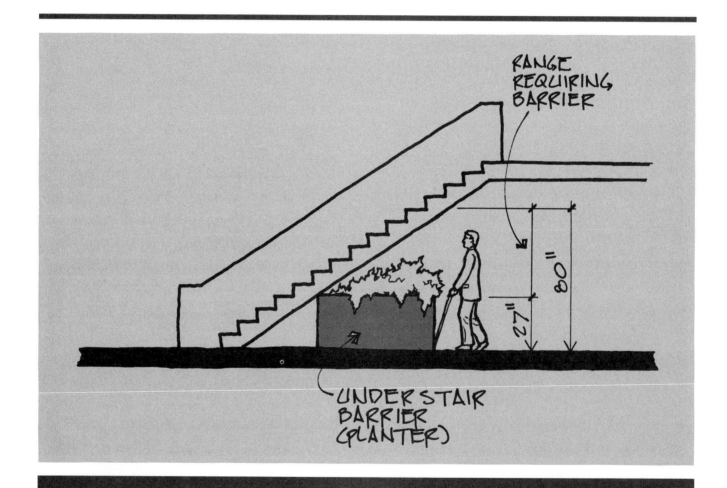

A lack of cane-detectable barriers under freestanding stairs can be extremely dangerous for people with visual impairments. When someone sweeps a cane in front of them, their head can hit an object below 80" before the cane touches it. Installing a barrier under a stair is vital to creating a safe, accessible facility.

ADAAG Reference
4.4 Protruding Objects

Where Applicable
Under freestanding staircases and any area along a public route of travel where the headroom is reduced below 80".

Design Requirements
- Barrier located 27" a.f.f. or lower, where headroom is less than 80" a.f.f.

Design Suggestions
The requirements for protection from protruding objects apply to all circulation paths, not just an accessible

route. It should not be assumed that an area is not within a route of travel just because it is not the intended route. Install barriers under all freestanding stairs and similar hazards that are apparent only to sighted people. The barrier could be functional, such as seating, a planter, or a trash receptacle. If space permits, the area under the stairs could be enclosed with partitions and used for storage. Avoid using movable objects that are not anchored in place of barriers. Check local code requirements for under-stair use.

Key Items
A barrier or barriers under each low-headroom hazard stair.

Level of Difficulty
Low.

Estimate

Install barrier under stairs

Description	Quantity	Unit	Work Hours	Material
Square wood planter, 48" x 48" x 24"	1.000	Ea.	1.067	575.00
Totals			1.067	575.00

Total per each including general contractor's overhead and profit: $1,027

Copyright R. S. Means Company, Inc, 1994

20 Install New Door: Drywall

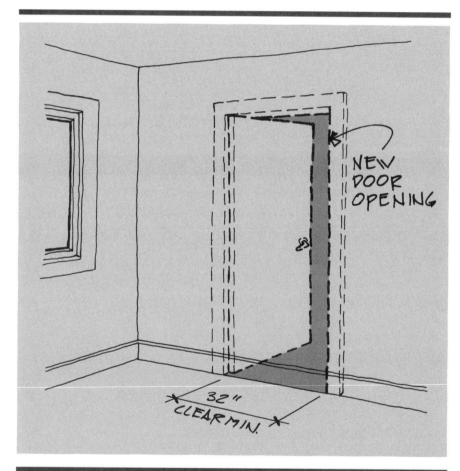

ADAAG Reference
4.13 Doors

Where Applicable
Interior doors to accessible areas on accessible routes placed in walls constructed of studs and gypsum wallboard.

Design Requirements
- 32" clear opening width.
- 18" clearance adjacent to the latch on the pull side of the door, 12" on the push side.
- 1/2" maximum beveled threshold.
- 5 lbs. maximum pull or push weight on interior doors (no ADAAG reference for exterior doors).
- Level maneuvering space on both sides of the door, depending on approach. Minimum required dimensions in front of door between 42" to 60", depending on approach and whether door has a closer (see ADAAG Fig. 25).
- 60" level surface, inside and outside of door at entrances.
- Accessible hardware (acceptable if operable with a closed fist).
- With door closer, 3 seconds minimum closing time to a point 3" from the latch.

Design Suggestions
Because the 32" clear opening is measured from the face of the door in a 90° open position to the stop on the opposite jamb, the door itself has to be wider (usually 36") in order to comply. 2'-10" doors are the smallest that can be used to comply, but might not meet the requirement. There are several accessible hardware options: a loop (allow at least 3" between inside of loop and face of door), lever handles, push plate, or panic bar. Where opening force is necessarily high or where adequate maneuvering space cannot be provided, installation of an automatic opener may be a solution.

Maneuvering space on each side of the door is determined by how it is approached. A 60" × 60" minimum clear space is best and complies with all approaches cited in ADAAG. Where only a straight-on approach is available, a 60" deep space measured from the face of the door is required on the pull side, and a 48" deep space is required on the push side. Where only a side approach is available, the required depth of the clear area in front of the door varies from 42" to 60", and the width is affected by the latch edge clearances and presence or absence of a door opener. Consult ADAAG 4.13 and ADAAG Fig. 25 for exact requirements.

Key Items

Door, new structural members for opening, wall/floor finishes to match existing.

Level of Difficulty

Moderate to high. Higher for load-bearing walls. Involves demolition, structural framing, finish work. Requires a building permit and possibly design drawings.

Estimate

Install solid core wood door in metal stud/drywall partition

Description	Quantity	Unit	Work Hours	Material
Remove metal studs/gypsum board	9.000	S.F.	0.414	0.00
Steel door frame, 3'-0" x 6'-8"	1.000	Ea.	1.000	65.50
Solid core wood door, flush birch face	1.000	Ea.	1.143	89.50
Lever-handled lockset	2.000	Ea.	0.000	159.56
Hinges	1.500	Pr.	0.000	69.00
Threshold	1.000	Ea.	0.400	27.50
Paint door	1.000	Ea.	0.941	4.96
Painting	76.000	S.F.	1.368	12.92
Totals			5.266	428.94

Total per each including general contractor's overhead and profit: $1,093

Copyright R. S. Means Company, Inc, 1994

21
Install New Door: Masonry

A new accessible door can often prevent the need for a long circuitous route of travel to an existing accessible entrance, or for the modification of an existing entrance to create access. In some cases, an existing window opening can be enlarged to create an accessible entrance to a facility or space.

ADAAG Reference
4.13 Doors

Where Applicable
Doors to accessible areas on accessible routes placed in walls constructed of brick, block, or stone.

Design Requirements
- 32" clear opening width.
- 18" clearance adjacent to the latch on the pull side of the door, 12" on the push side.
- 1/2" maximum beveled threshold.
- 5 lbs. maximum pull or push weight on interior doors (no ADAAG reference for exterior doors).
- Level maneuvering space on both sides of the door, depending on approach. Minimum required dimensions in front of door between 42" to 60", depending on approach and whether door has a closer (see ADAAG Fig. 25).
- 60" level surface, inside and outside of door at entrances.
- Accessible hardware (acceptable if operable with a closed fist).
- With door closer, 3 seconds minimum closing time to a point 3" from the latch.

Design Suggestions
Because the 32" clear opening is measured from the face of the door in a 90° open position to the stop on the opposite jamb, the door itself has to be wider (usually 36") in order to comply. 2'-10" doors are the smallest that can be used to comply, but might not meet the requirement. There are several accessible hardware options: a loop (allow at least 3" between inside of loop and face of door), lever handles, push plate, or panic bar. It is recommended

32" CLEAR OPENING REQUIRED — MASONRY OPENING DETERMINED BY FIELD CONDITIONS (but not required) that the maximum opening force for exterior doors be 8 lbs. This may not be possible due to wind pressure or interior air circulation systems. Where opening force is necessarily high or where adequate maneuvering space cannot be provided, installation of an automatic opener may be a solution.

Maneuvering space on each side of the door is determined by how it is approached. A 60" × 60" minimum clear space is best and complies with all approaches cited in ADAAG. Where only a straight-on approach is available, a 60" deep space measured from the face of the door is required on the pull side, and a 48" deep space is required on the push side. Where only a side approach is available, the required depth of the clear area in front of the door varies from 42" to 60", and the width is affected by the latch edge clearances and presence or absence of a door opener. Consult ADAAG 4.13 and ADAAG Fig. 25 for exact requirements.

Key Items
Door, new structural members for opening, wall/floor finishes to match existing.

Level of Difficulty
Moderate to high. Higher for load-bearing walls. Involves demolition, bracing of new opening, structural framing, finish work. Requires a building permit and possibly design drawings.

Estimates

Install solid core wood door in 12" thick brick exterior wall

Description	Quantity	Unit	Work Hours	Material
Saw cutting brick wall, per inch of depth	288.000	L.F.	19.008	66.24
Brick wall demolition	34.000	C.F.	6.188	0.00
Remove window	1.000	Ea.	0.615	0.00
Labor minimum for masonry repairs	1.000	Job	2.000	0.00
Steel lintel, 4" × 3-1/2" × 1/4", 5'-0" long	1.000	Ea.	0.200	15.50
Steel door frame, 3'-0" × 6'-8"	1.000	Ea.	1.000	65.50
Solid core wood door, flush birch face	1.000	Ea.	1.143	89.50
Door closer	1.000	Ea.	1.333	95.50
Lever-handled lockset	1.000	Ea.	0.000	79.78
Hinges	1.500	Pr.	0.000	69.00
Threshold	1.000	Ea.	0.400	27.50
Paint door	1.000	Ea.	0.941	4.96
Silicone sealant	17.000	L.F.	0.578	4.08
Totals			33.406	517.56

Total per each including general contractor's overhead and profit **$2,929**

Copyright R. S. Means Company, Inc., 1994

21. Install New Door: Masonry (continued)

Install solid core wood door in masonry veneer exterior wall

Description	Quantity	Unit	Work Hours	Material
Saw cutting brick wall, per inch of depth	96.000	L.F.	6.336	22.08
Brick wall demolition	12.000	C.F.	2.184	0.00
Remove window	1.000	Ea.	0.615	0.00
Remove metal studs, interior & exterior gypsum board	34.000	S.F.	1.564	0.00
Labor minimum for masonry repairs	1.000	Job	2.000	0.00
Steel lintel, 4" x 3-1/2" x 1/4", 5'-0" long	1.000	Ea.	0.200	15.50
Steel door frame, 3'-0" x 6'-8"	1.000	Ea.	1.000	65.50
Solid core wood door, flush birch face	1.000	Ea.	1.143	89.50
Door closer	1.000	Ea.	1.333	95.50
Lever-handled lockset	1.000	Ea.	0.000	79.78
Hinges	1.500	Pr.	0.000	69.00
Threshold	1.000	Ea.	0.400	27.50
Paint door	1.000	Ea.	0.941	4.96
Silicone sealant	17.000	L.F.	0.578	4.08
Totals			18.294	473.40

Total per each including general contractor's overhead and profit: $2,095

Install solid core wood door in 8" block exterior wall

Description	Quantity	Unit	Work Hours	Material
Saw cutting block wall, per inch of depth	192.000	L.F.	12.672	44.16
Concrete block demolition	34.000	S.F.	1.666	0.00
Remove window	1.000	Ea.	0.615	0.00
Labor minimum for masonry repairs	1.000	Job	2.000	0.00
Steel lintel, 4" x 3-1/2" x 1/4", 5'-0" long	1.000	Ea.	0.200	15.50
Steel door frame, 3'-0" x 6'-8"	1.000	Ea.	1.000	65.50
Solid core wood door, flush birch face	1.000	Ea.	1.143	89.50
Door closer	1.000	Ea.	1.333	95.50
Lever-handled lockset	1.000	Ea.	0.000	79.78
Hinges	1.500	Pr.	0.000	69.00
Threshold	1.000	Ea.	0.400	27.50
Paint door	1.000	Ea.	0.941	4.96
Silicone sealant	17.000	L.F.	0.578	4.08
Totals			22.548	495.48

Total per each including general contractor's overhead and profit: $2,341

Copyright R. S. Means Company, Inc., 1994

Install hollow metal door in 12" thick brick exterior wall

Description	Quantity	Unit	Work Hours	Material
Saw cutting brick wall, per inch of depth	288.000	L.F.	19.008	66.24
Brick wall demolition	34.000	C.F.	6.188	0.00
Remove window	1.000	Ea.	0.615	0.00
Labor minimum for masonry repairs	1.000	Job	2.000	0.00
Steel lintel, 4" × 3-1/2" × 1/4", 5'-0" long	1.000	Ea.	0.200	15.50
Steel door frame, 3'-0" × 6'-8"	1.000	Ea.	1.000	65.50
Hollow metal door	1.000	Ea.	0.941	147.00
Door closer	1.000	Ea.	1.333	95.50
Lever-handled lockset	1.000	Ea.	0.000	79.78
Hinges	1.500	Pr.	0.000	69.00
Threshold	1.000	Ea.	0.400	27.50
Paint door	1.000	Ea.	0.941	4.96
Silicone sealant	17.000	L.F.	0.578	4.08
Totals			33.204	575.06

Total per each including general contractor's overhead and profit $3,013

Install hollow metal door in masonry veneer exterior wall

Description	Quantity	Unit	Work Hours	Material
Saw cutting brick wall, per inch of depth	96.000	L.F.	6.336	22.08
Brick wall demolition	12.000	C.F.	2.184	0.00
Remove window	1.000	Ea.	0.615	0.00
Remove metal studs, interior & exterior gypsum board	34.000	S.F.	1.564	0.00
Labor minimum for masonry repairs	1.000	Job	2.000	0.00
Steel lintel, 4" × 3-1/2" × 1/4", 5'-0" long	1.000	Ea.	0.200	15.50
Steel door frame, 3'-0" × 6'-8"	1.000	Ea.	1.000	65.50
Hollow metal door	1.000	Ea.	0.941	147.00
Door closer	1.000	Ea.	1.333	95.50
Lever-handled lockset	1.000	Ea.	0.000	79.78
Hinges	1.500	Pr.	0.000	69.00
Threshold	1.000	Ea.	0.400	27.50
Paint door	1.000	Ea.	0.941	4.96
Silicone sealant	17.000	L.F.	0.578	4.08
Totals			18.092	530.90

Total per each including general contractor's overhead and profit $2,179

Copyright R. S. Means Company, Inc., 1994

21. Install New Door: Masonry (continued)

Install hollow metal door in 8" block exterior wall

Description	Quantity	Unit	Work Hours	Material
Saw cutting block wall, per inch of depth	192.000	L.F.	12.672	44.16
Concrete block demolition	34.000	S.F.	1.666	0.00
Remove window	1.000	Ea.	0.615	0.00
Labor minimum for masonry repairs	1.000	Job	2.000	0.00
Steel lintel, 4" x 3-1/2" x 1/4", 5'-0" long	1.000	Ea.	0.200	15.50
Steel door frame, 3'-0" x 6'-8"	1.000	Ea.	1.000	65.50
Hollow metal door	1.000	Ea.	0.941	147.00
Door closer	1.000	Ea.	1.333	95.50
Lever-handled lockset	1.000	Ea.	0.000	79.78
Hinges	1.500	Pr.	0.000	69.00
Threshold	1.000	Ea.	0.400	27.50
Paint door	1.000	Ea.	0.941	4.96
Silicone sealant	17.000	L.F.	0.578	4.08
Totals			22.346	552.98

Total per each including general contractor's overhead and profit $2,425

Copyright R. S. Means Company, Inc, 1994

Notes

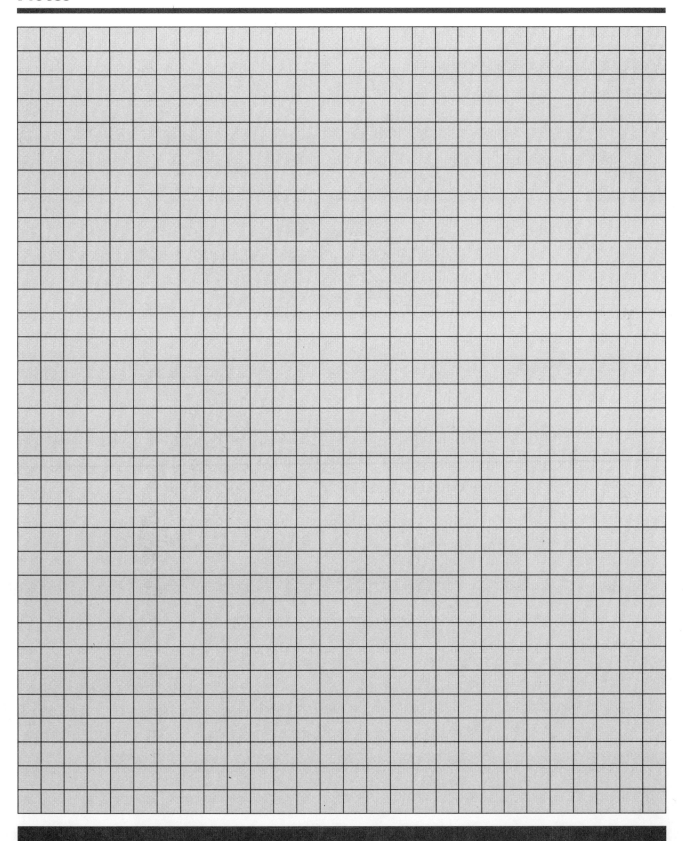

22
Install New Door: Glass Storefront

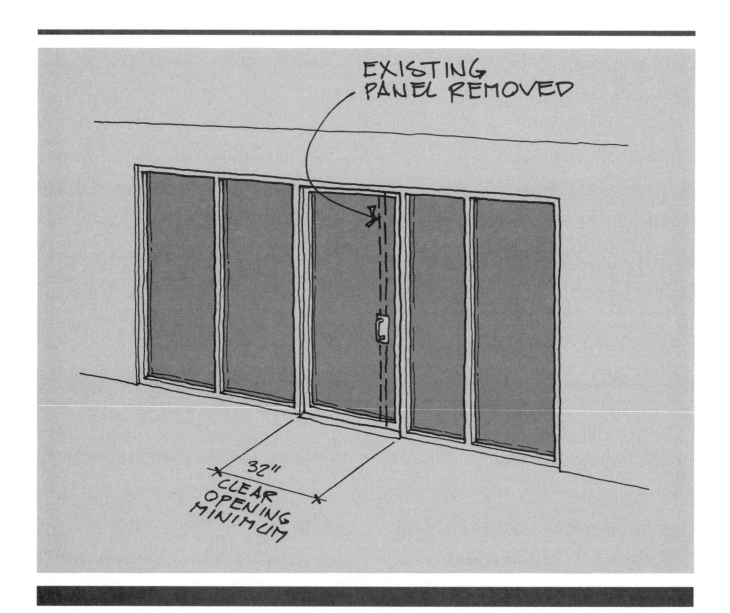

Many public facilities have glass storefront facades. Although systems and materials vary widely, many storefronts are made of modules or partitions that can be removed and replaced relatively simply. This allows for the installation of an accessible door in an existing building while maintaining the building design.

ADAAG Reference

4.13 Doors

Where Applicable

Doors to accessible areas on accessible routes of travel placed in walls constructed of metal framed glass.

Design Requirements

- 32" clear opening width.
- 18" clearance adjacent to the latch on the pull side of the door, 12" on the push side.
- 1/2" maximum beveled threshold.
- 5 lbs. maximum pull or push weight on interior doors (no ADAAG reference for exterior doors).
- Level maneuvering space on both sides of the door, depending on approach. Minimum required dimensions in front of door between 42" to 60", depending on approach and whether door has a closer (see ADAAG Fig. 25).
- 60" level surface, inside and outside of door at entrances.
- Accessible hardware (acceptable if operable with a closed fist).
- With door closer, 3 seconds minimum closing time to a point 3" from the latch.

Design Suggestions

Because the 32" clear opening is measured from the face of the door in a 90° open position to the stop on the opposite jamb, the door itself has to be wider (usually 36") in order to comply. 2'-10" doors are the smallest that can be used to comply, but might not meet the requirement. There are several accessible hardware options: a loop (allow at least 3" between inside of loop and face of door), lever handles, push plate, or panic bar. It is recommended (but not required) that the maximum opening force for exterior doors be 8 lbs. This may not be possible due to wind pressure or interior air circulation systems. Where opening force is necessarily high or where adequate maneuvering space cannot be provided, installation of an automatic opener may be a solution.

Maneuvering space on each side of the door is determined by how it is approached. A 60" × 60" minimum clear space is best and complies with all approaches cited in ADAAG. Where only a straight-on approach is available, a 60" deep space measured from the face of the door is required on the pull side, and a 48" deep space is required on the push side. Where only a side approach is available, the required depth of the clear area in front of the door varies from 42" to 60", and the width is affected by the latch edge clearances and presence or absence of a door opener. Consult ADAAG 4.13 and ADAAG Fig. 25 for exact requirements.

Key Items

Door, new structural members for opening, wall/floor finishes to match existing. Installation of an accessible door in an existing metal and glass storefront wall involves removal of adjacent panels, and replacing with narrower panels to accommodate the wider door.

Level of Difficulty

Moderate to high. Involves demolition, bracing of new opening, and finish work. If storefront is one integral unit, installing or widening a door can require removal of the wall. Requires a building permit and possibly design drawings.

22. Install New Door: Glass Storefront *(continued)*

Estimates

Install new storefront door

Description	Quantity	Unit	Work Hours	Material
Remove glass	60.000	S.F.	3.180	0.00
Cut back tube frame sills	1.000	Ea.	1.000	0.00
New tube frame jambs	20.000	L.F.	4.580	153.00
Install new tube frame header	3.000	L.F.	0.615	31.95
Door stop (snap in)	23.000	L.F.	0.966	43.24
Install new insulated glass	39.000	S.F	8.307	528.45
Install new storefront door	1.000	Ea.	8.000	605.00
Totals			26.648	1361.64

Total per each including general contractor's overhead and profit **$3,828**

Remove glass panels

Description	Quantity	Unit	Work Hours	Material
Remove glass, 3'-0" × 8'-0"	24.000	S.F.	1.272	0.00
Totals			1.272	0.00

Total per each including general contractor's overhead and profit **$55**

Copyright R. S. Means Company, Inc, 1994

Notes

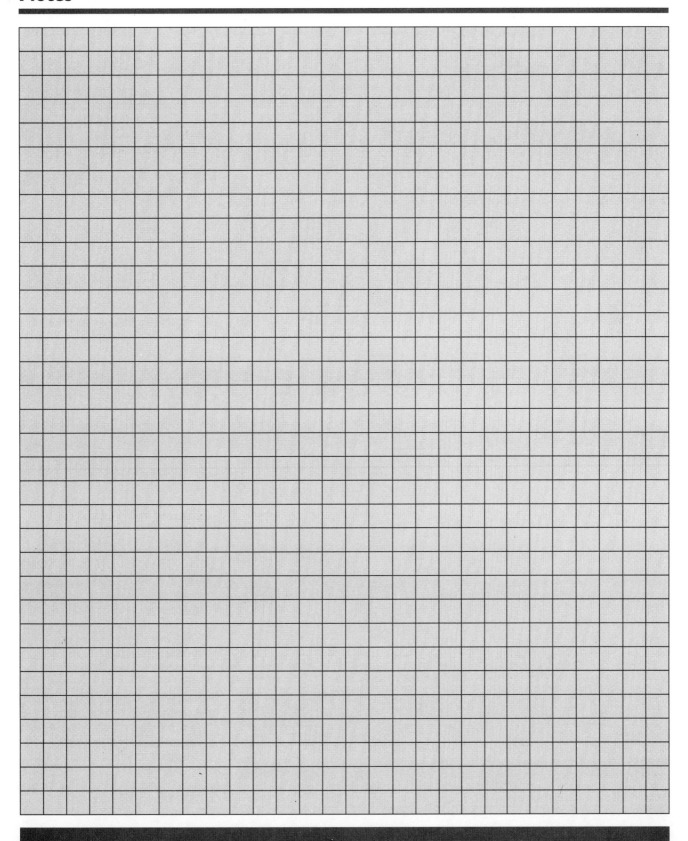

23
Install Automatic Door Opener

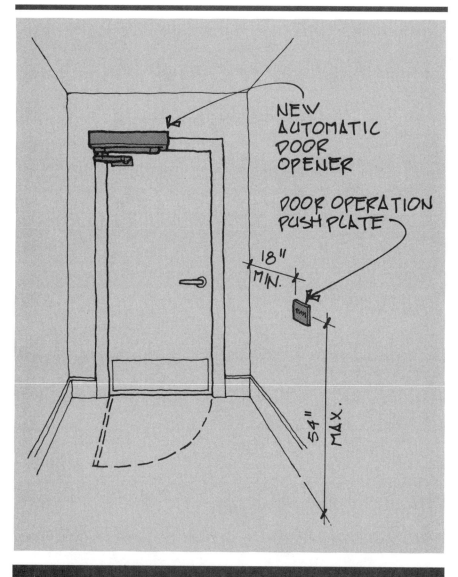

ADAAG Reference
4.13 Doors

Where Applicable
Doors to accessible areas on accessible routes of travel which are either heavy to open or don't have sufficient space in front of, or adjacent to, the door to allow wheelchair access.

Design Requirements
- 32" clear opening width.
- 18" clearance adjacent to the latch on the pull side of the door, 12" on the push side.
- 1/2" maximum beveled threshold.
- 5 lbs. maximum pull or push weight on interior doors (no ADAAG reference for exterior doors).
- Level maneuvering space on both sides of the door, depending on approach. Minimum required dimensions in front of door between 42" to 60", depending on approach and whether door has a closer *(see ADAAG Fig. 25)*.
- 60" level surface, inside and outside of door at entrances.
- Accessible hardware (acceptable if operable with a closed fist).
- With door closer, 3 seconds minimum closing time to a point 3" from the latch.

Design Suggestions
Automatic openers are useful for making doors accessible where the door opening pressure is excessive or there is insufficient maneuvering clearance on one or both sides of the door. Some automatic openers must be used when opening the door, but others have an arm which rests against the door allowing the option of automatic or manual use and these are preferred. It is recommended that the maximum opening force for exterior doors be 8 lbs. of pressure. This may not be possible due to wind pressure or interior air circulation systems, but if operating force exceeds 8 lbs. for exterior doors, consider installing an automatic door opener or a power assist.

Key Items
Door opener and electrical supply, possible wall demo and finish work.

Level of Difficulty
Moderate to high. Higher for load-bearing walls. Involves demolition, bracing of new opening, structural framing, finish work.

Estimates

Install automatic exterior door opener

Description	Quantity	Unit	Work Hours	Material
Cutout demolition of ext. wall (12" thick brick wall)	1.000	Ea.	4.000	0.00
Conductor	0.100	C.L.F.	0.296	1.55
Install junction box	1.000	Ea.	0.400	4.65
Install outlet	1.000	Ea.	0.296	5.25
Install plate	1.000	Ea.	0.100	2.30
Automatic opener, button operation	1.000	Ea.	2.000	1025.00
Totals			7.092	1038.75

Total per each including general contractor's overhead and profit: $2,149

Install automatic interior door opener

Description	Quantity	Unit	Work Hours	Material
Cutout demolition of partition	1.000	Ea.	0.333	0.00
Conductor	0.100	C.L.F.	0.296	1.55
Install junction box	1.000	Ea.	0.400	4.65
Install outlet	1.000	Ea.	0.296	5.25
Install plate	1.000	Ea.	0.100	2.30
Repair gypsum board	1.000	Job	2.000	0.00
Paint gypsum board—minimum	1.000	Job	2.000	0.00
Automatic opener, button operation	1.000	Ea.	2.000	1025.00
Totals			7.425	1038.75

Total per each including general contractor's overhead and profit: $2,289

Install door with infrared activated automatic opener

Description	Quantity	Unit	Work Hours	Material
Single swinging door automatic opener	1.000	Ea.	20.000	2000.00
Infrared detector	1.000	Ea.	3.478	232.00
Totals			23.478	2232.00

Total per each including general contractor's overhead and profit: $5,194

Install power assist interior door opener/closer

Description	Quantity	Unit	Work Hours	Material
Cast iron power assist door opener/closer	1.000	Ea.	1.333	134.00
Totals			1.333	134.00

Total per each including general contractor's overhead and profit: $305

Copyright R. S. Means Company, Inc, 1994

24 Modify Existing Double-Leaf Doors

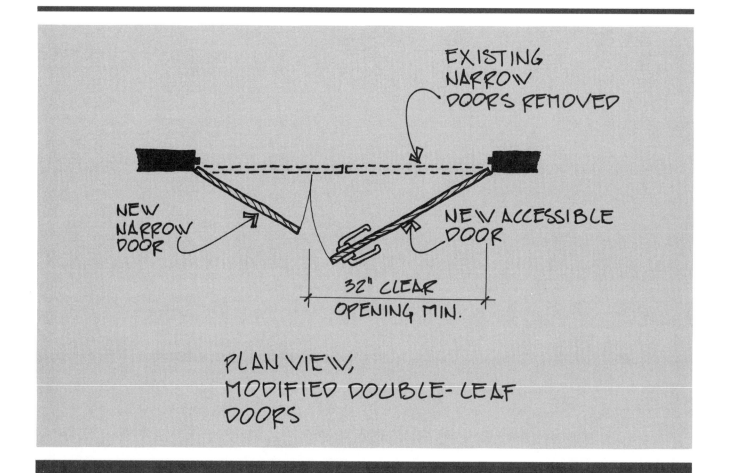

The entire opening of a double-leaf doorway is almost always sufficient for complying with the necessary clear width for an accessible door, so modifying the doors to comply can be a very useful, and often simple, access modification.

ADAAG Reference
4.13 Doors

Where Applicable
All double doorways along accessible routes of travel where at least one leaf does not have a 32" clear opening.

Design Requirements
- At least one leaf of a pair with 32" clear opening.
- All door requirements.

Design Suggestions

Depending on existing conditions, there may be several options for modifying non-compliant double doors:

1. Widen doorway to allow two 34" doors (very difficult if doors span the width of a hallway). As with all doors, 34" is a minimum; 36" is recommended.
2. Install one 34" door (minimum) and fixed sidelight.
3. Install uneven doors, with one leaf 34" minimum.
4. Install automatic door openers (both leaves).
5. Remove doors, if possible without sacrificing privacy, security, or fire safety.
6. Keep doors open during public operating hours (if permitted).

Key Items

Replacement doors and hardware as required, finish materials, possibly structural bracing. Button-operated or infrared automatic opener, hold-open device.

Level of Difficulty

- Option 1: High; requires structural and finish work.
- Options 2 & 3: Moderate to high; more difficult for metal doors.
- Option 4: Varies; some automatic openers can be installed by maintenance staff, while others require electrical and finish work.
- Options 5 & 6: Low.

Estimates

Widen existing brick/block opening & replace w/solid wood door

Description	Quantity	Unit	Work Hours	Material
Remove interior door	1.000	Ea.	0.400	0.00
Remove door frame	1.000	Ea.	0.571	0.00
Saw cutting block wall, per inch of depth	64.000	L.F.	4.224	14.72
Remove masonry partition	9.000	S.F.	0.441	0.00
Labor minimum for masonry repairs	1.000	Job	2.000	0.00
Steel lintel, 4" × 3-1/2" × 1/4", 5'-0" long	1.000	Ea.	0.200	15.50
Interior door frame	18.000	L.F.	0.774	57.60
Interior wood door, birch face, 3'-0" × 6'-8"	1.000	Ea.	1.143	65.00
Add for solid core door	1.000	Ea.	0.000	27.50
Hinges	1.500	Pr.	0.000	69.00
Lever-handled lockset	1.000	Ea.	0.000	79.78
Threshold	1.000	Ea.	0.400	27.50
Paint door	1.000	Ea.	0.941	4.96
Painting	68.000	S.F.	1.224	11.56
Totals			12.318	373.12

Total per each including general contractor's overhead and profit: $1,615

Copyright R. S. Means Company, Inc., 1994

24. Modify Existing Double-Leaf Doors (continued)

Widen existing brick/block opening & replace w/hollow wood door

Description	Quantity	Unit	Work Hours	Material
Remove interior door	1.000	Ea.	0.400	0.00
Remove door frame	1.000	Ea.	0.571	0.00
Saw cutting block wall, per inch of depth	64.000	L.F.	4.224	14.72
Remove masonry partition	9.000	S.F.	0.441	0.00
Labor minimum for masonry repairs	1.000	Job	2.000	0.00
Steel lintel, 4" x 3-1/2" x 1/4", 5'-0" long	1.000	Ea.	0.200	15.50
Interior door frame	18.000	L.F.	0.774	57.60
Interior wood door, birch face, 3'-0" x 6'-8"	1.000	Ea.	1.143	65.00
Hinges	1.500	Pr.	0.000	69.00
Lever-handled lockset	1.000	Ea.	0.000	79.78
Threshold	1.000	Ea.	0.400	27.50
Paint door	1.000	Ea.	0.941	4.96
Painting	68.000	S.F.	1.224	11.56
Totals			12.318	345.62

Total per each including general contractor's overhead and profit: $1,568

Widen existing brick/block opening & replace w/metal door

Description	Quantity	Unit	Work Hours	Material
Remove interior door	1.000	Ea.	0.400	0.00
Remove door frame	1.000	Ea.	0.571	0.00
Saw cutting block wall, per inch of depth	64.000	L.F.	4.224	14.72
Remove masonry partition	9.000	S.F.	0.441	0.00
Labor minimum for masonry repairs	1.000	Job	2.000	0.00
Steel lintel, 4" x 3-1/2" x 1/4", 5'-0" long	1.000	Ea.	0.200	15.50
Interior door frame	1.000	Ea.	1.000	65.50
Hollow metal flush door, 3'-0" x 6'-8"	1.000	Ea.	0.941	147.00
Hinges	1.500	Pr.	0.000	69.00
Lever-handled lockset	1.000	Ea.	0.000	79.78
Threshold	1.000	Ea.	0.400	27.50
Paint door	1.000	Ea.	0.941	4.96
Painting	68.000	S.F.	1.224	11.56
Totals			12.342	435.52

Total per each including general contractor's overhead and profit: $1,723

Copyright R. S. Means Company, Inc., 1994

Widen existing brick/block opening & replace w/double solid wood door

Description	Quantity	Unit	Work Hours	Material
Remove interior door	1.000	Ea.	0.400	0.00
Remove door frame	1.000	Ea.	0.571	0.00
Saw cutting block wall, per inch of depth	80.000	L.F.	5.280	18.40
Remove masonry partition	23.000	S.F.	1.127	0.00
Labor minimum for masonry repairs	1.000	Job	2.000	0.00
Steel lintel, 4" x 3-1/2" x 1/4", 9'-0" long	1.000	Ea.	0.229	28.00
Interior door frame	20.000	L.F.	0.860	64.00
Interior wood door, birch face, 3'-0" x 6'-8"	1.000	Ea.	1.143	65.00
Interior wood door, birch face, 2'-6" x 6'-8"	1.000	Ea.	1.067	58.50
Add for solid core door	2.000	Ea.	0.000	55.00
Hinges	3.000	Pr.	0.000	138.00
Lever-handled lockset	1.000	Ea.	0.000	79.78
Threshold	2.000	Ea.	0.800	55.00
Paint door	2.000	Ea.	1.882	9.92
Painting	80.000	S.F.	1.440	13.60
Totals			16.799	585.20

Total per each including general contractor's overhead and profit $2,197

Widen existing brick/block opening & replace w/double hollow wood door

Description	Quantity	Unit	Work Hours	Material
Remove interior door	1.000	Ea.	0.400	0.00
Remove door frame	1.000	Ea.	0.571	0.00
Saw cutting block wall, per inch of depth	80.000	L.F.	5.280	18.40
Remove masonry partition	23.000	S.F.	1.127	0.00
Labor minimum for masonry repairs	1.000	Job	2.000	0.00
Steel lintel, 4" x 3-1/2" x 1/4", 9'-0" long	1.000	Ea.	0.229	28.00
Interior door frame	20.000	L.F.	0.860	64.00
Interior wood door, birch face, 3'-0" x 6'-8"	1.000	Ea.	1.143	65.00
Interior wood door, birch face, 2'-6" x 6'-8"	1.000	Ea.	1.067	58.50
Hinges	3.000	Pr.	0.000	138.00
Lever-handled lockset	1.000	Ea.	0.000	79.78
Threshold	2.000	Ea.	0.800	55.00
Paint door	2.000	Ea.	1.882	9.92
Painting	80.000	S.F.	1.440	13.60
Totals			16.799	530.20

Total per each including general contractor's overhead and profit $2,102

Copyright R. S. Means Company, Inc., 1994

24. Modify Existing Double-Leaf Doors (continued)

Widen existing brick/block opening & replace w/double hollow metal door

Description	Quantity	Unit	Work Hours	Material
Remove interior door	1.000	Ea.	0.400	0.00
Remove door frame	1.000	Ea.	0.571	0.00
Saw cutting block wall, per inch of depth	80.000	L.F.	5.280	18.40
Remove masonry partition	23.000	S.F.	1.127	0.00
Labor minimum for masonry repairs	1.000	Job	2.000	0.00
Steel lintel, 4" x 3-1/2" x 1/4", 9'-0" long	1.000	Ea.	0.229	28.00
Interior door frame	1.000	Ea.	1.143	85.50
Hollow metal flush door, 3'-0" x 6'-8"	1.000	Ea.	0.941	147.00
Hollow metal flush door, 2'-0" x 6'-8"	1.000	Ea.	0.800	130.00
Hinges	3.000	Pr.	0.000	138.00
Lever-handled lockset	1.000	Ea.	0.000	79.78
Threshold	2.000	Ea.	0.800	55.00
Paint door	2.000	Ea.	1.882	9.92
Painting	80.000	S.F.	1.440	13.60
Totals			16.613	705.20

Total per each including general contractor's overhead and profit: $2,389

Widen existing stud wall opening & replace w/solid wood door

Description	Quantity	Unit	Work Hours	Material
Remove interior door	1.000	Ea.	0.400	0.00
Remove door frame	1.000	Ea.	0.571	0.00
Remove stud wall partition	7.000	S.F.	0.322	0.00
Interior door frame	18.000	L.F.	0.774	57.60
Interior wood door, birch face, 3'-0" x 6'-8"	1.000	Ea.	1.143	65.00
Add for solid core door	1.000	Ea.	0.000	27.50
Hinges	1.500	Pr.	0.000	69.00
Lever-handled lockset	1.000	Ea.	0.000	79.78
Threshold	1.000	Ea.	0.400	27.50
Paint door	1.000	Ea.	0.941	4.96
Painting	68.000	S.F.	1.224	11.56
Totals			5.775	342.90

Total per each including general contractor's overhead and profit: $923

Copyright R. S. Means Company, Inc., 1994

Widen existing stud wall opening & replace w/hollow wood door

Description	Quantity	Unit	Work Hours	Material
Remove interior door	1.000	Ea.	0.400	0.00
Remove door frame	1.000	Ea.	0.571	0.00
Remove stud wall partition	7.000	S.F.	0.322	0.00
Interior door frame	18.000	L.F.	0.774	57.60
Interior wood door, birch face, 3'-0" × 6'-8"	1.000	Ea.	1.143	65.00
Hinges	1.500	Pr.	0.000	69.00
Lever-handled lockset	1.000	Ea.	0.000	79.78
Threshold	1.000	Ea.	0.400	27.50
Paint door	1.000	Ea.	0.941	4.96
Painting	68.000	S.F.	1.224	11.56
Totals			5.775	315.40

Total per each including general contractor's overhead and profit $876

Widen existing stud wall opening & replace w/metal door

Description	Quantity	Unit	Work Hours	Material
Remove interior door	1.000	Ea.	0.400	0.00
Remove door frame	1.000	Ea.	0.571	0.00
Remove stud wall partition	7.000	S.F.	0.322	0.00
Interior door frame	1.000	Ea.	1.000	65.50
Hollow metal flush door, 3'-0" × 6'-8"	1.000	Ea.	0.941	147.00
Hinges	1.500	Pr.	0.000	69.00
Lever-handled lockset	1.000	Ea.	0.000	79.78
Threshold	1.000	Ea.	0.400	27.50
Paint door	1.000	Ea.	0.941	4.96
Painting	68.000	S.F.	1.224	11.56
Totals			5.799	405.30

Total per each including general contractor's overhead and profit $1,031

Widen existing stud wall opening & replace w/double solid wood door

Description	Quantity	Unit	Work Hours	Material
Remove interior door	1.000	Ea.	0.400	0.00
Remove door frame	1.000	Ea.	0.571	0.00
Remove stud wall partition	21.000	S.F.	0.966	0.00
Interior door frame	20.000	L.F.	0.860	64.00
Interior wood door, birch face, 3'-0" × 6'-8"	1.000	Ea.	1.143	65.00
Interior wood door, birch face, 2'-6" × 6'-8"	1.000	Ea.	1.067	58.50
Add for solid core door	2.000	Ea.	0.000	55.00
Hinges	3.000	Pr.	0.000	138.00
Lever-handled lockset	1.000	Ea.	0.000	79.78
Threshold	2.000	Ea.	0.800	55.00
Paint door	2.000	Ea.	1.882	9.92
Painting	80.000	S.F.	1.440	13.60
Totals			9.129	538.80

Total per each including general contractor's overhead and profit $1,419

Copyright R. S. Means Company, Inc., 1994

24. Modify Existing Double-Leaf Doors *(continued)*

Widen existing stud wall opening & replace w/double hollow wood door

Description	Quantity	Unit	Work Hours	Material
Remove interior door	1.000	Ea.	0.400	0.00
Remove door frame	1.000	Ea.	0.571	0.00
Remove stud wall partition	21.000	S.F.	0.966	0.00
Interior door frame	20.000	L.F.	0.860	64.00
Interior wood door, birch face, 3'-0" × 6'-8"	1.000	Ea.	1.143	65.00
Interior wood door, birch face, 2'-6" × 6'-8"	1.000	Ea.	1.067	58.50
Hinges	3.000	Pr.	0.000	138.00
Lever-handled lockset	1.000	Ea.	0.000	79.78
Threshold	2.000	Ea.	0.800	55.00
Paint door	2.000	Ea.	1.882	9.92
Painting	80.000	S.F.	1.440	13.60
Totals			9.129	483.80

Total per each including general contractor's overhead and profit   **$1,325**

Widen existing stud wall opening & replace w/double hollow metal door

Description	Quantity	Unit	Work Hours	Material
Remove interior door	1.000	Ea.	0.400	0.00
Remove door frame	1.000	Ea.	0.571	0.00
Remove stud wall partition	21.000	S.F.	0.966	0.00
Interior door frame	1.000	Ea.	1.143	85.50
Hollow metal flush door, 3'-0" × 6'-8"	1.000	Ea.	0.941	147.00
Hollow metal flush door, 2'-0" × 6'-8"	1.000	Ea.	0.800	130.00
Hinges	3.000	Pr.	0.000	138.00
Lever-handled lockset	1.000	Ea.	0.000	79.78
Threshold	2.000	Ea.	0.800	55.00
Paint door	2.000	Ea.	1.882	9.92
Painting	80.000	S.F.	1.440	13.60
Totals			8.943	658.80

Total per each including general contractor's overhead and profit   **$1,611**

Install automatic door opener

Description	Quantity	Unit	Work Hours	Material
Cutout demolition of partition	1.000	Ea.	0.333	0.00
Conductor fished to nearby junction box	0.100	C.L.F.	0.333	2.90
Junction box	1.000	Ea.	0.400	1.54
Automatic opener, button operation	1.000	Ea.	2.000	1025.00
Totals			3.066	1029.44

Total per each including general contractor's overhead and profit   **$1,921**

Copyright R. S. Means Company, Inc., 1994

Remove doors

Description	Quantity	Unit	Work Hours	Material
Remove interior door	1.000	Ea.	0.400	0.00
Totals			**0.400**	**0.00**

Total per each including general contractor's overhead and profit — **$18**

Install magnetic hold-open devices

Description	Quantity	Unit	Work Hours	Material
Magnetic door holder	1.000	Ea.	2.000	71.00
Cutout demolition of partition	1.000	Ea.	0.333	0.00
Conductor fished to nearby junction box	0.100	C.L.F.	0.333	2.90
Junction box	1.000	Ea.	0.400	1.54
Totals			**3.066**	**75.44**

Total per each including general contractor's overhead and profit — **$308**

Copyright R. S. Means Company, Inc, 1994

25
Modify Existing Door

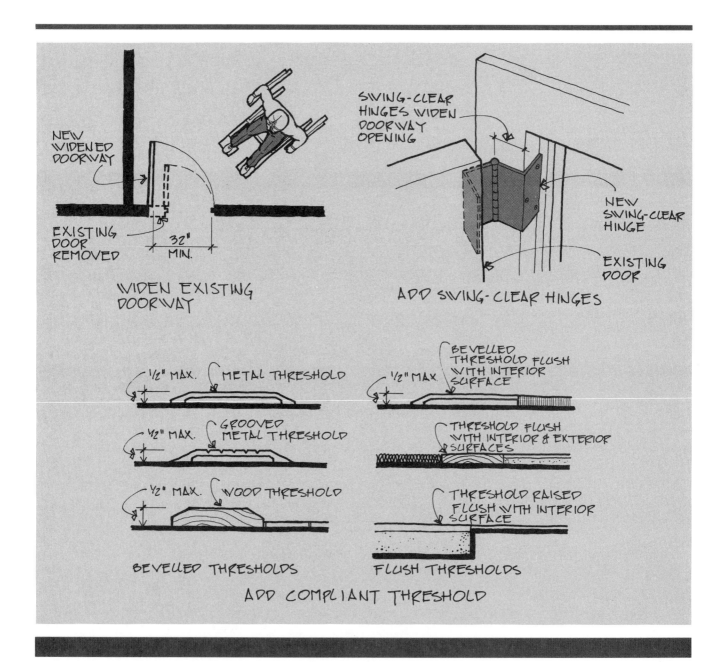

Some modifications are relatively simple on any door. Examples include replacing or modifying existing hardware, beveling an existing threshold, or removing an existing threshold. Reversing a door swing can solve maneuvering clearance problems, but can be difficult with metal doors because hinges and hardware have to be changed and the frame sometimes has to be reversed in the opening. Swing-clear hinges free the space normally occupied by the door (when the door is open), although they can also be difficult to install on metal doors.

ADAAG Reference
4.13 Doors

Where Applicable
Non-compliant doors along an accessible route.

Design Requirements
- 32" clear opening width.
- Accessible hardware (acceptable if operable with a closed fist).
- Threshold 1/2" high maximum, beveled.
- Interior pull weight 5 lbs. maximum (no ADAAG reference for exterior doors).
- 18" clearance adjacent to the latch on the pull side of the door, 12" on the push side, with level clearances on both sides of the door as required (consult ADAAG for exact dimensions).
- 3 second minimum closing time to a point 3" from the latch.

Design Suggestions
If there is a closer on the door, the closing speed can usually be adjusted which improves accessibility. If there are hinges, oiling them is another simple way to make a door easier to open. It is almost always possible (but sometimes expensive) to install an automatic opener to compensate for insufficient latchside clearance or heavy door opening weights.

Key Items
Door hardware, hinges, thresholds, automatic opener.

Level of Difficulty
Low to moderate. Some finish work required for threshold removal. Automatic openers require preliminary electrical work, and can be expensive.

Estimates

Reverse door swing, wood door

Description	Quantity	Unit	Work Hours	Material
Remove door	1.000	Ea.	0.400	0.00
Remove door frame & trim	1.000	Ea.	0.571	0.00
Interior door frame	18.000	L.F.	0.774	57.60
Door trim set, 1 head, 2 sides, 2-1/2" pine	2.000	Opng.	2.712	22.00
Remove door hinge	3.000	Ea.	0.501	0.00
Install door hinge	3.000	Ea.	0.501	0.00
Re-install door	1.000	Ea.	1.143	0.00
Remove lockset	1.000	Ea.	0.040	0.00
Re-install lockset	1.000	Ea.	0.667	0.00
Paint door frame & trim	18.000	L.F.	0.450	0.72
Totals			7.759	80.32

Total per each including general contractor's overhead and profit: $571

Copyright R. S. Means Company, Inc., 1994

25. Modify Existing Door (continued)

Replace existing hinges with swing-clear hinges

Description	Quantity	Unit	Work Hours	Material
Remove interior door	1.000	Ea.	0.400	0.00
Remove door hinge	3.000	Ea.	0.501	0.00
Swing clear hinges	1.500	Pr.	0.000	136.50
Install door hinge	3.000	Ea.	0.501	0.00
Re-install door	1.000	Ea.	1.143	0.00
Totals			2.545	136.50

Total per set of hinges including general contractor's overhead and profit — **$368**

Replace existing lockset with lever-handled lockset

Description	Quantity	Unit	Work Hours	Material
Remove lockset	1.000	Ea.	0.040	0.00
Lever-handled lockset	1.000	Ea.	0.000	79.78
Totals			0.040	79.78

Total per each including general contractor's overhead and profit — **$204**

Install push plates on wood door

Description	Quantity	Unit	Work Hours	Material
Aluminum push plates, both sides of door	1.000	Ea.	0.571	11.90
Totals			0.571	11.90

Total per each including general contractor's overhead and profit — **$51**

Install push plates and panic bar

Description	Quantity	Unit	Work Hours	Material
Aluminum push plates, both sides of door	1.000	Ea.	0.571	11.90
Panic bar and verticle rod, for exit only	1.000	Ea.	1.600	410.00
Totals			2.171	421.90

Total per each including general contractor's overhead and profit — **$834**

Remove threshold, fill floor flush with existing flooring (vinyl tile)

Description	Quantity	Unit	Work Hours	Material
Remove threshold	1.000	Ea.	0.200	0.00
Vinyl tile	3.000	S.F.	0.048	8.25
Minimum labor to patch subfloor and lay new tile	1.000	Job	2.000	0.00
Totals			2.248	8.25

Total per each including general contractor's overhead and profit — **$136**

Copyright R. S. Means Company, Inc., 1994

Bevel 2 existing wood thresholds

Description	Quantity	Unit	Work Hours	Material
Minimum labor to bevel 2 wood thresholds	1.000	Job	2.000	0.00
Totals			2.000	0.00

Total per each including general contractor's overhead and profit	**$57**
Total per 2 thresholds including general contractor's overhead and profit	**$114**

Remove 2 existing thresholds

Description	Quantity	Unit	Work Hours	Material
Min. labor to remove 2 existing thresholds (no patching incl.)	1.000	Job	2.000	0.00
Totals			2.000	0.00

Total per each including general contractor's overhead and profit	**$57**
Total per 2 thresholds including general contractor's overhead and profit	**$114**

Adjust door-closing speed (5 doors)

Description	Quantity	Unit	Work Hours	Material
Minimum labor to adjust 5 door closers	1.000	Job	2.000	0.00
Totals			2.000	0.00

Total per each including general contractor's overhead and profit	**$23**
Total per 5 closers including general contractor's overhead and profit	**$114**

Install automatic door opener

Description	Quantity	Unit	Work Hours	Material
Cutout demolition of partition	1.000	Ea.	0.333	0.00
Conductor fished to nearby junction box	0.100	C.L.F.	0.333	2.90
Junction box	1.000	Ea.	0.400	1.54
Automatic opener, button operation	1.000	Ea.	2.000	1025.00
Totals			3.066	1029.44

Total per each including general contractor's overhead and profit	**$1,921**

Copyright R. S. Means Company, Inc, 1994

26
Install Sliding Door

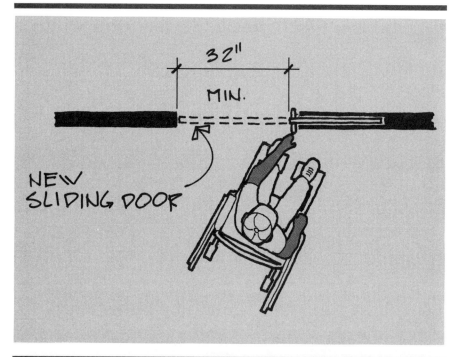

Installation of an accessible sliding door can create an accessible doorway where door swings might otherwise prevent access. Sliding doors can fit within the width of a standard stud wall, and can be a useful and creative method of creating an accessible route between two spaces.

ADAAG Reference
4.13 Doors

Where Applicable
Doors to areas on accessible routes of travel where swinging doors block required maneuvering space.

Design Requirements
- 32" minimum clear opening width.
- 5 lbs. maximum pull weight on interior doors (no ADAAG reference for exterior doors).
- Thresholds at maximum height of 3/4" for exterior sliding doors, 1/2" maximum for others. All thresholds beveled at a maximum slope of 1:2 maximum.
- 48" clearance from face of door for front approach, 42" for side approach.
- Accessible hardware (acceptable if operable with a closed fist).

Design Suggestions
The 32" clear opening requirement is a minimum. Wider sliding doors are only marginally more expensive than narrow doors, and construction costs are the same. Finding compliant hardware for sliding doors is usually more difficult than for swinging doors (sliding doors usually just have a small latch or button), so nonstandard hardware might be needed.

Key Items
Door, door frame, track hardware, new structural members for opening, and wall/floor finishes to match existing.

Level of Difficulty
Moderate to high. Involves demolition, bracing of new opening, structural framing, finish work.

Estimates

Install solid core wood pocket door in metal stud wall

Description	Quantity	Unit	Work Hours	Material
Remove interior door	1.000	Ea.	0.400	0.00
Remove door frame	1.000	Ea.	0.571	0.00
Remove stud wall partition	21.000	S.F.	0.966	0.00
Pocket door frame	1.000	Ea.	1.000	79.00
Gypsum board finish on pocket door frame	64.000	S.F.	1.088	12.80
Cased opening jamb and header	10.000	L.F.	0.400	5.30
Door trim set, 1 head, 2 sides, 2-1/2" pine	2.000	Opng.	2.712	22.00
Interior wood door, birch face, 3'-0" x 6'-8"	1.000	Ea.	1.143	65.00
Add for solid core door	1.000	Ea.	0.000	27.50
Lever-handled lockset	1.000	Ea.	0.000	79.78
Totals			8.280	291.38

Total per each including general contractor's overhead and profit — **$980**

Install solid core wood pocket door in wood stud wall

Description	Quantity	Unit	Work Hours	Material
Remove interior door	1.000	Ea.	0.400	0.00
Remove door frame	1.000	Ea.	0.571	0.00
Remove stud wall partition	21.000	S.F.	0.966	0.00
Pocket door frame	1.000	Ea.	1.000	79.00
Gypsum board finish on pocket door frame	64.000	S.F.	1.088	12.80
Cased opening jamb and header	10.000	L.F.	0.400	5.30
Door trim set, 1 head, 2 sides, 2-1/2" pine	2.000	Opng.	2.712	22.00
Interior wood door, birch face, 3'-0" x 6'-8"	1.000	Ea.	1.143	65.00
Add for solid core door	1.000	Ea.	0.000	27.50
Lever-handled lockset	1.000	Ea.	0.000	79.78
Totals			8.280	291.38

Total per each including general contractor's overhead and profit — **$980**

Install hollow core wood pocket door in metal stud wall

Description	Quantity	Unit	Work Hours	Material
Remove interior door	1.000	Ea.	0.400	0.00
Remove door frame	1.000	Ea.	0.571	0.00
Remove stud wall partition	21.000	S.F.	0.966	0.00
Pocket door frame	1.000	Ea.	1.000	79.00
Gypsum board finish on pocket door frame	64.000	S.F.	1.088	12.80
Cased opening jamb and header	10.000	L.F.	0.400	5.30
Door trim set, 1 head, 2 sides, 2-1/2" pine	2.000	Opng.	2.712	22.00
Interior wood door, birch face, 3'-0" x 6'-8"	1.000	Ea.	1.143	65.00
Lever-handled lockset	1.000	Ea.	0.000	79.78
Totals			8.280	263.88

Total per each including general contractor's overhead and profit — **$933**

Copyright R. S. Means Company, Inc., 1994

26. Install Sliding Door (continued)

Install hollow core wood pocket door in wood stud wall

Description	Quantity	Unit	Work Hours	Material
Remove interior door	1.000	Ea.	0.400	0.00
Remove door frame	1.000	Ea.	0.571	0.00
Remove stud wall partition	21.000	S.F.	0.966	0.00
Pocket door frame	1.000	Ea.	1.000	79.00
Gypsum board finish on pocket door frame	64.000	S.F.	1.088	12.80
Cased opening jamb and header	10.000	L.F.	0.400	5.30
Door trim set, 1 head, 2 sides, 2-1/2" pine	2.000	Opng.	2.712	22.00
Interior wood door, birch face, 3'-0" x 6'-8"	1.000	Ea.	1.143	65.00
Lever-handled lockset	1.000	Ea.	0.000	79.78
Totals			8.280	263.88

Total per each including general contractor's overhead and profit: $933

Copyright R. S. Means Company, Inc, 1994

Notes

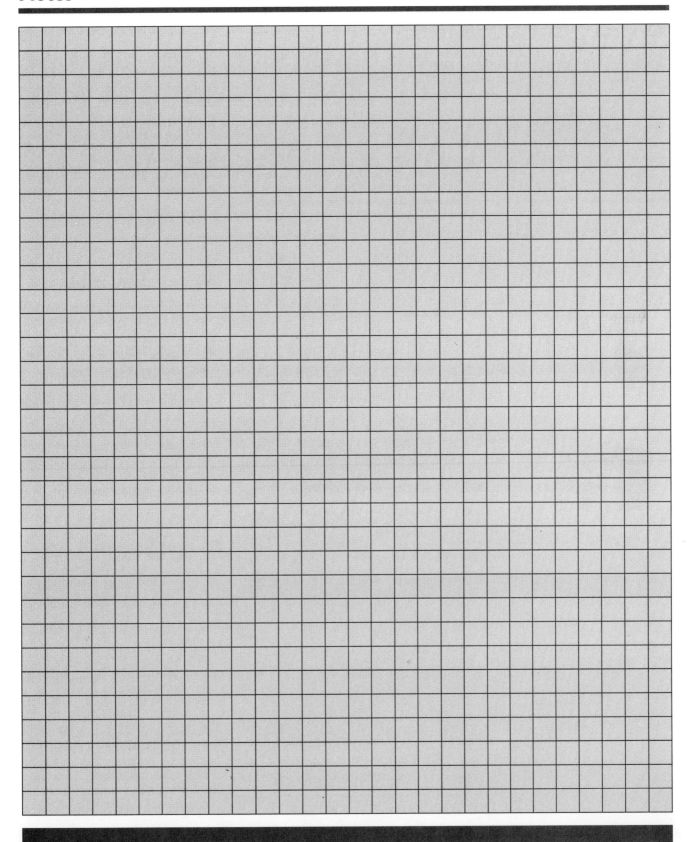

27
Enlarge Vestibule

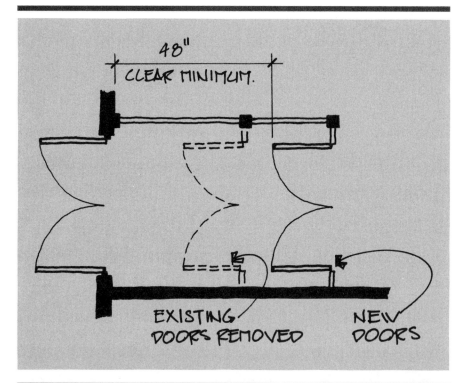

Vestibules at building entrances serve the vital function of conserving energy, and vestibules at rest rooms create visual privacy. If these are the only design considerations used, however, a small vestibule can create a barrier to a building or space for people who need larger maneuvering space. A vestibule with compliant maneuvering space between the doors not only fulfills the energy and visual functions of a vestibule, but also makes it accessible for all users.

ADAAG References
4.13 Doors
4.14 Entrances

Where Applicable
All vestibules on accessible routes with insufficient maneuvering clearances.

Design Requirements
- 48" clear minimum between door swings.
- At least one door of each pair must provide a 32" clear opening.
- Doors compliant with hardware, threshold, and pull weight requirements.
- Level maneuvering spaces on both sides of each door as required (between 42" to 60" from the door depending on approach).

Design Suggestions
Vestibules are built most often at building entrances and toilet rooms. It might be possible to remove the inner set of doors of a vestibule, to enlarge it. However, this is not recommended for a building entrance from an energy standpoint (and might possibly be prohibited by state or local building code). Reversing the door swings at either set of doors might solve wheelchair maneuvering clearance issues, but this is likely to be prohibited by local fire egress codes. Depending on the vestibule's configuration, it could be cheaper to extend the vestibule into the building, rather than add to the existing face of the vestibule, which

would require new foundation work. Even less costly might be adding synchronized door openers operable only on demand.

Key Items
New vestibule and doors, usually interior finish work on floor and wall surfaces, possibly new foundation work.

Level of Difficulty
High.

Estimates

Enlarge existing vestibule by adding on

Description	Quantity	Unit	Work Hours	Material
Excavation	8.000	C.Y.	1.184	0.00
Footing formwork	24.000	SFCA	1.584	9.12
Wall forms	72.000	SFCA	5.688	37.44
Reinforcing (@ 50 lbs./C.Y.)	0.100	Ton	1.067	48.50
Concrete, material only	4.000	C.Y.	0.000	220.00
Placing concrete	4.000	C.Y.	2.132	0.00
Backfill	4.000	C.Y.	2.284	0.00
Slab on grade, 4"	16.000	S.F.	0.352	18.24
Storefront system, commercial grade (with door)	96.000	S.F.	10.272	1056.00
Flat roof framing	35.000	L.F.	0.455	19.25
Sheathing	25.000	S.F.	0.275	9.25
Fascia	18.000	L.F.	0.648	10.62
Cornice	18.000	L.F.	0.720	9.00
Drip edge, 8" girth aluminum	18.000	L.F.	0.360	4.14
Roll roofing	0.250	Sq.	0.519	9.25
Lead flashing	5.000	S.F.	0.295	13.50
Reglet	5.000	L.F.	0.180	4.30
Polyethylene backer rod	0.420	L.F.	0.730	2.90
Silicone sealant	42.000	L.F.	1.428	10.08
Totals			30.173	1481.59

Total per square foot including general contractor's overhead and profit	**$258**
Total per each 16 square foot vestibule including general contractor's overhead and profit	**$4,132**

Remove vestibule doors

Description	Quantity	Unit	Work Hours	Material
Remove double doors	1.000	Ea.	0.500	0.00
Remove glazing	64.000	S.F.	3.392	0.00
Totals			3.892	0.00

Total per set including general contractor's overhead and profit	**$168**

Copyright R. S. Means Company, Inc., 1994

27. Enlarge Vestibule (continued)

Extend vestibule inside building

Description	Quantity	Unit	Work Hours	Material
Remove double doors	1.000	Ea.	0.500	0.00
Remove glazing	64.000	S.F.	3.392	0.00
Storefront system, commercial grade	24.000	S.F.	2.568	264.00
6' × 7' door with hardware	42.000	S.F.	4.998	472.50
Totals			11.458	736.50

Total per set including general contractor's overhead and profit: $1,813

Copyright R. S. Means Company, Inc, 1994

Notes

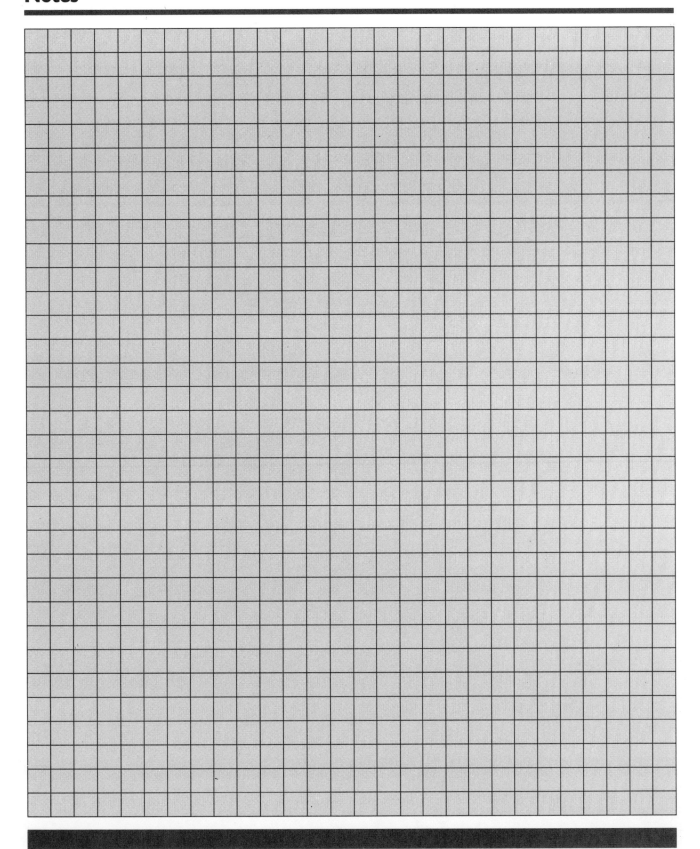

28
Modify Buzzer or Intercom

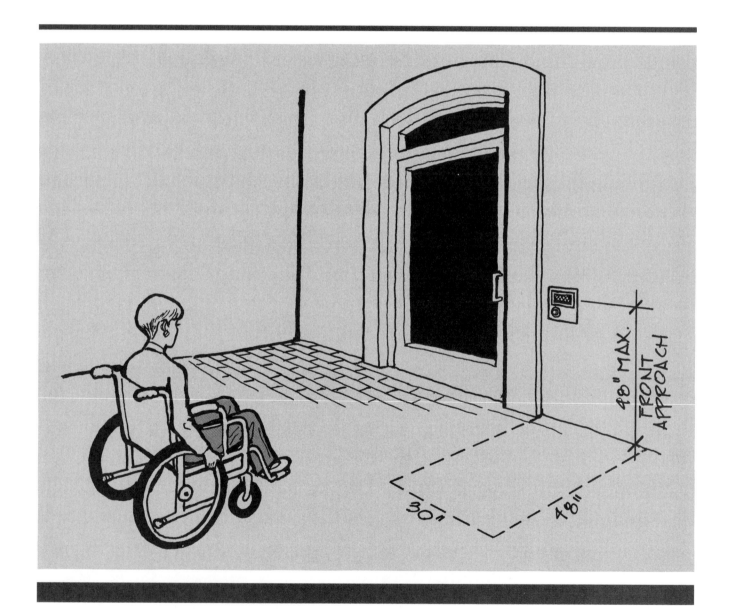

Even where a facility's entrance has compliant door widths and maneuvering spaces, if an individual is unable to reach the communication panel to gain entry, as far as they are concerned, the building is essentially inaccessible. Locating the building's buzzers or intercom at accessible heights, in accessible locations, and with accessible controls creates an entrance that is usable by a much larger group of people with a much wider range of abilities.

ADAAG Reference
4.27 Controls and Operating Mechanisms

Where Applicable
All accessible entrances with buzzers or intercoms.

Design Requirements
- Intercom, buzzer, or control within reach range (48" a.f.f. for front reach, 54" a.f.f. for side reach).
- Accessible controls (if operable with a closed fist).
- 30" × 48" minimum clear floor space in front of buzzer.
- No objects (plants, ashtrays, etc.) below blocking access.

Design Suggestions
If the panel is being lowered, it might be possible to make other alterations at the same time, such as adding large buttons or characters. The voice panel should also be lower than 48". It may be possible to surface-mount the controls to reduce costs. Large push-plates are easier to operate than buttons. If an intercom panel is being replaced as well as lowered, nonvoice communication could be installed as well.

Key Items
Electrical and finish work to match existing.

Level of Difficulty
Moderate to high, depending on the wall material and whether the panel is flush-mounted or installed on surface.

Estimates

Lower existing buzzer/intercom panel in 8" block wall (painted)

Description	Quantity	Unit	Work Hours	Material
Remove panel and fixture	1.000	Ea.	0.031	0.00
Remove junction box	1.000	Ea.	0.100	0.00
Cutout demolition of 8" block	1.000	Ea.	1.481	0.00
Re-install junction box	1.000	Ea.	0.400	0.00
Install outlet	1.000	Ea.	0.296	5.25
Install plate	1.000	Ea.	0.100	2.30
Miscellaneous materials for block repair and painting	1.000	Job	0.000	100.00
Repair block	1.000	Job	2.000	0.00
Paint block—minimum	1.000	Job	2.000	0.00
Totals			6.408	107.55

Total per each including general contractor's overhead and profit: $506

Copyright R. S. Means Company, Inc., 1994

28. Modify Buzzer or Intercom (continued)

Lower existing buzzer/intercom panel in sheet rock/metal stud wall

Description	Quantity	Unit	Work Hours	Material
Remove panel and fixture	1.000	Ea.	0.031	0.00
Remove junction box	1.000	Ea.	0.100	0.00
Cutout demolition of partition	1.000	Ea.	0.333	0.00
Re-install junction box	1.000	Ea.	0.400	0.00
Install outlet	1.000	Ea.	0.296	5.25
Install plate	1.000	Ea.	0.100	2.30
Misc. materials for gypsum board painting and repairs	1.000	Job	0.000	75.00
Repair gypsum board	1.000	Job	2.000	0.00
Paint gypsum board—minimum	1.000	Job	2.000	0.00
Totals			5.260	82.55

Total per each including general contractor's overhead and profit: $545

Copyright R. S. Means Company, Inc, 1994

Notes

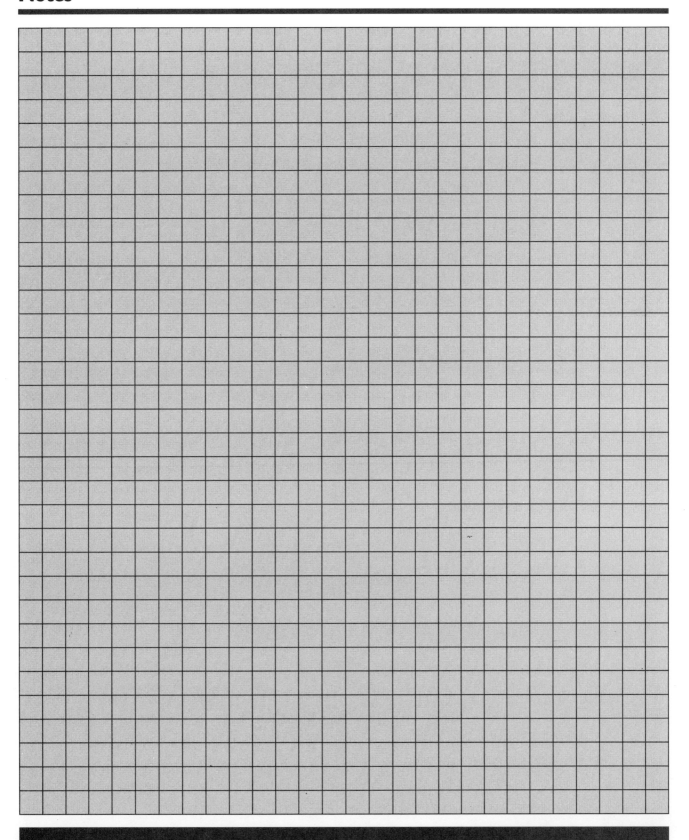

29
Remove or Relocate Partitions

Many accessibility problems can occur within an interior route of travel: Changes in level, protruding objects, slippery surfaces, or insufficient lighting, but tight maneuvering spaces can make the route especially difficult to use. Removing or relocating a few linear feet of partition can often make all the difference in creating an accessible (as well as generally less constrained) route.

ADAAG References
4.3 Accessible Route
4.13 Doors (if alterations to door(s) involved)

Where Applicable
Interior routes of travel with insufficient maneuvering space.

Design Requirements
- 36" wide accessible route.
- Level maneuvering spaces on both sides of each door as required (between 42" to 60" from the door depending on approach).
- If an accessible route makes a U-turn either: (1) a 48" minimum wide space is required to separate two parallel 36" wide pathways with a 36" minimum wide space at the turn, or (2) if the parallel pathways are closer together than 48", they are required to be at least 42" wide with a minimum 48" wide space at the turn (see ADAAG 4.3.3 Accessible Route Width, Figures 7a and 7b).

Design Suggestions
Although the width of an accessible route may be decreased to 32" at a doorway, level maneuvering space always must be provided. For ease of use, a 48" wide route of travel is recommended at 90° turns.

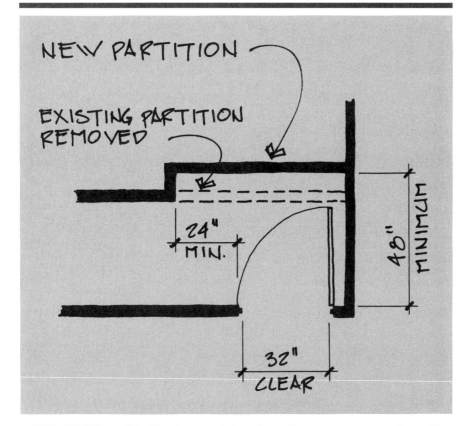

Key Items
New wall materials (studs, gypsum wallboard), new wall/ceiling finish materials to match existing.

Level of Difficulty
Moderate to high. Requires carpentry, electrical, finish work.

Estimate

Remove 8-foot-high wood stud/gypsum board partition and replace (no ceiling work)

Description	Quantity	Unit	Work Hours	Material
Remove stud wall partition	1.000	S.F.	0.046	0.00
1/2" gypsum board on both sides of 2" × 4" wall	1.000	S.F.	0.052	0.78
Painting, 3 coats by roller	2.000	S.F.	0.020	0.20
Vinyl tile	0.125	S.F.	0.002	0.34
Totals			0.120	1.32

Total per square foot including general contractor's overhead and profit: $8

Copyright R. S. Means Company, Inc, 1994

30
Install Wing Walls

Where a building element such as a drinking fountain protrudes into a path of travel, it can present a hazard to people with visual impairments. Building wing walls on either side of it can be an effective and permanent solution to preventing the danger without having to remove or relocate the object.

ADAAG Reference
4.3 Accessible Route
4.4 Protruding Objects

Where Applicable
Where an object protrudes more than 4" into the route of travel and is higher than 27" above the floor and walls are built on either side of the object.

Design Requirements
- Wing walls must protrude at least 12" from wall, and be at least 27" high.

Design Suggestions
Wing walls can provide a permanent means of ensuring that a protruding object complies with ADAAG without replacing or removing the object. A 36" wide route of travel must be maintained; ensure that wing walls do not intrude into 36" clear dimension.

Key Items
Studs/gypsum wallboard, finishes to match existing.

Level of Difficulty
Moderate.

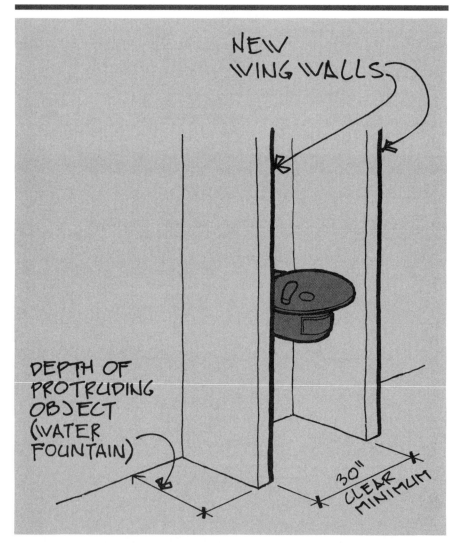

Estimate

Install wing walls

Description	Quantity	Unit	Work Hours	Material
1/2" gypsum board on both sides of 2" × 4" wall	1.000	S.F.	0.052	0.78
Painting, 3 coats by roller	2.000	S.F.	0.020	0.20
Totals			0.072	0.98

Total per square foot including general contractor's overhead and profit: $5

Copyright R. S. Means Company, Inc, 1994

31
Install Slip-Resistant Flooring Materials

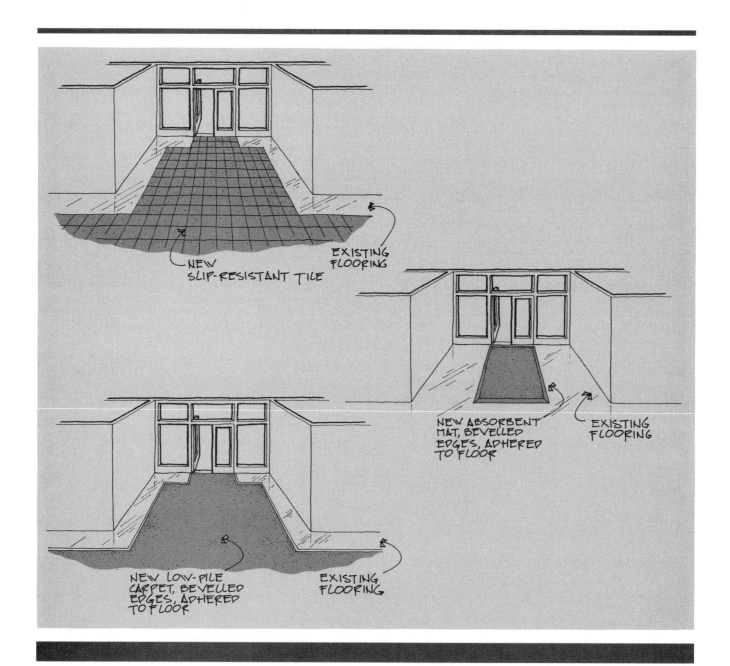

There are two main difficult situations in creating a slip-resistant surface: (1) treating slippery surfaces that are already in place, and (2) use of materials such as vinyl tile or unglazed quarry tile which are of uncertain or varying slip-resistance. For both situations, there are at least four types of solutions:

1. Replace the flooring. This can be expensive, especially with stone or tile.
2. Cover the flooring along the entire route of travel. This is not as expensive as replacing the flooring, but can be difficult at doorways.
3. Cover the floor with carpets or mats at key areas, such as entrance doors (both inside and outside), main interior routes (to information desks or elevators), elevator lobbies, and other highly used areas. Mats can be installed permanently, which is preferred, or only during inclement weather. The mats must be firmly attached to the floor, and must either have beveled edges or be lower than 1/4" in height.
4. Apply a slip-resistant coating to the floor. Terrazzo and vinyl tile can be coated with a non-glossy finish, but the actual slip-resistance of such coatings is unknown. They must also be applied on a regular basis.

Manufacturer's specifications should be checked, and it is recommended that this solution be used in conjunction with slip-resistant mats or carpeting.

ADAAG Reference
4.3 Accessible Route

Where Applicable
All public routes of travel.

Design Requirements
- Flooring to be stable, firm, and slip-resistant. No absolute definition of slip-resistance given; ADAAG Appendix, which is advisory only, recommends using a 0.6 coefficient of friction (check manufacturer's specifications of surfacing materials for level surfaces prior to installation, if available).
- Carpet securely attached, with a firm cushion or no pad, maximum pile thickness 1/2".
- Floor mats securely attached.

Design Suggestions/Considerations
There is no real list of slip-resistant materials, since slip resistance is affected by too many factors, such as the presence of water or slush on the surface and the slipperiness of the sole of a person's shoe. Many designated exterior slip-resistant materials (broom-finish concrete, rough stone, brick or asphalt pavers) might be inappropriate for interior applications. Some materials are usually considered *not* to be slip-resistant: Polished stone, polished terrazzo, and glazed tile with a glossy finish. Others, such as some untextured vinyl tiles designated as slip-resistant by the manufacturer, are slippery when wet. If new flooring is being installed along a public route of travel, a material should be chosen which is designated by the manufacturer as slip-resistant. It is also useful, however, to examine similar installations and obtain input from people with mobility impairments to help determine the actual slip-resistance of a specific material in use.

Key Items
New flooring material (rough stone, unglazed tile, carpet). Carpet, carpet runners, slip-resistant mats.

Level of Difficulty
Low, except for installing new stone or ceramic tile, which is moderate to high.

Estimates

Remove vinyl tile and replace with unglazed quarry tile

Description	Quantity	Unit	Work Hours	Material
Remove 12" × 12" tiles	1.000	S.F.	0.016	0.00
6" × 6" red quarry tile, mud set, 1/2" thick	1.000	S.F.	0.114	2.74
Totals			0.130	2.74

Total per square foot including general contractor's overhead and profit	$11

Copyright R. S. Means Company, Inc., 1994

31. Install Slip-Resistant Flooring Materials *(continued)*

Remove vinyl tile and replace with carpet

Description	Quantity	Unit	Work Hours	Material
Remove 12" x 12" tiles	9.000	S.F.	0.144	0.00
40 oz. nylon carpet	1.000	S.Y.	0.140	18.50
Totals			0.284	18.50

Total per square yard including general contractor's overhead and profit — **$45**

Install carpet runner with beveled edges

Description	Quantity	Unit	Work Hours	Material
Olefin carpet runner (3' x 60')	1.000	Ea.	0.250	380.00
Totals			0.250	380.00

Total per each including general contractor's overhead and profit — **$659**

Install absorbent mat

Description	Quantity	Unit	Work Hours	Material
Black rubber mat, with nosing, 3/8" thick	1.000	S.F.	0.052	11.70
Totals			0.052	11.70

Total per square foot including general contractor's overhead and profit — **$22**

Copyright R. S. Means Company, Inc, 1994

Notes

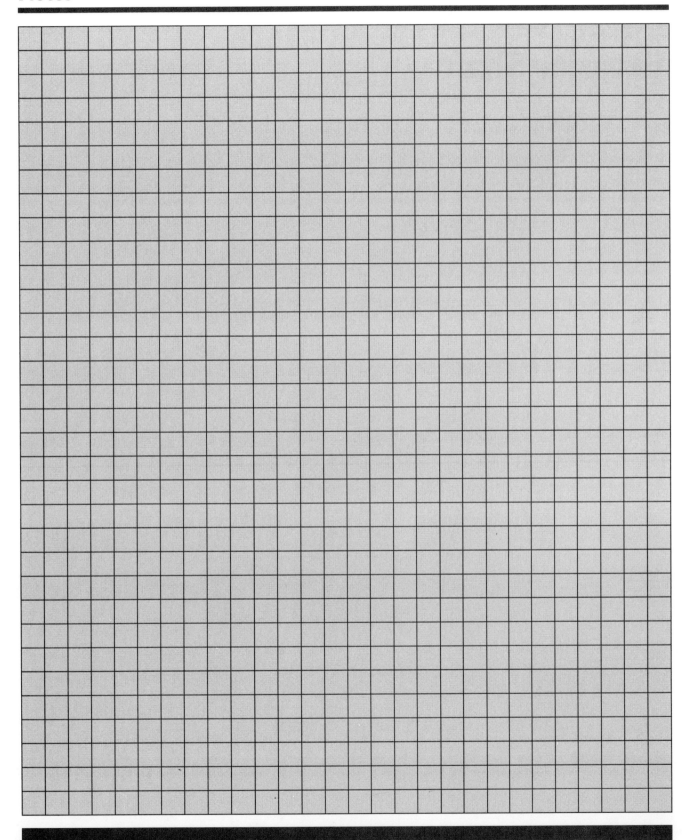

32
Install New Elevator: Exterior Shaft

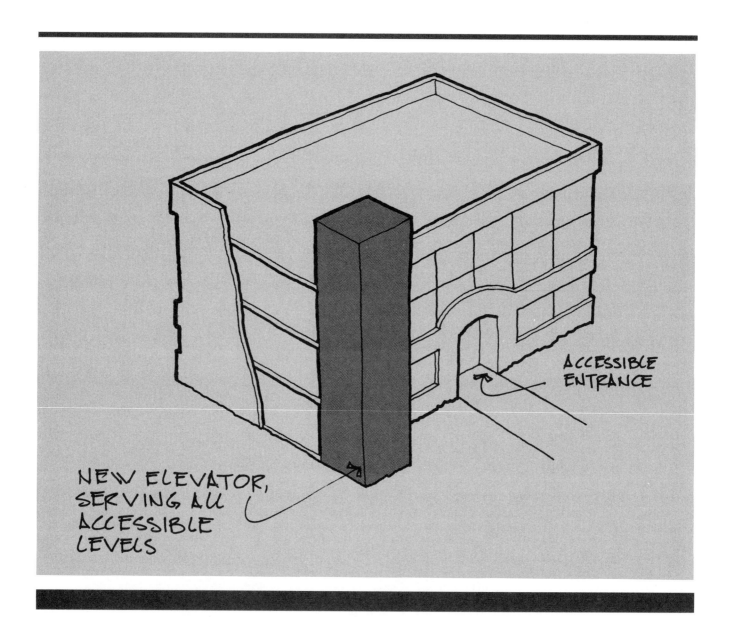

No architectural modification for creating accessible vertical circulation over large height distances is as effective as an elevator. Unlike a ramp, it is capable of bridging distances over several floors; unlike a wheelchair platform lift, it can be used by everyone. Installing an elevator against an existing building exterior allows for minimal disruption of the building's structure, can sometimes accommodate a larger shaft than an interior application, and can include an accessible exterior door at grade level.

ADAAG References

4.1.3 (5) **Accessible Buildings, New Construction**
4.1.6 (3) (c) **Accessible Buildings, Alterations**
4.3 **Accessible Route**
4.10 **Elevators**

Where Applicable

Required in new construction in buildings over two stories or 3,000 square feet per floor and any shopping centers or health care providers, or additions to existing buildings conforming to the same conditions. For Title II (State and Local Governments) facilities using ADAAG as the standard, elevators might be necessary to create program access. Under Title III (Public Accommodations and Commercial Facilities), elevators are only required in renovations if an area of primary function is being modified, and alterations to the route of travel do not exceed 20% of total renovation costs.

Design Requirements
(Access requirements only: strict building code requirements also apply)

- Accessible route to the elevator and at all stops.
- Cab 54" × 68" clear inside dimensions for side-opening doors, 54" × 80" for centered doors.
- Compliance with all elevator cab signal, control, communication, and reach requirements *(not cited here; consult ADAAG 4.10 for details)*.

Design Suggestions

Careful consideration must be given to elevator placement. Elevators should create accessible vertical circulation where it is most needed, but should serve all floors if possible. If there are level changes within a building, the elevator should serve the most heavily used levels, and the possibility of creating accessible routes at level changes (such as ramps or lifts) should be examined. Installation of an exterior elevator allows the building interior framing to remain intact, but will require bracing of the building facade and spandrel (edge) framing.

The most efficient door configuration for a wheelchair user is a two-door cab, with the doors at opposite ends. This prevents the need for turns. A two-door cab with doors on adjacent walls works almost as well. No matter what configuration is used, the cab size must still conform to ADA standards.

Key Items

Elevator shaft, cab, pit, mechanical room, accessible route to elevator and all stops.

Level of Difficulty

Very high. Requires major structural renovation, foundation and roof modifications, electrical preparation, and finish work. Necessitates design drawings, specifications, and shop drawings, approved by local building departments and/or elevator board.

32. Install New Elevator: Exterior Shaft (continued)

Estimate

Add new elevator and shaft to building exterior

Description	Quantity	Unit	Work Hours	Material
Saw cutting exterior walls (per inch of depth)	360.000	L.F.	19.800	82.80
Exterior wall demolition	54.000	S.F.	2.646	0.00
Excavation	15.000	C.Y.	2.220	0.00
Footing formwork	52.000	SFCA	3.432	19.76
Wall forms	156.000	SFCA	12.324	81.12
Reinforcing (@ 50 lbs./C.Y.)	0.150	Ton	1.600	72.75
Concrete, material only	6.000	C.Y.	0.000	330.00
Placing concrete	6.000	C.Y.	3.198	0.00
Backfill	5.000	C.Y.	2.855	0.00
Slab on grade, 4"	65.000	S.F.	1.430	74.10
Structural steel frame, plates, angles, and fasteners	3.000	Ton	23.763	2745.00
Shaft wall, including light-gauge framing	1110.000	S.F.	81.030	1554.00
Face brick, running bond	1110.000	S.F.	202.020	2419.80
Scaffolding	11.100	C.S.F.	15.862	260.85
Elevator pit ladder	4.000	V.L.F.	1.504	100.00
3" deep, 22-gauge metal decking	65.000	S.F.	0.650	52.00
6" aluminum gravel stop	40.000	L.F.	2.360	80.80
2" thick polyisocyanurate roof insulation	65.000	S.F.	0.455	32.50
EPDM roofing with ballast	65.000	S.F.	0.520	50.05
Roofing minimum	1.000	Job	4.000	0.00
Hydraulic passenger elevator (3 stop)	1.000	Ea.	800.000	20,800.00
Totals			1181.669	28,755.53

Total per each including general contractor's overhead and profit $120,229

Copyright R. S. Means Company, Inc, 1994

Notes

33
Install New Elevator: Interior Shaft

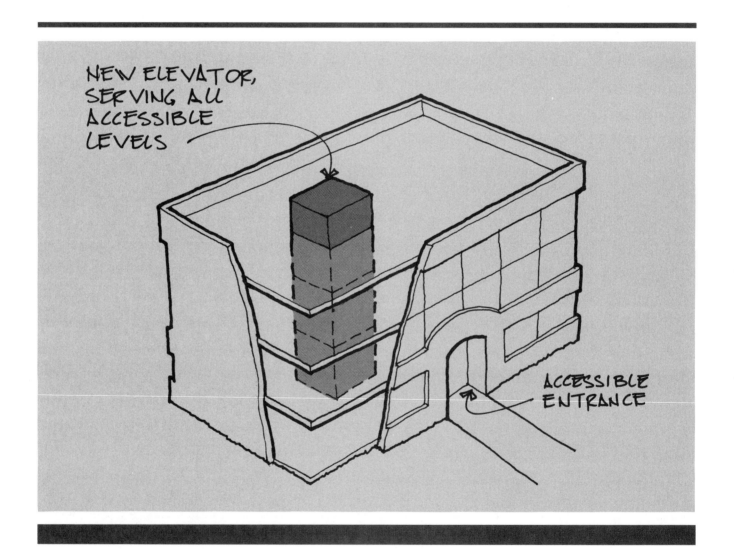

The same benefits for installing an elevator on the exterior of a building apply to an interior installation: virtually unlimited height travel and unrestricted access use. In addition, installation of an elevator in the interior of an existing building does not require a weatherproof shaft and would not block an exterior route of travel. Interior elevators can be monitored more easily where security is an issue.

ADAAG References

4.1.3 (5) Accessible Buildings, New Construction
4.1.6 (3) (c) Accessible Buildings, Alterations
4.3 Accessible Route
4.10 Elevators

Where Applicable

Required in new construction in buildings over two stories or 3,000 square feet per floor and any shopping centers or health care providers, or additions to existing buildings conforming to the same conditions. For existing facilities under Title II (State and Local Governments) using ADAAG as the design standard, elevators might be necessary to create program access. Under Title III (Public Accommodations and Commercial Facilities), elevators are only required in renovations if an area of primary function is being modified, and alterations to the route of travel do not exceed 20% of total renovation costs.

Design Requirements
(Access requirements only: strict building code requirements also apply)

- Accessible route to the elevator and at all stops.
- Cab 54" × 68" clear inside dimensions for side-opening doors, 54" × 80" for centered doors.
- Compliance with all elevator cab signal, control, communication, and reach requirements *(not cited here; consult ADAAG 4.10 for details)*.

Design Suggestions

Careful consideration must be given to elevator placement. Elevators should create accessible vertical circulation where it is most needed, but should serve all floors if possible. If there are level changes within a building, the elevator should serve the most heavily used levels, and the possibility of creating accessible routes at level changes (such as ramps or lifts) should be examined. Installation of an interior elevator may require extensive reinforcing of the building frame.

The most efficient door configuration for a wheelchair user is a two-door cab, with the doors at opposite ends. This prevents the need for turns. A two-door cab with doors on adjacent walls works almost as well. No matter what configuration is used, the cab size must still conform to ADA standards.

Key Items

Elevator shaft, cab, pit, mechanical room, accessible route to elevator and all stops.

Level of Difficulty

Very high. Requires structural renovation, foundation and roof modifications, electrical preparation, and finish work. Necessitates design drawings, specifications, and shop drawings, approved by local building departments and/or elevator board.

33. Install New Elevator: Interior Shaft (continued)

Estimate

Add new elevator and shaft to building interior

Description	Quantity	Unit	Work Hours	Material
Cut out demolition, 6" thick slab on grade	65.000	S.F.	24.765	0.00
Cut out demolition, floor slabs	195.000	S.F.	156.000	0.00
Shoring of floor openings	0.720	M.B.F.	15.709	342.00
Hand excavation	15.000	C.Y.	30.000	0.00
Footing formwork	26.000	SFCA	1.716	9.88
Wall forms	78.000	SFCA	6.162	40.56
Reinforcing (@ 50 lbs./C.Y.)	0.150	Ton	1.600	72.75
Concrete, material only	7.000	C.Y.	0.000	385.00
Placing concrete (pumped)	6.000	C.Y.	4.518	0.00
Backfill	5.000	C.Y.	2.855	0.00
Slab on grade, 4"	1.000	C.Y.	0.533	0.00
Structural steel frame, plates, angles, and fasteners	5.000	Ton	39.605	4575.00
Shaft wall, including light-gauge framing	1320.000	S.F.	96.360	1848.00
Face brick, running bond	330.000	S.F.	60.060	719.40
Scaffolding	0.330	C.S.F.	0.472	7.76
Elevator pit ladder	4.000	V.L.F.	1.504	100.00
3" deep, 22-gauge metal decking	65.000	S.F.	0.650	52.00
6" aluminum gravel stop	40.000	L.F.	2.360	80.80
2" thick polyisocyanurate roof insulation	65.000	S.F.	0.455	32.50
EPDM roofing with ballast	65.000	S.F.	0.520	50.05
Roofing labor minimum	1.000	Job	4.000	0.00
Painting, 3 coats by roller	990.000	S.F.	9.900	99.00
Electric passenger elevator (3 stop)	1.000	Ea.	458.000	40,825.00
Totals			**917.744**	**49,239.70**

Total per each including general contractor's overhead and profit $136,864

Copyright R. S. Means Company, Inc, 1994

Notes

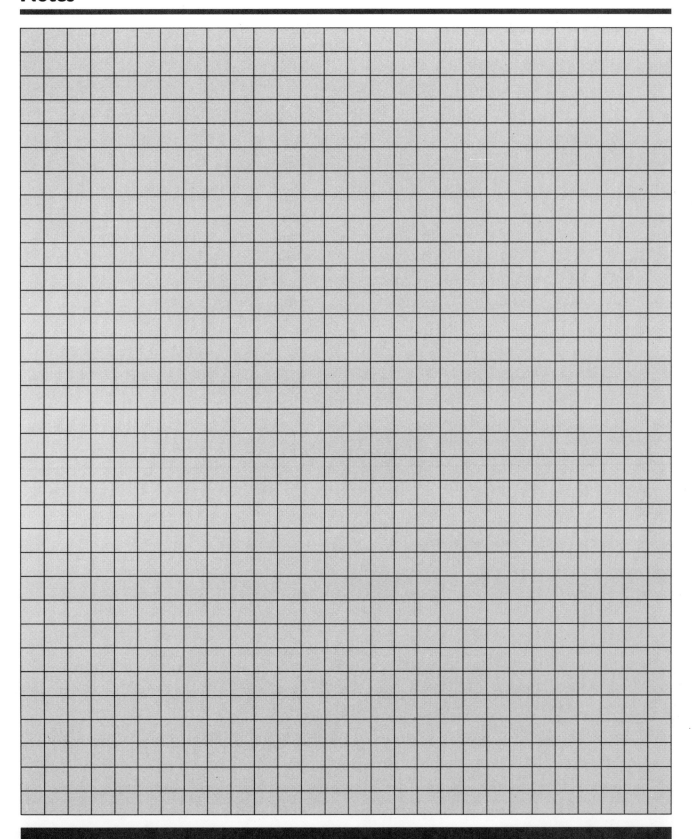

34
Modify Existing Elevator Cab

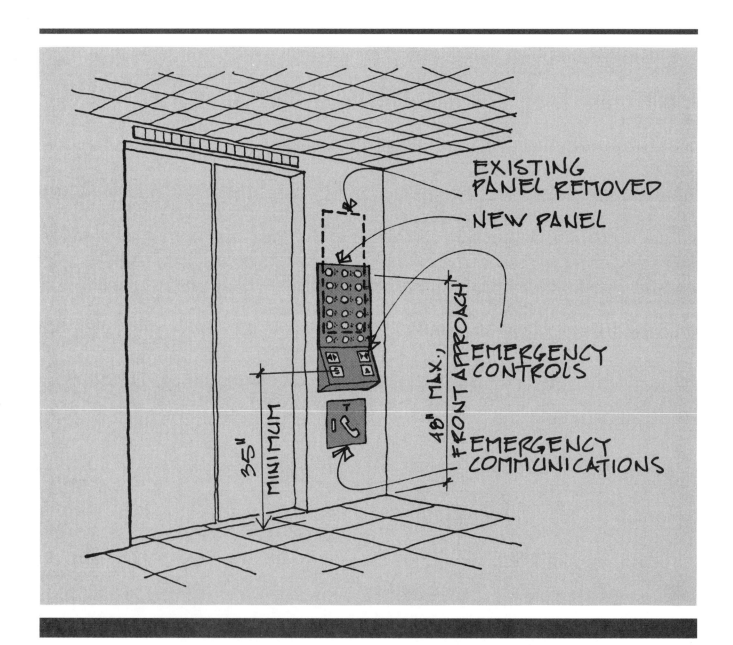

If there were no numbers on the panel in the elevator, how would you know what button to push? Without some basic modifications, such as raised characters and Braille or visual and audible signals, an elevator cab can be just as difficult to use for some people as if there were no numbers. These kinds of additions, and other simple modifications like increased lighting and handrails, can greatly increase the usability and compliance of an elevator cab for all users.

ADAAG Reference
4.10 Elevators

Where Applicable
All passenger elevators.

Design Requirements
- Visual car position signals above the door or call panel.
- Doors open for a minimum of three seconds when called; must reopen automatically.
- Audible signals indicating direction and passing floors, two signals for up, one for down.
- Control panel at front or side of cab.
- Panel within reach range (54" maximum, side reach, 48" maximum front reach).
- Buttons raised or flush with panel, 3/4" diameter minimum, visual indication of call registration.
- Raised character and Braille next to each button.
- Emergency controls on bottom of panel.
- Emergency communication system (if provided), not voice dependent, 48" a.f.f. maximum, operable with a closed fist.
- Slip-resistant flooring surface.
- Adequate lighting (5 footcandles minimum).

Design Suggestions
Reduce interior reflective surfaces. They can make controls difficult to see, and can be disorienting for people with low vision or cognitive impairments. Increased lighting can make call button and characters easier to see for people with low vision. If the panel is both too high and has non-compliant call buttons, consider replacing the existing control panel with a more compliant panel since it must be lowered or moved anyway. A handrail, although not required for ADA compliance, is welcomed by many people because it helps to maintain balance.

Key Items
Signals, control panel, communications system, floor surfaces, tactile and Braille characters, possibly handrails.

Level of Difficulty
Low for installation of tactile and Braille numbers and floor surfacing, high for all electrical work. Trades involved: Electricians, possibly manufacturer's installers. Production of shop drawings could be necessary.

Estimates

Convert freight elevator to passenger elevator (5-stop, 3000 lb. cap., 100 F.P.M.)

Description	Quantity	Unit	Work Hours	Material
Self-leveling device (hoist motor and controller change)	1.000	Ea.		
Lower control panel, add audible and visual signals	1.000	Ea.	Price includes an elevator	
Add railings to cab interior	20.000	L.F.	subcontractor's materials and labor	
Totals				

Total per each including general contractor's overhead and profit **$38,357**

Replace self-leveler (5-stop, 3500 lb. capacity, 100 F.P.M.)

Description	Quantity	Unit	Work Hours	Material
Self-leveling device (hoist motor and controller change)	1.000	Ea.	Price includes an elevator subcontractor's materials and labor	
Totals				

Total per each including general contractor's overhead and profit **$31,277**

Copyright R. S. Means Company, Inc., 1994

34. Modify Existing Elevator Cab (continued)

Install/replace non-compliant doors, including opening device (5-stop passenger elevator)

Description	Quantity	Unit	Work Hours	Material
New doors, frames, sills, headers	1.000	Floor		Price includes an elevator subcontractor's materials and labor
Totals				

Total per floor including general contractor's overhead and profit: $6,398

Add visual signal

Description	Quantity	Unit	Work Hours	Material
Add visual signal	1.000	Ea.		Price includes an elevator subcontractor's materials and labor
Totals				

Total per each including general contractor's overhead and profit: $398

Add audible signal

Description	Quantity	Unit	Work Hours	Material
Add audible signal	1.000	Ea.		Price includes an elevator subcontractor's materials and labor
Totals				

Total per each including general contractor's overhead and profit: $398

Lower existing elevator panel

Description	Quantity	Unit	Work Hours	Material
Existing panel relocation	1.000	Ea.		Price includes an elevator subcontractor's materials and labor
Totals				

Total per each including general contractor's overhead and profit: $6,398

Replace existing elevator panel

Description	Quantity	Unit	Work Hours	Material
New control panel	1.000	Ea.		Price includes an elevator subcontractor's materials and labor
Totals				

Total per each including general contractor's overhead and profit: $10,663

Copyright R. S. Means Company, Inc., 1994

Add aluminum railing in cab (48" long, one wall)

Description	Quantity	Unit	Work Hours	Material
1-1/2" diameter aluminum railing	4.000	L.F.	0.600	25.20
Totals			0.600	25.20

Total per each including general contractor's overhead and profit: $88

Add stainless steel railing in cab (48" long, one wall)

Description	Quantity	Unit	Work Hours	Material
Stainless steel railing, 2" × 3/8"	4.000	L.F.	0.600	34.00
Totals			0.600	34.00

Total per each including general contractor's overhead and profit: $141

Add wood railing in cab (48" long, one wall)

Description	Quantity	Unit	Work Hours	Material
Wood railing	4.000	L.F.	0.532	20.00
Totals			0.532	20.00

Total per each including general contractor's overhead and profit: $63

Add raised character/Braille signage to control panel

Description	Quantity	Unit	Work Hours	Material
Adhesive-mounted signage (for 9-stop elevator)	1.000	Ea.	0.000	135.00
Labor minimum	1.000	Job	2.000	0.00
Totals			2.000	135.00

Total per each including general contractor's overhead and profit: $325

Replace emergency communications system

Description	Quantity	Unit	Work Hours	Material
Emergency system	1.000	Ea.	Price includes an elevator subcontractor's materials and labor	
Totals				

Total per each including general contractor's overhead and profit: $2,559

Copyright R. S. Means Company, Inc, 1994

35 Modify Existing Elevator Hall Signals

Often a noisy elevator door is the only clue people with visual impairments have to tell them that an elevator has arrived. This does not indicate which direction the elevator is traveling, and the sound of the doors is of no use to someone with hearing and visual impairments. Both visual and audible hall signals are vital to let all elevator users know where an elevator is and which direction it is going, especially at elevator banks with multiple cars.

ADAAG Reference
4.10 Elevators

Where Applicable
All passenger elevators.

Design Requirements
- Visual hall signals, minimum 72" a.f.f., at least 2-1/2" in the smallest dimension, visible from the call button area.
- Audible signals indicating car arrival, sounding once for up, twice for down.
- Call buttons 42" high to center line, at least 3/4" in diameter, button for "up" on top.
- Call buttons raised or flush with the surrounding surface.
- No obstructions below call buttons projecting more than 4".
- Hoistway entrances, floor designations in raised characters and Braille.

Design Suggestions
In addition to visual elevator signals, ensure that lighting outside the elevator is sufficient to see all signage.

Key Items
Visual signal(s), audible signal(s), call buttons.

Level of Difficulty
Moderate. Signal and call button alterations must be done by elevator installer. Finish work to wall is also required.

Estimates

Add hallway visual signal

Description	Quantity	Unit	Work Hours	Material
Add visual signal	1.000	Ea.	Price includes an elevator subcontractor's materials and labor	

Totals

			Total per each including general contractor's overhead and profit	**$398**

Add hallway audible signal

Description	Quantity	Unit	Work Hours	Material
Add audible signal	1.000	Ea.	Price includes an elevator subcontractor's materials and labor	

Totals

			Total per each including general contractor's overhead and profit	**$398**

Lower hallway call buttons

Description	Quantity	Unit	Work Hours	Material
Remove panel and receptacle/switch/fixture	1.000	Ea.	0.031	0.00
Remove junction box	1.000	Ea.	0.100	0.00
Cutout demolition of partition	1.000	Ea.	0.333	0.00
Re-install junction box	1.000	Ea.	0.400	0.00
Re-install call button fixture	1.000	Ea.	1.000	0.00
Re-install call button panel	1.000	Ea.	0.100	0.00
Misc. materials for gypsum board painting and repair	1.000	Job	0.000	75.00
Repair gypsum board	1.000	Job	2.000	0.00
Paint gypsum board—minimum	1.000	Job	2.000	0.00
Totals			5.964	75.00

			Total per each including general contractor's overhead and profit	**$574**

Replace hallway call buttons

Description	Quantity	Unit	Work Hours	Material
Remove panel and receptacle/switch/fixture	1.000	Ea.	0.031	0.00
Remove junction box	1.000	Ea.	0.100	0.00
Cutout demolition of partition	1.000	Ea.	0.333	0.00
Install new call button fixture	1.000	Ea.	0.400	280.00
Misc. materials for gypsum board painting and repair	1.000	Job	0.000	75.00
Repair gypsum board	1.000	Job	2.000	0.00
Paint gypsum board—minimum	1.000	Job	2.000	0.00
Totals			4.864	355.00

			Total per each including general contractor's overhead and profit	**$922**

Copyright R. S. Means Company, Inc, 1994

36
Elevator Raised and Braille Characters

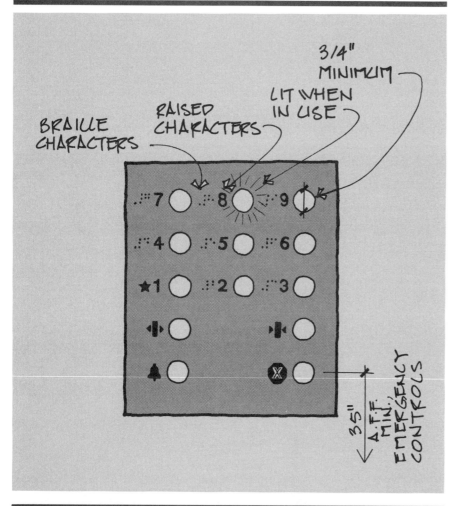

Installing raised and Braille characters is one of the easiest modifications to help make an elevator cab more accessible. Without these features, people who are blind or have severe vision impairments have no way of locating what button to push. Many of the standardized symbols, such as the star for the ground floor, can also be useful for sighted people in situations where the first floor, ground floor, or main floors are different.

ADAAG References
4.10 Elevators (4.10.5 Raised and Braille Characters on Hoistway Entrances, 4.10.12 Car Controls.)
4.30 Signage

Where Applicable
All public elevators.

Design Requirements
- Raised characters and Braille on both sides of all elevator door jambs at 60" a.f.f. to the sign centerline.
- Raised character and Braille adjacent to the left of all buttons in the cab, 5/8" to 2" high, with a star indicating the first (ground) floor (regardless of other signaling devices).
- Raised symbol on emergency communication device door.

Design Suggestions
Except in those rare instances where an elevator has an audible voice system calling out every floor, raised and Braille characters on the door jambs and call panels can be the only means that a blind person has of using the elevator unassisted. Also, the availability, easy installation and low cost of adding raised and Braille characters usually makes this a readily achievable modification in a public accommodation. The requirements listed in ADAAG should be followed as closely as possible, since the dimensions and locations given in ADAAG for raised and Braille characters are the same as those cited in the American National Standards

Institute's *Accessible and Usable Buildings and Facilities* and the national model building codes. This is part of a national attempt to institute a standard location for the characters to help ensure that a person using the characters will know where to look for them.

Key Item

Signage, either bolted or adhered to surface.

Level of Difficulty

Low. Providing raised-character or Braille signage, installation may require only adhering new signage to existing panels.

Estimates

Add 5 adhesive-mounted floor designation signs

Description	Quantity	Unit	Work Hours	Material
Floor designation jamb signage	5.000	Ea.	0.000	80.00
Labor minimum	1.000	Job	2.000	0.00
Totals			2.000	80.00

Total per each sign including general contractor's overhead and profit **$50**

Total per 5 signs including general contractor's overhead and profit **$250**

Add adhesive-mounted metal elevator control panel signage

Description	Quantity	Unit	Work Hours	Material
Adhesive-mounted signage (for 3-stop elevator)	1.000	Ea.	0.000	105.00
Labor minimum	1.000	Job	2.000	0.00
Totals			2.000	105.00

Total per each including general contractor's overhead and profit **$293**

Copyright R. S. Means Company, Inc, 1994

37
Install or Modify Public Telephones

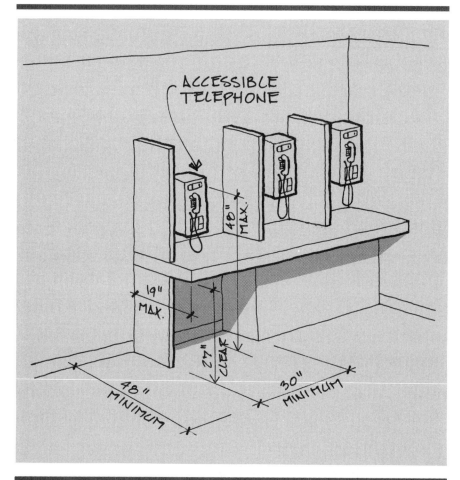

Telephones are one of the easiest building elements to make accessible, or to make inaccessible. Since they are one of the most used building elements, it is vital to avoid the latter. Very often, inaccessible phones are located on accessible routes. Locating phones too high, installing them in narrow booths, or not including volume controls generally make the phones difficult to use for everyone and impossible to use for people with mobility or hearing impairments. Simple modifications like removing a phone booth, lowering a phone, or installing a volume control phone can go a long way in making a facility more accessible for a wide range of users.

ADAAG References
4.1.3 (17), Accessible Buildings, New Construction
4.31 Telephones

Where Applicable
At least one phone at all public phone banks.

Design Requirements
- On an accessible route.
- One per floor, and one bank per floor if there are more than two telephone banks per floor, required to comply with the following:
  — within reach range—54" maximum to the highest operable part for side reach, except where frontal approach is required and 48" is the maximum;
  — push buttons;
  — hearing aid compatibility;
  — volume control;
  — signs indicating accessibility features.
- 25% of all other public phones are required to have volume controls.
- If there are more than four public phones with one phone inside, a TTY required.
- A TTY shelf and outlet are required at interior banks of three or more phones.
- Signage is required indicating the location of the nearest accessible phone/TTY.

Design Suggestions
Locate accessible phones adjacent to existing phones, in the most visible location. Do not locate phones in a corner unless frontal approach is possible and maneuvering space requirements are met.

Key Items
Phone equipment, usually supplied by phone company. Signage as appropriate. Possible finish work if phones are installed or relocated on an existing surface.

Level of Difficulty
Most phone work done by the phone company, often at no charge. Installation of signage can be done by others. Trades involved: electrical, possible finish work.

Estimates

Lower public telephones

Description	Quantity	Unit	Work Hours	Material
Misc. materials for gypsum board painting and repair	1.000	Job	0.000	75.00
Gypsum board patching labor minimum	1.000	Job	2.000	0.00
Paint gypsum board—minimum	1.000	Job	2.000	0.00
Totals			4.000	75.00

Total per each including general contractor's overhead and profit: $458

Provide interior text telephone

Description	Quantity	Unit	Work Hours	Material
Telecommunications display device	1.000	Ea.	0.000	265.00
Totals			0.000	265.00

Total per each including general contractor's overhead and profit: $452

Provide volume control

Description	Quantity	Unit	Work Hours	Material
Volume control handset	1.000	Ea.	0.000	45.00
Totals			0.000	45.00

Total per each including general contractor's overhead and profit: $77

Remove phone booth

Description	Quantity	Unit	Work Hours	Material
Demolition crew minimum	1.000	Job	8.000	0.00
Totals			8.000	0.00

Total per each including general contractor's overhead and profit: $455

Copyright R. S. Means Company, Inc., 1994

37. Install or Modify Public Telephones (continued)

Add signage to phone

Description	Quantity	Unit	Work Hours	Material
Plastic, adhesive-mounted signage	2.000	Ea.	0.000	88.00
Labor minimum	1.000	Job	2.000	0.00
Totals			2.000	88.00

Total per each including general contractor's overhead and profit **$264**

Provide new accessible telephone

Description	Quantity	Unit	Work Hours	Material
Telephone company labor minimum	1.000	Job	4.000	0.00
Totals			4.000	0.00

Total per each including general contractor's overhead and profit **$313**

Add counter and outlet for portable TDD

Description	Quantity	Unit	Work Hours	Material
Telephone enclosure	1.000	Ea.	3.200	115.00
Telephone receptacle w/20' of 4/C phone wire	1.000	Ea.	0.308	5.95
Totals			3.508	120.95

Total per each including general contractor's overhead and profit **$403**

Copyright R. S. Means Company, Inc, 1994

Notes

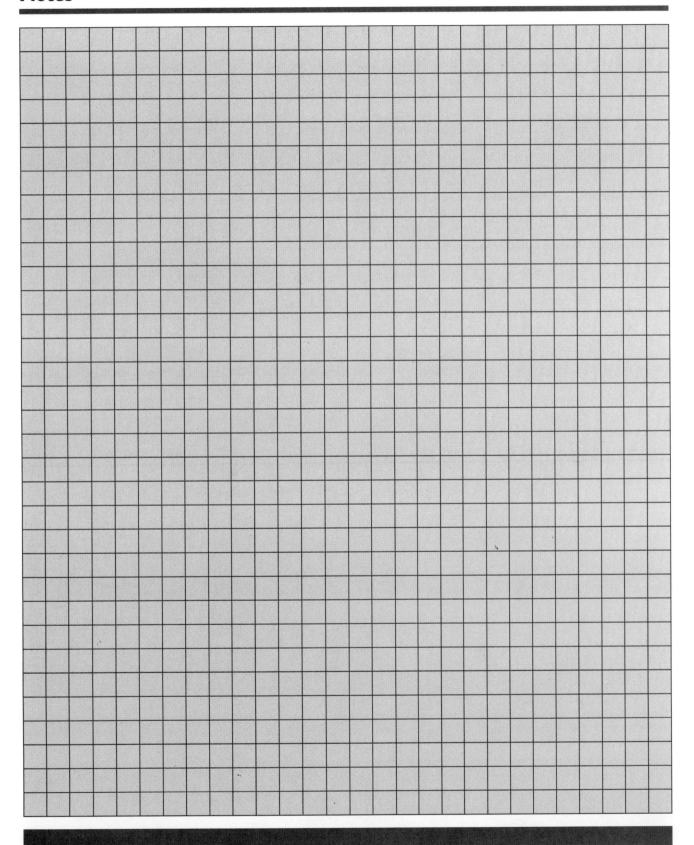

38
Install Public Text Telephones (TTYs/TDDs)

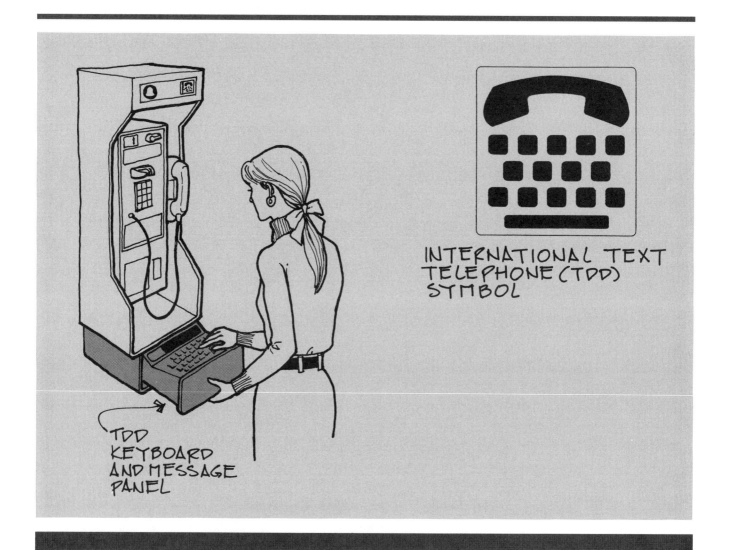

Accessibility is often thought of only in terms of people who use wheelchairs, but the public text telephone is one of the most important advances in access technology. It enables someone with a hearing impairment to communicate using the phone, something most people take for granted. A public text telephone which uses a keyboard and digital display (or TDD outlet where required) is important not only as a building amenity, but can be vital in an emergency situation and is an integral part of an accessible facility.

ADAAG References
4.1.3 (17), **Accessible Buildings, New Construction**
4.31 Telephones

Where Applicable
At least one phone at all public phone banks.

Design Requirements
- On an accessible route.
- Within reach range—54" maximum to the highest operable part for side reach, except where frontal approach is required and 48" is the maximum, and push buttons.
- If there are more than four public phones with one inside, a TTY is required.
- A TTY shelf and outlet are required at interior banks of three or more phones.
- Signage is required indicating location of nearest accessible phone/TTY.

Design Suggestions
Locate text telephones adjacent to existing phones, in the most visible location.

Key Items
Phone equipment, usually supplied by phone company. Signage as appropriate. Possible finish work if phones are installed or relocated on an existing surface.

Level of Difficulty
Most phone work done by the phone company. Installation of signage can be done by others. Other trades involved: electrical, possible finish work.

Note: This includes the cost of the public text telephone unit itself. In some instances, the text telephone is leased or supplied by the local phone company. In this case, the construction costs would involve electrical work to supply the phone lines and outlet(s) and finish costs only.

Estimate

Install public text telephone

Description	Quantity	Unit	Work Hours	Material
Telecom. display device in stainless steel enclosure	1.000	Ea.	0.000	875.00
Telephone company labor minimum	1.000	Job	4.000	0.00
Totals			4.000	875.00

Total per each including general contractor's overhead and profit: $1,806

Copyright R. S. Means Company, Inc, 1994

39
Install or Modify Drinking Fountains

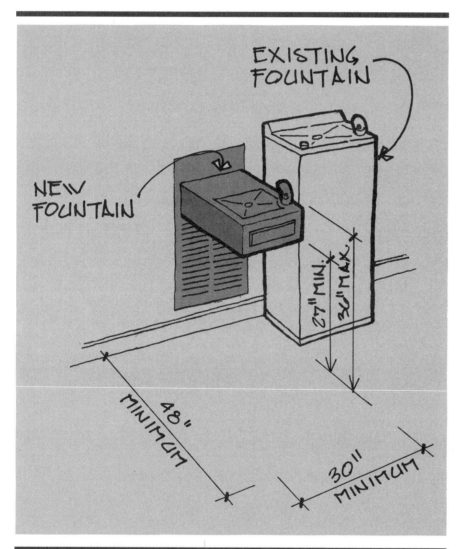

As with telephones, inaccessible drinking fountains are often located on accessible routes. Replacing an inaccessible fountain can often use existing piping and electrical supply. Even in cases where this is not possible, installing an accessible fountain may be a relatively simple construction operation that adds an important accessible element to a facility.

ADAAG References
4.1.3 (10) New Construction
4.15 Drinking Fountains and Water Coolers

Where Applicable
50% of drinking fountains.

Design Requirements
- Spout 36" a.f.f. maximum.
- Spout at the front of the fountain.
- Accessible control (acceptable if operable with a closed fist).
- Water flow at least 4" high.
- 30" × 48" clear floor space minimum when facing the fountain at wall- and post-mounted units.
- 30" × 48" clear floor space minimum parallel to the fountain at freestanding or built-in units.
- Knee space below, 27" a.f.f. minimum., if wall-mounted.
- Cannot intrude more than 4" into the circulation path, if higher than 27" a.f.f. and not otherwise protected (as with wing walls).

Design Suggestions
- Fountains with a push-bar control across the front are easier to use for a wider range of people.
- High/low fountains work well for a variety of users.
- Install cup dispenser as a barrier removal modification if installing a new or additional fountain is not readily achievable.

Key Items
Drinking fountain, plumbing/water supply, electricity source, possible finish materials.

Level of Difficulty
High; modifying or installing fountains requires plumbing, electrical, and finish work as well as structural reinforcement if the fountain is wall-mounted.

Estimates

Install new high/low fountain

Description	Quantity	Unit	Work Hours	Material
Remove existing freestanding water fountain	1.000	Ea.	1.000	0.00
Drinking fountain, wall-mounted, accessible type	1.000	Ea.	4.000	955.00
Water fountain supply, waste, and vent	1.000	Ea.	3.620	49.50
Totals			8.620	1004.50

Total per each including general contractor's overhead and profit $2,216

Install new low, wall-mounted fountain

Description	Quantity	Unit	Work Hours	Material
Water fountain supply, waste, and vent	1.000	Ea.	3.620	49.50
Drinking fountain, wall-mounted, accessible type	1.000	Ea.	4.000	955.00
Totals			7.620	1004.50

Total per each including general contractor's overhead and profit $2,154

Lower existing fountain

Description	Quantity	Unit	Work Hours	Material
Remove existing water fountain	1.000	Ea.	1.000	0.00
Water fountain supply, waste, and vent	1.000	Ea.	3.620	49.50
Re-install water fountain, labor minimum	1.000	Job	4.000	0.00
Totals			8.620	49.50

Total per each including general contractor's overhead and profit $627

Copyright R. S. Means Company, Inc, 1994

40 Install or Modify Controls

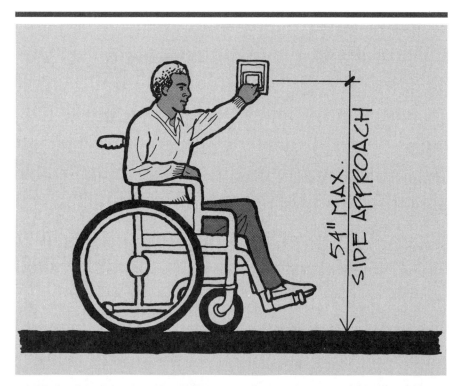

Controls in public places are often vital for public safety. Examples include fire alarm pull boxes and emergency assistance boxes. Placing public controls in accessible locations or making the control devices accessible not only makes them usable by people with mobility or grasping impairments, it also makes them easier to reach and use for everyone in the facility.

ADAAG References
4.2 Reach Ranges
4.27 Controls and Operating Mechanisms

Where Applicable
All controls and alarms operated by building occupants.

Design Requirements
- Within reach range: 54" a.f.f. for side reach, 48" a.f.f. for front reach.
- All controls operable without tight grasping, pinching, or twisting of the wrist.
- 30" × 48" clear floor space in front of control.
- Operating force no greater than 5 pounds/foot.

Design Suggestions
- Install controls at 48" a.f.f. maximum even if side reach is possible. Fire alarms can be covered with plastic shields to help prevent false alarms, but shields must comply with 4.17 and be operable without tight grasping, pinching, or twisting of the wrist.
- Locate all controls at least 18" from an inside corner.

Key Items
Controls, wiring, finishes to match existing.

Level of Difficulty
Moderate. Involves electrical and finish work.

Estimate

Lower 5 fire alarm boxes (surface-mounted conduit)

Description	Quantity	Unit	Work Hours	Material
No. 700 wiremold raceway	10.000	L.F.	0.800	6.30
#18 fire alarm conductor	0.100	C.L.F.	0.100	6.40
Labor minimum	1.000	Job	2.000	0.00
Totals			2.900	12.70

Total per each including general contractor's overhead and profit	**$39**
Total per 5 alarm boxes including general contractor's overhead and profit	**$196**

Copyright R. S. Means Company, Inc, 1994

41
Install or Modify Outlets

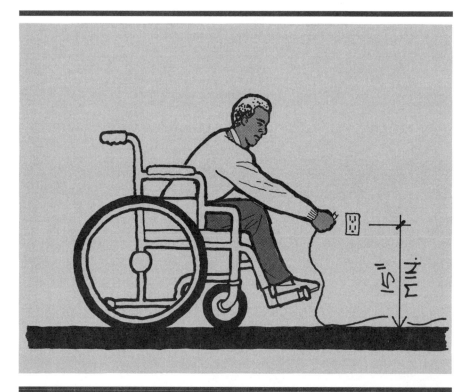

While placing outlets in accessible locations is essential for people who use wheelchairs, it makes them easier to use for everyone. Outlets located close to the floor require stooping even for people who do not use wheelchairs. Installing or relocating outlets to accessible heights is a relatively simple modification that may only require minor electrical and finish work to complete.

ADAAG References
4.2 Reach Ranges
4.27 Controls and Operating Mechanisms

Where Applicable
All outlets operated by building occupants.

Design Requirements
- Within reach range: 54" a.f.f. for side reach, 48" a.f.f. for front reach, 15" minimum a.f.f. for electrical or convenience outlets and telephone and data jacks.
- 30" × 48" clear floor space in front of control.

Design Suggestions
- Locate all controls at least 18" from an inside corner.

Key Items
Outlets, wiring, finishes to match existing.

Level of Difficulty
Low to moderate. Involves electrical and finish work.

Estimates

Raise 5 outlets, gypsum/metal stud wall

Description	Quantity	Unit	Work Hours	Material
Remove plate and receptacle/switch/fixture	5.000	Ea.	0.155	0.00
Cutout demolition of partition	5.000	Ea.	1.665	0.00
Conductor	0.100	C.L.F.	0.296	1.55
Install junction box	5.000	Ea.	2.000	23.25
Re-install outlet	5.000	Ea.	1.480	0.00
Install plate	5.000	Ea.	0.500	11.50
Misc. materials for gypsum board painting and repair	1.000	Job	0.000	75.00
Repair gypsum board	1.000	Job	2.000	0.00
Paint gypsum board—minimum	1.000	Job	2.000	0.00
Totals			10.096	111.300

Total per each including general contractor's overhead and profit	**$176**
Total per 5 outlets including general contractor's overhead and profit	**$878**

Install 5 outlets, gypsum/metal stud wall

Description	Quantity	Unit	Work Hours	Material
Cutout demolition of partition	5.000	Ea.	1.665	0.00
Conductor	0.100	C.L.F.	0.296	1.55
Install junction box	5.000	Ea.	2.000	23.25
Install outlet	5.000	Ea.	1.480	26.25
Install plate	5.000	Ea.	0.500	11.50
Misc. materials for gypsum board painting and repair	1.000	Job	0.000	75.00
Repair gypsum board	1.000	Job	2.000	0.00
Paint gypsum board—minimum	1.000	Job	2.000	0.00
Totals			9.941	137.550

Total per each including general contractor's overhead and profit	**$182**
Total per 5 outlets including general contractor's overhead and profit	**$909**

Copyright R. S. Means Company, Inc., 1994

42
Install or Modify Switches

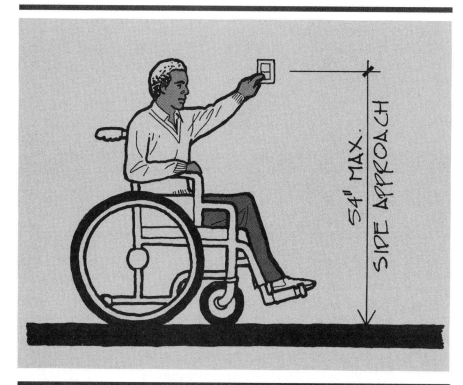

It is now common in new construction to locate light switches within accessible reach range, but in older facilities and in some new construction, switches are placed too high or too close to a corner, making them unusable by people who use wheelchairs and others with limited range of arm movement. As with outlets, locating a switch in a compliant location in new construction or renovations costs the same as placing it in an inaccessible location, and relocating a switch might involve only minor electrical and finish work.

ADAAG References
4.2 Reach Ranges
4.27 Controls and Operating Mechanisms

Where Applicable
All switches operated by building occupants.

Design Requirements
- Within reach range: 54" a.f.f. for side reach, 48" a.f.f. for front reach, 15" minimum a.f.f. for outlets and jacks.
- All controls operable without tight grasping, pinching, or twisting of the wrist.
- 30" × 48" clear floor space in front of control.

Design Suggestions
- Install controls at 48" a.f.f. maximum even if side reach is possible.
- Install rocker switches (easier to use for people with limited fine motor control).
- Locate all controls at least 18" from an inside corner.

Key Items
Switches, wiring, finishes to match existing.

Level of Difficulty
Low to moderate. Involves electrical and finish work.

Estimates

Lower 5 light switches, gypsum/metal stud wall

Description	Quantity	Unit	Work Hours	Material
Remove plate and receptacle/switch/fixture	5.000	Ea.	0.155	0.00
Cutout demolition of partition	5.000	Ea.	1.665	0.00
Conductor	0.100	C.L.F.	0.296	1.55
Install junction box	5.000	Ea.	2.000	23.25
Re-install switch	5.000	Ea.	3.635	0.00
Install plate	5.000	Ea.	0.500	11.50
Misc. materials for gypsum board painting and repair	1.000	Job	0.000	75.00
Repair gypsum board—minimum	1.000	Job	2.000	0.00
Paint gypsum board—minimum	1.000	Job	2.000	0.00
Totals			12.251	111.30

Total per each including general contractor's overhead and profit	**$201**
Total per 5 switches including general contractor's overhead and profit	**$1,004**

Install 5 rocker switches, gypsum/metal stud wall

Description	Quantity	Unit	Work Hours	Material
Cutout demolition of partition	5.000	Ea.	1.665	0.00
Conductor	0.100	C.L.F.	0.296	1.55
Install junction box	5.000	Ea.	2.000	23.25
20 amp rocker switch	5.000	Ea.	1.480	66.25
Install plate	5.000	Ea.	0.500	9.25
Misc. materials for gypsum board painting and repair	1.000	Job	0.000	75.00
Repair gypsum board	1.000	Job	2.000	0.00
Paint gypsum board—minimum	1.000	Job	2.000	0.00
Totals			9.941	175.300

Total per each including general contractor's overhead and profit	**$194**
Total per 5 switches including general contractor's overhead and profit	**$971**

Copyright R. S. Means Company, Inc, 1994

43
Install Audible and Visual Fire Alarms

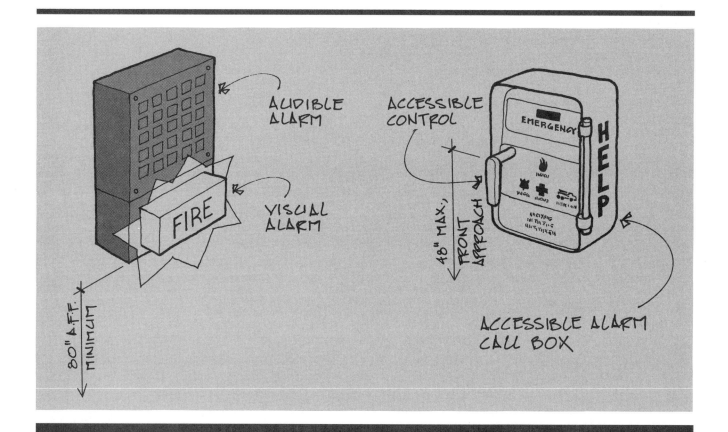

Maximum fire safety is based on assuming the worst case scenario and taking steps to prevent it. Accessible fire safety includes providing a means of warning people with visual and hearing impairments of an emergency wherever they may be in an accessible facility. This means that both visual and audible alarms are necessary. Installing them has the added benefit of increasing the level of fire safety for all of a facility's users.

ADAAG Reference
4.28 Alarms

Where Applicable
All common use areas where fire alarms are provided.

Design Requirements
- Alarms to be audible and visual.
- Visible from all locations within the space.
- No more than 50' from any point, 100' apart along corridors or in large rooms.
- Clear strobe lamp, 75 candela minimum, 3-Hz flash rate.

- 80" above the highest floor level, or 6" below the ceiling, whichever is lower.

Design Suggestions
- Install visual alarms in rooms (such as toilet rooms) where audible alarms are provided. Even if not required as part of a renovation, consider installing audible and visual alarms in high-use spaces, such as rest rooms, where people might be alone.

Key Items
Audible or visual components added to existing alarm, new audible and visual alarm, electrical conduit to main wiring system.

Level of Difficulty
High. Requires hard-wiring alarms to existing system, and may substantially increase power requirements. May require new wiring system.

Estimates

Install new visual alarm

Description	Quantity	Unit	Work Hours	Material
Fire alarm light	1.000	Ea.	1.509	96.00
#18 fire alarm conductor	0.100	C.L.F.	0.100	6.40
Totals			1.609	102.40

Total per each including general contractor's overhead and profit: $271

Add visual alarm to existing audible alarm

Description	Quantity	Unit	Work Hours	Material
Fire alarm light	1.000	Ea.	1.509	96.00
#18 fire alarm conductor	0.100	C.L.F.	0.100	6.40
Totals			1.609	102.40

Total per each including general contractor's overhead and profit: $271

Install new audible/visual alarm

Description	Quantity	Unit	Work Hours	Material
Fire alarm light and horn	1.000	Ea.	1.509	96.00
#18 fire alarm conductor	0.200	C.L.F.	0.200	12.80
Totals			1.709	108.80

Total per each including general contractor's overhead and profit: $288

Copyright R. S. Means Company, Inc, 1994

44
Install Signage

SIGNAGE INSTALLED AT INACCESSIBLE ENTRANCE

Signage has proven to be one of the most carefully watched areas for ADA compliance. Providing information is a vital component of any facility, and installing signs is often considered a readily achievable modification in businesses. Signage and wayfinding are very often insufficient even for people with full vision, and such deficiencies are increased dramatically for those with visual or cognitive impairments (as well as for people not fully literate or fluent in English). Each facility should be carefully studied in order to provide the clearest directional and informational signage. It should also be remembered that even where signs themselves are not required to comply with section 4.29 of ADAAG, any information provided to the public must be available in some form to anyone who needs it.

ADAAG Reference
4.30 Signage

Where Applicable
Signs are required at all entrances and rest rooms when not all are accessible, and at parking spaces, volume control telephones, text telephones, and assistive listening systems. Any signs designating permanent spaces (such as room numbers, exit signs, and bathroom designations) must have tactile and Braille lettering and be mounted at the door. If provided, signage conveying directions to a facility and information about spaces inside must comply with ADA requirements for contrast, character height, font, and proportion. Temporary signage, such as building directories and menus, is not required to comply.

Design Requirements
For all permanent signs:
- High contrast between characters/pictures and background (dark on light, or light on dark).
- Simple-serif or sans-serif font.
- Matte or other non-reflective surface.
- Character width-to-height ratio between 3:5 and 1:1.

- Stroke width-to-height ratio between 1:5 and 1:10.
- Overhead signs that are above 80" a.f.f. must have letters no less than 3" high.
- International symbol of accessibility at entrances if all are not accessible, with directions to the nearest accessible entrance; similar for rest rooms. All symbols (pictograms) to have 6" border height minimum (no size requirement for the symbol itself).

For signs at permanently identified rooms and spaces:
- Characters raised at least 1/32", with upper case characters 5/8" to 2" high.
- Grade II Braille (an abbreviated version).
- Signs to be located on wall adjacent to the door, on the latch side, 60" a.f.f. A person must be able to approach within 3" of the sign.

Design Suggestions

Regarding contrast, evidence suggests that light characters on a dark background are more easily read. Since there is no firm definition of high contrast, choose colors that leave as little doubt as possible, e.g., white characters on a red background. Although room signs have been customarily placed on doors, this practice is not effective for people who must read the sign by touch or be able to approach within inches of the sign to see the letters. If a sign exists on a door, keeping it and adding an accessible sign adjacent to the door can be effective for both groups of users. Raised characters and Braille are not required for informational or directional signs, but if possible, some method of wayfinding should be provided for people with severe visual impairments, such as audible signs at decision points or information booths. Temporary signs (e.g., menus or prices) are not required to comply with ADA Standards, but the *effective communications* requirement of ADA does require some method of conveying the information, such as a Braille menu or having the choices read to a customer. The Appendix of ADAAG provides some very useful information on making signage as helpful as possible.

Key Items

Signs, mounting or adhering mechanism.

Level of Difficulty

Low to moderate. While interior signs can be quite easy to install with self-adhesive backings, attaching exterior signs or signs on masonry may require difficult anchoring.

Estimates

Anchor plastic exterior signage to masonry wall

Description	Quantity	Unit	Work Hours	Material
Plastic signage	1.000	Ea.	0.000	20.00
Layout & drilling of holes, per inch (4 holes, 2" deep)	8.000	Ea.	0.856	0.64
Wedge anchors	4.000	Ea.	0.212	1.36
Totals			1.068	22.00

Total per each including general contractor's overhead and profit **$95**

Anchor metal exterior signage to masonry

Description	Quantity	Unit	Work Hours	Material
Metal signage	1.000	Ea.	0.457	8.75
Layout & drilling of holes, per inch (4 holes, 2" deep)	8.000	Ea.	0.856	0.64
Wedge anchors	4.000	Ea.	0.212	1.36
Totals			1.525	10.75

Total per each including general contractor's overhead and profit **$107**

Copyright R. S. Means Company, Inc., 1994

44. Install Signage (continued)

Anchor metal interior sign to gypsum board/stud wall

Description	Quantity	Unit	Work Hours	Material
Signage	1.000	Ea.	0.457	8.75
Layout & drilling of anchor holes	4.000	Ea.	0.212	0.16
Plastic shields and screws	4.000	Ea.	0.000	0.12
Totals			0.669	9.03

Total per each including general contractor's overhead and profit: $58

Anchor metal room sign to gypsum board/stud wall

Description	Quantity	Unit	Work Hours	Material
Signage	1.000	Ea.	0.457	8.75
Layout & drilling of anchor holes	4.000	Ea.	0.212	0.16
Plastic shields and screws	4.000	Ea.	0.000	0.12
Totals			0.669	9.03

Total per each including general contractor's overhead and profit: $58

Copyright R. S. Means Company, Inc, 1994

Notes

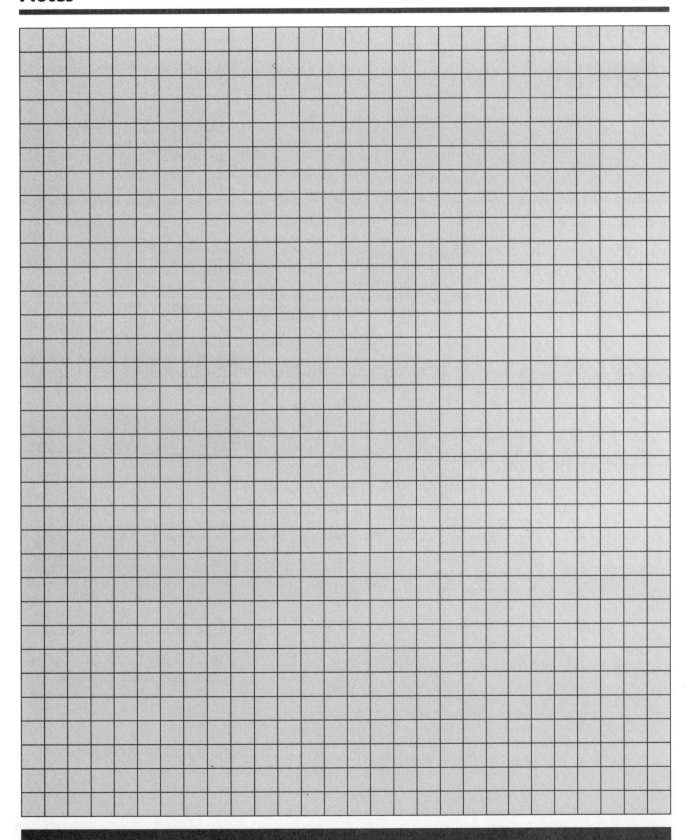

45
Install Electronic Display Signage

In facilities where changing information is conveyed to the public on a regular basis, electric display signage assists all users, especially those with hearing impairments. The ADA Standards require that all spoken information is available to all users in some form, and electric signage is a very effective way of meeting that requirement. Such signs can usually be retrofitted into an existing facility, often without using wiring recessed into the walls or ceiling.

ADAAG References
10.3.1(14) Fixed Facilities and Stations
10.4.1(6) Airports

Where Applicable
At transit facilities that broadcast information to the general public.

Design Requirements
- At fixed transit facilities and airports, information broadcast to the general public must also be available to people with auditory impairments. One possible method is a visual paging system.

Design Suggestions
ADAAG does not give any design requirements for visual paging systems, and ADAAG 4.30 Signage only covers signs that provide information about specific spaces. It is recommended that any electronic signs comply with ADAAG 4.30.2, 3, 5, and 7, the standards for permanent signs:
- High contrast between characters/pictures and background (dark on light, or light on dark).
- Simple serif or sans serif font.
- Matte or other non-reflective surface.
- Character width-to-height ratio between 3:5 and 1:1.
- Stroke width-to-height ratio between 1:5 and 1:10.
- Overhead signs with 80" clearance below minimum, letters no less than 3" high when measured by a lower-case X.

These standards should be applied at all other installations where such information is conveyed, such as convention centers or hotels, which also have to provide information to people with visual and hearing impairments.

Key Items
Signs, Light-Emitting Diode boxes, mounting or adhering mechanism.

Level of Difficulty
Low to moderate. Involves electrical work and finish work.

Estimate

Install electric signage (2 lines of text, 3" high characters, indoor application)

Description	Quantity	Unit	Work Hours	Material
Electric display signage (2 lines of text, 3" high)	1.000	Ea.	0.000	5000.00
Cutout demolition of ceiling	1.000	Ea.	0.333	0.00
Support for and installation of sign	1.000	Ea.	8.000	205.00
Conductor	0.400	C.L.F.	0.800	24.40
Install junction box	1.000	Ea.	0.400	4.65
Install outlet	1.000	Ea.	0.296	5.25
Install plate	1.000	Ea.	0.100	2.30
Misc. materials for gypsum board painting and repair	1.000	Job	0.000	75.00
Repair gypsum board	1.000	Job	2.000	0.00
Paint gypsum board—minimum	1.000	Job	2.000	0.00
Totals			13.929	5316.60

Total per each including general contractor's overhead and profit **$9,460**

Copyright R. S. Means Company, Inc, 1994

46 Construct Single-User Toilet Rooms

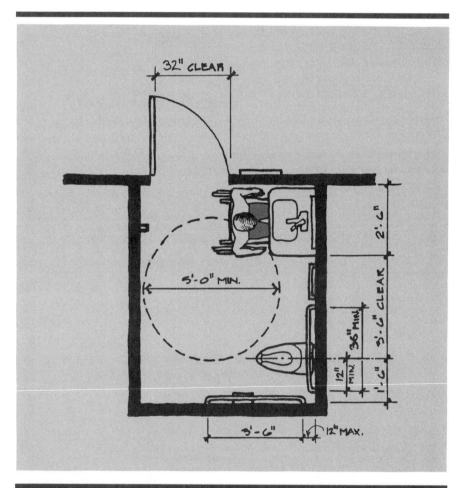

As an access modification, single-use accessible rest rooms are most often considered in cases where existing, multiple-stall toilet rooms are inaccessible, and creating access to both is technically infeasible. If the toilet room is to be unisex, verify that local plumbing codes allow unisex accessible toilet rooms.

ADAAG Reference
4.22 Toilet Rooms

Where Applicable
All public and common-use, single-user, toilet rooms.

Design Requirements
- Located on an accessible route.
- Door compliant with all width, hardware, pull weight, and maneuvering space requirements.
- Raised character, Braille, and international symbol of accessibility signage if not all rest rooms are accessible, adjacent to door, latch side, 60" a.f.f.
- 60" diameter clear turning space inside (can extend 19" under wall-mounted sink).
- 30" × 48" clear space in front of sink, extends 19" maximum under sink.
- Slip-resistant flooring.
- Sink with paddle faucets or other accessible faucets, 34" maximum to top, 29" clear knee space below apron, 27" clear below bowl measured 8" from front.
- Exposed pipes wrapped with insulation or protected from contact.
- Mirror 40" a.f.f. maximum to bottom.
- Toilet: seat 17" to 19" a.f.f., centerline 18" from wall, flush valve on open side.
- Grab bars 33" to 36" a.f.f., 1-1/2" diameter, 1-1/2" from wall, 42" long minimum at side wall, 36" long minimum at back wall.
- Toilet paper dispenser below with grab bar within 36" of rear wall, 19" a.f.f. minimum.
- All dispensers within reach range (48" front reach, 54" side reach) and accessible in operation.

Design Suggestions
It is possible to use standard sinks that are not designated "HP" units to meet the apron height and knee space requirements; they will be less costly, more attractive and easier to install and use. The same is true of tilted "HP" mirrors; if a mirror above the sink cannot be lowered to the correct height, add another mirror in a different

location. Having tilted mirrors does not meet accessibility requirements if they are too high. Be sure that grab bars are anchored to either framing or blocking, so that they meet the minimum weight resistance requirements.

Key Items

Plumbing fixtures/hook-ups: minimum toilet, sink/paddle faucets, door with privacy lock, framing, grab bars, finishes, mirror, dispensers, ventilation.

Level of Difficulty

High. Requires plumbing, framing, electrical, and finish (usually tile) work.

Estimate

Install single-use toilet room

Description	Quantity	Unit	Work Hours	Material
Remove metal stud/gypsum board partition	21.000	S.F.	0.966	0.00
Remove gypsum board from stud face at room interior	55.000	S.F.	0.440	0.00
Remove vinyl flooring	60.000	S.F.	0.960	0.00
Ceiling demolition (suspended A.C.T.)	60.000	S.F.	0.660	0.00
Metal stud partition (3-5/8" wide, 16" O.C.)	230.000	S.F.	4.370	66.70
1/2" gypsum board, taped and finished	230.000	S.F.	3.910	46.00
2" x 4" blocking	0.025	M.B.F.	1.429	13.63
Water-resistant 1/2" gypsum board	300.000	S.F.	2.400	69.00
Ceramic tile cove base	30.000	L.F.	5.280	78.30
Ceramic tile floor, thin-set 4" x 4" tiles	60.000	S.F.	7.980	213.00
Ceramic tile walls, thin-set 4-1/4" x 4-1/4"	300.000	S.F.	25.200	570.00
Mineral fiber suspended ceiling	60.000	S.F.	1.380	74.40
Vinyl tile flooring	23.000	S.F.	0.368	32.20
Vinyl base, 6" high	23.000	L.F.	0.575	18.17
Painting	230.000	S.F.	4.140	39.10
Interior door frame	1.000	Ea.	1.000	65.50
Hollow metal flush door, 3'-0" x 6'-8"	1.000	Ea.	0.941	147.00
Hinges	1.500	Pr.	0.000	69.00
Lever-handled lockset	1.000	Ea.	0.000	79.78
Signage	1.000	Ea.	0.457	8.75
Layout & drilling of anchor holes	4.000	Ea.	0.212	0.16
Plastic shields and screws	4.000	Ea.	0.000	0.12
Grab bars	3.000	Ea.	1.200	105.00
Combined soap/towel dispenser/mirror/shelf	1.000	Ea.	0.800	255.00
Double-roll toilet tissue dispenser	1.000	Ea.	0.333	21.00
13-gallon waste receptacle	1.000	Ea.	0.800	144.00
Accessible lavatory	1.000	Ea.	2.286	210.00
Rough in supply, waste and vent	1.000	Ea.	9.639	128.00
Water closet	1.000	Ea.	3.200	276.00
Rough in supply, waste and vent	1.000	Ea.	5.634	113.00
Totals			86.560	2842.81

Total per square foot including general contractor's overhead and profit	**$402**
Total per each including general contractor's overhead and profit	**$9,237**

Copyright R. S. Means Company, Inc, 1994

47
Modify Multiple-Stall Toilet Rooms

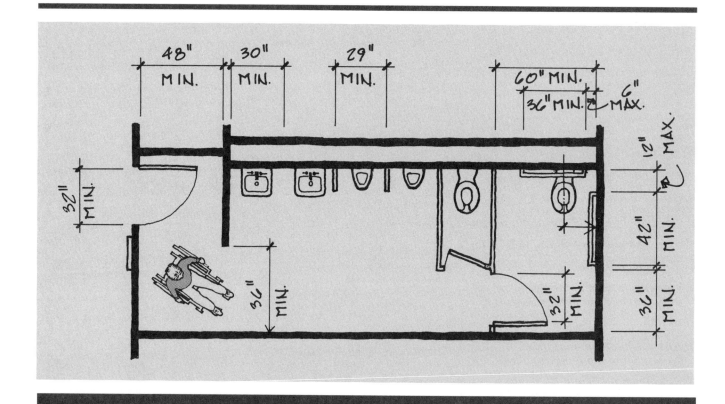

Privacy partitions at the entrances to toilet rooms can create difficult accessibility, even when the clear widths are compliant with ADA. If possible, try to locate the entrance door so there is no need for a privacy partition, or use an entry vestibule with doors on opposite walls so that tight turns are not necessary, but be certain to provide required maneuverability space at each door and between doors.

ADAAG Reference
4.22 Toilet Rooms

Where Applicable
All public and common-use, multiple-use, toilet rooms.

Design Requirements
- Located on an accessible route.
- Accessible doors compliant with all width, hardware, pull weight, and space requirements.
- Raised character, Braille, and international symbol of accessibility signage (if not all rest rooms are accessible) adjacent to door, latch side, 60" a.f.f.
- 60" diameter clear turning space (can be measured partially under sinks).
- 30" × 48" clear floor space in front of sink, extends 19" maximum under sink.
- Slip-resistant flooring.
- At least one sink with paddle faucets or other accessible faucets, 34"

maximum to top, 29" clear knee space below apron, 27" clear below bowl measured 8" from front.
- Mirror 40" a.f.f. maximum to bottom edge of reflective surface.
- Exposed pipes wrapped with insulation or protected from contact.
- At least one fully accessible toilet stall, 60" × 56" minimum with out-swinging door, 32" clear door opening, toilet seat 17" to 19" a.f.f., centerline 18" from wall, flush valve on open side, grab bars 33" to 36" a.f.f., 1-1/2" diameter, 1-1/2" from wall, 42" long minimum at side wall, 36" long minimum at rear wall, toilet paper dispenser below grab bar within 36" of rear wall, 19" a.f.f. minimum.
- If there are six or more stalls, one 36" wide stall with out-swinging door and grab bars on sides of stall.
- 9" toe clearance under partitions, if stalls are less than 60" deep.
- Urinal rim 17" a.f.f. maximum, flush valve 44" a.f.f. maximum, 29" minimum between privacy screens if they do not project beyond urinal.
- All other dispensers and coat hooks within reach range (48" a.f.f. for front reach, 54"a.f.f. for side reach).

Design Suggestions

It is possible to use standard sinks that are not designated "HP" units to meet the apron height and knee space requirements; they will be less costly and easier to install and use. The same is true of tilted "HP" mirrors; if a mirror above the sink cannot be lowered, add another mirror in a different location. Tilting in and of itself, does not create an accessible alternative. At least one mirror must be accessible with its bottom edge no more than 40" above the floor. Be sure that grab bars are anchored to either framing or blocking, so that they meet the minimum structural strength requirements. Stall door hardware must also be accessible. Most knobs are not; slide latches are usually easier to use.

Key Items

Plumbing fixtures/hookups: toilets, partitions, sinks with paddle faucets, door, framing, grab bars, finishes, mirror, dispensers, ventilation.

Level of Difficulty

High. Requires plumbing, framing, electrical, and finish (usually tile) work.

Estimate

Remodel multi-use toilet room

Description	Quantity	Unit	Work Hours	Material
Remove metal stud/gypsum board partition	21.000	S.F.	0.966	0.00
Paddle faucets with spout	1.000	Ea.	0.800	126.00
1" wall flexible closed cell foam insulation	10.000	L.F.	0.950	23.50
Remove hollow metal toilet partitions	1.000	Ea.	1.000	0.00
Remove wall-mounted water closet	1.000	Ea.	1.143	0.00
Labor minimum to disconnect plumbing	1.000	Job	4.000	0.00
Misc. materials for gypsum board and ceramic tile repair	1.000	Job	0.000	100.00
Labor minimum to repair gypsum board and ceramic tile	1.000	Job	4.923	0.00
Painted metal toilet partitions	1.000	Ea.	2.667	330.00
Grab bars	3.000	Ea.	1.200	105.00
Combined soap/towel dispenser/mirror/shelf	1.000	Ea.	0.800	255.00
Water closet	1.000	Ea.	3.200	276.00
Rough in supply, waste and vent	1.000	Ea.	5.634	113.00
Interior door frame	1.000	Ea.	1.000	65.50
Hollow metal flush door, 3'-0" × 6'-8"	1.000	Ea.	0.941	147.00
Totals			29.224	1541.00

Total per each including general contractor's overhead and profit **$4,215**

Copyright R. S. Means Company, Inc, 1994

48
Create Accessible Stall

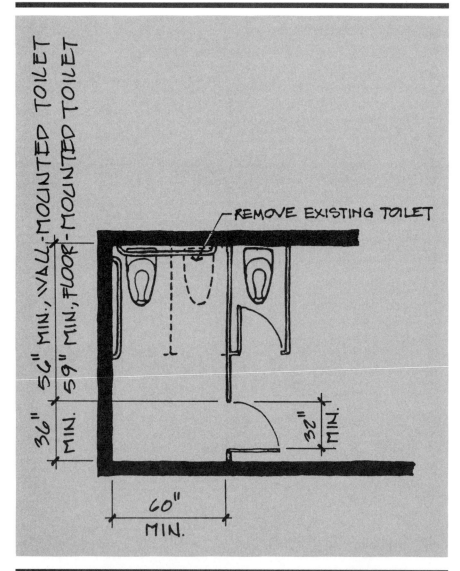

Many toilet stalls are at the end of a path of travel between the row of stalls and the wall. It is sometimes possible to create an accessible stall by extending the stall across the path of travel to the opposite wall and entering straight on, but only if the stall has compliant maneuvering space. In a multiple-stall toilet room with typical inaccessible stalls, it is often possible to create an accessible stall by removing one toilet and combining the space of two stalls, although this might be prohibited by local plumbing codes requiring a certain number of fixtures for a given building population. In such cases, one of the alternative accessible stall configurations should be used or an accessible unisex bathroom added.

ADAAG Reference
4.17 Toilet Stalls

Where Applicable
All accessible toilet rooms with at least one stall.

Design Requirements
- Located on an accessible route.
- Slip-resistant flooring.
- Stall door with 32" clear opening, accessible hardware.
- 60" × 56" minimum clear inside dimension for wall-mounted toilet (60" × 59" for floor-mounted toilet), with outswinging door where possible. Alternate configurations for smaller spaces allowable for compliance: 48" × 54" stall or a 36" × 69" stall if only a 42" wide approach is possible from the latch side of the door, or if a 48" wide approach is possible from the hinged side of the door (see ADAAG 4.17, Figure 30, for details).
- 9" toe clearance under partitions, if stall is less than 60" deep.
- Toilet: seat 17" to 19" a.f.f., centerline 18" from wall, flush valve on open side.
- Grab bars 33" to 36" a.f.f., 1-1/2" diameter, 1-1/2" from wall, 42" long minimum at side wall, 36" long minimum at rear wall.

- Toilet paper dispenser below grab bar, within 36" of rear wall, 19" a.f.f. minimum, freely dispenses paper.
- Coat hook within reach range (48" a.f.f. for forward reach).

Design Suggestions

It is usually recommended that the end toilet be used for the accessible stall, so grab bars can be attached to the wall rather than to a partition. Be sure that grab bars are anchored to either framing or blocking, so that they meet the minimum structural strength requirements (250 lbs. pressure minimum).

Key Items

Minimum toilet, grab bars, partitions, door hardware.

Level of Difficulty

Moderate to high. Requires plumbing and finish work.

Estimates

Install accessible stall

Description	Quantity	Unit	Work Hours	Material
Remove hollow metal toilet partitions	1.000	Ea.	1.000	0.00
Remove wall-mounted water closet	1.000	Ea.	1.143	0.00
Labor minimum to disconnect plumbing	1.000	Job	4.000	0.00
Water closet	1.000	Ea.	3.200	276.00
Rough in supply, waste and vent	1.000	Ea.	5.634	113.00
Misc. materials for gypsum board and ceramic tile repair	1.000	Job	0.000	100.00
Labor minimum to repair gypsum board and ceramic tile	1.000	Job	4.923	0.00
Painted metal toilet partitions	1.000	Ea.	2.667	330.00
Grab bars	3.000	Ea.	1.200	105.00
Totals			23.767	924.00

Total per each including general contractor's overhead and profit: $2,860

Install grab bar, gypsum/metal stud wall with ceramic tile

Description	Quantity	Unit	Work Hours	Material
Cutout demolition of partition	1.000	Ea.	0.333	0.00
2" x 4" blocking	0.005	M.B.F.	0.306	2.92
Misc. materials for gypsum board and ceramic tile repair	1.000	Job	0.000	100.00
Labor minimum to repair gypsum board and ceramic tile	1.000	Job	4.923	0.00
Grab bars	1.000	Ea.	0.400	35.00
Totals			5.962	137.92

Total per each including general contractor's overhead and profit: $531

Add new toilet partition with 36" door

Description	Quantity	Unit	Work Hours	Material
Ceiling-hung painted metal cubicle	1.000	Ea.	4.000	350.00
Add for accessible options	1.000	Ea.	0.000	189.00
Totals			4.000	539.00

Total per each including general contractor's overhead and profit: $1,106

Copyright R. S. Means Company, Inc, 1994

49
Install New Toilet

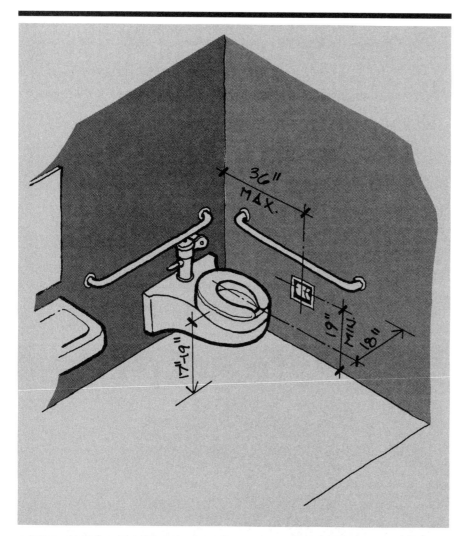

In some facilities which are being modified for access, an old toilet might not be compliant, or its location might not be accessible. In such situations, it might be easier (and cheaper) to install a new toilet compliant with the ADA Standards rather than modify the existing toilet. At the same time, other upgrade features can be added such as a low-flush water-saving model and other general upgrade features.

ADAAG Reference
4.16 Water Closets

Where Applicable
Toilets in accessible stalls or in accessible single-use toilet rooms.

Design Requirements
- Toilet centerline 18" from the wall.
- Toilet seat 17" to 19" a.f.f. Seats that spring automatically to upright position are not acceptable.
- Flush valve on the approach side of the toilet.

Design Suggestions
If the toilet is very far from the side wall and is being replaced but not relocated, blocking can be added to the wall surface to bring the grab bar closer to the toilet, or a new low wall can be added against the existing wall to provide a mounting surface for the grab bar. Wall-hung toilets provide more maneuvering space than floor-mounted toilets.

Key Items
Toilet, piping, finish materials.

Level of Difficulty
Moderate to high. Requires plumber, can involve finish/tile work.

Estimates

Replace toilet in an existing location

Description	Quantity	Unit	Work Hours	Material
Remove wall-mounted water closet	1.000	Ea.	1.143	0.00
Labor minimum to disconnect and reconnect plumbing	1.000	Job	4.000	0.00
Water closet	1.000	Ea.	3.200	276.00
Totals			8.343	276.00

Total per each including general contractor's overhead and profit — **$951**

Relocate toilet to new location where plumbing is available

Description	Quantity	Unit	Work Hours	Material
Labor minimum for toilet relocation	1.000	Job	4.000	0.00
Rough in supply, waste and vent	1.000	Ea.	5.634	113.00
Misc. materials for gypsum board and ceramic tile repair	1.000	Job	0.000	100.00
Labor minimum to repair gypsum board and ceramic tile	1.000	Job	4.923	0.00
Totals			14.557	213.00

Total per each including general contractor's overhead and profit — **$1,143**

Relocate flush valve from wall side to approach side

Description	Quantity	Unit	Work Hours	Material
Labor minimum to relocate flush valve	1.000	Job	2.000	0.00
Totals			2.000	0.00

Total per each including general contractor's overhead and profit — **$127**

Lower urinal

Description	Quantity	Unit	Work Hours	Material
Labor minimum for urinal relocation	1.000	Job	4.000	0.00
Rough in supply, waste, and vent	1.000	Ea.	5.654	84.50
Misc. materials for gypsum board and ceramic tile repair	1.000	Job	0.000	100.00
Labor minimum to repair gypsum board and ceramic tile	1.000	Job	4.923	0.00
Totals			14.577	184.5

Total per each including general contractor's overhead and profit — **$1,097**

Copyright R. S. Means Company, Inc, 1994

50
Modify Existing Toilet

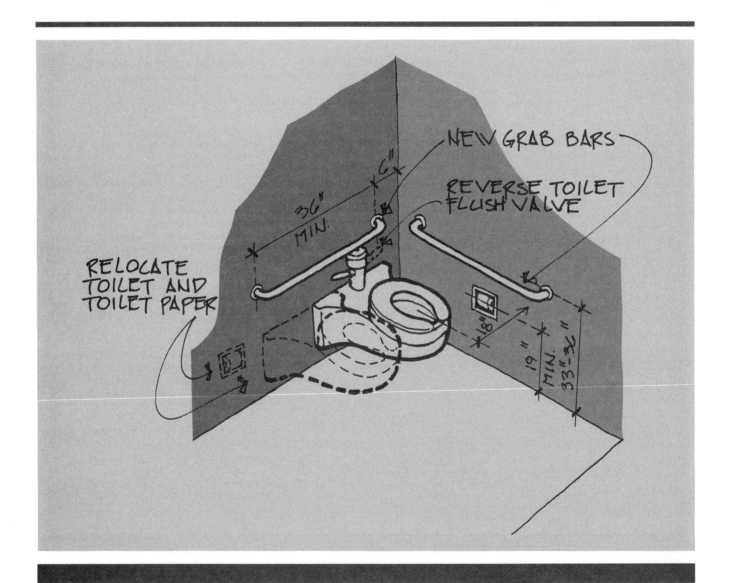

If an existing toilet is in an accessible location but has inaccessible features, or would be too expensive or difficult to move, several elements can still be modified to increase accessibility. Reversing the flush valve, adding a seat height extender, or relocating the toilet paper holder and grab bars can all increase compliance and ease of use.

ADAAG Reference

4.16 Water Closets

Where Applicable

Toilets in accessible stalls or in accessible single-use toilet rooms.

Design Requirements

- Toilet centerline 18" from the wall.
- Toilet seat 17" to 19" a.f.f. Cannot spring automatically to upright position.
- Flush valve on the approach side of the toilet.
- Urinal rim 17" a.f.f. max., flush valve 44" a.f.f. max.

Design Suggestions

Several features of an existing toilet can be modified to make it accessible, including the toilet location, seat height, flush valve location, and toilet paper location. Grab bars can also be added, using one of several approaches.

It might be possible to add a permanently installed seat height extender to an existing toilet if the seat is low. If the toilet is very far from the side wall and it would be expensive to move it (typically the case), blocking can be added to the wall surface to bring the grab bar closer to the toilet; or a new low wall can be added against the existing wall to provide a mounting surface for the grab bar. (It might be possible to add floor-mounted grab bars 18" from the centerline of the toilet, but it is very difficult to meet ADAAG's structural requirements for grab bars with these and they are not recommended. If installed, floor-mounted grab bars should be braced to the side wall to prevent wobbling.) Lowering a urinal is possible, but expensive, since it involves both plumbing and finish work. If a toilet is accessible, modifying the urinal is usually unnecessary. A floor-mounted urinal easily complies with the rim height requirement, but care must be taken to ensure that the flush valve is within reach range.

Key Items

Toilet, piping, finish materials.

Level of Difficulty

Moderate to high. Requires plumber and may involve finish/tile work. Lowering the urinal requires structural modification as well.

Estimate

Modify toilet

Description	Quantity	Unit	Work Hours	Material
Labor minimum for toilet and flush valve relocation	1.000	Job	4.000	0.00
Rough in supply, waste and vent	1.000	Ea.	5.634	113.00
Misc. materials for gypsum board and ceramic tile repair	1.000	Job	0.000	100.00
Labor min. to repair gypsum board and ceramic tile	1.000	Job	4.923	0.00
Grab bars	3.000	Ea.	1.200	105.00
Double-roll toilet tissue dispenser	6.000	Ea.	1.998	126.00
Totals			17.755	444.00

Total per each including general contractor's overhead and profit: $1,707

Copyright R. S. Means Company, Inc, 1994

51
Install New Sink

In situations where a sink is inaccessible but is too difficult to modify (such as an old pedestal sink, a sink with an exceptionally deep bowl, or a sink resting on a vanity), replacing it with an accessible sink might be the easiest solution to achieve accessibility and compliance. In many situations, the sink can be replaced using the existing plumbing and perhaps the structural bracing, but replacing the sink also allows the possibility of moving it to a more accessible location. Replacing the sink does not require using a designated "handicap" sink. Attractive standard models that comply with ADA Standards can be installed.

ADAAG Reference
4.19 Lavatories and Mirrors
(Note: ADAAG makes a distinction between lavatories, which are basins for hand washing, and sinks, which are other types of basins. ADAAG 4.19 refers to lavatories only.)

Where Applicable
At least one sink in all public toilet rooms.

Design Requirements
- 34" maximum to rim, 29" minimum clear knee space below rim, 27" clear below bowl.
- Bowl 6-1/2" deep maximum.
- Pipes wrapped with insulation.
- No sharp or abrasive surfaces under sink.
- Faucets operable with closed fist (electronic sensor faucets acceptable); self-closing faucets to remain open for at least 10 seconds.
- Mirror 40" a.f.f. maximum.
- Dispensers (such as soap dispensers) operable with a closed fist, and within reach range (48" a.f.f. for front approach, 54" for side approach).

Design Suggestions
If a sink is located in the only accessible bathroom, or in a high-use bathroom, and is not accessible due to insufficient

knee clearance below, it might be cheaper to replace it than to modify it or lower it. It is possible to use standard sinks to meet apron height and knee space requirements; they are not only less costly, but also easier to install and minimize the stigma and visual disruption of having one "accessible" sink in a row of standard sinks. (This is also true for tilted "HP" mirrors.) At least one mirror has to be installed with its bottom edge no more than 40" a.f.f., but the accessible height mirror does not have to be above the accessible sink.

Key Items

Sink, piping, faucets, insulation, and under-sink or wall bracing.

Level of Difficulty

Moderate to high. Involves plumbing, finish, possibly structural work.

Estimate

Install new sink, faucet, handles and shelf w/tilted mirror (plumbing available)

Description	Quantity	Unit	Work Hours	Material
Remove lavatory	1.000	Ea.	0.800	0.00
Rough in supply, waste and vent	1.000	Ea.	9.639	128.00
Wall-hung porcelain enamel lavatory (22" x 19")	1.000	Ea.	2.000	315.00
Faucet, handles and drain	1.000	Ea.	0.800	126.00
Misc. materials for gypsum board and ceramic tile repair	1.000	Job	0.000	100.00
Labor minimum to repair gypsum board and ceramic tile	1.000	Job	4.923	0.00
Mirror with stainless steel shelf	1.000	Ea.	0.400	67.50
Totals			18.562	736.50

Total per each including general contractor's overhead and profit: $2,261

Copyright R. S. Means Company, Inc, 1994

52
Modify Existing Sink

A sink on an accessible route can almost always be modified to increase accessibility. Removing a vanity or apron, replacing the existing faucet with a single-lever or paddle handles, or even lowering the bowl can all help make a sink compliant and more accessible for all users. (Ask anyone who's ever tried to use a ball faucet with soapy hands.)

ADAAG Reference
4.19 Lavatories and Mirrors *(Note: ADAAG makes a distinction between lavatories, which are basins for hand washing, and sinks, which are other types of basins. ADAAG 4.19 refers to lavatories only.)*

Where Applicable
At least one sink in all public toilet rooms.

Design Requirements
- 34" maximum to rim, 29" minimum clear knee space below rim, 27" clear below bowl.
- Bowl 6-1/2" deep maximum.
- Pipes wrapped with insulation.
- No sharp or abrasive surfaces under sink.
- Faucets operable with closed fist (electronic sensor faucets acceptable); self-closing faucets to remain open for at least 10 seconds.
- Mirror 40" a.f.f. maximum.
- Dispensers (such as soap dispensers) operable with a closed fist, and within reach range (48" a.f.f. for front approach, 54" for side approach).

Design Suggestions
If a sink is required to be accessible, there are many modifications that can make it more compliant and easier to use, such as lowering the sink, replacing inaccessible faucets with paddle faucets, or modifying the apron below to create sufficient knee space. At least one mirror has to be installed with its bottom edge no more than 40" a.f.f., but that mirror is not required to be above the accessible sink.

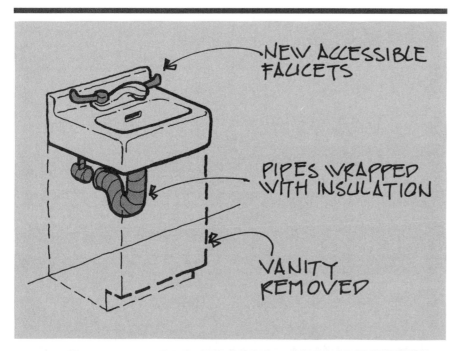

Key Items
Sink, piping, faucets, insulation, mirror, and under-sink or wall bracing.

Level of Difficulty
Varies. Involves insulation, plumbing, finish, and possible structural work.

Estimates

Lower existing sink

Description	Quantity	Unit	Work Hours	Material
Labor minimum for lowering sink	1.000	Ea.	2.000	0.00
Rough in supply, waste and vent	1.000	Ea.	9.639	128.00
Misc. materials for gypsum board and ceramic tile repair	1.000	Job	0.000	100.00
Labor minimum to repair gypsum board and ceramic tile	1.000	Job	4.923	0.00
Totals			16.562	228.00

Total per each including general contractor's overhead and profit: $1,286

Replace knob faucets with paddle faucets

Description	Quantity	Unit	Work Hours	Material
Labor minimum to remove faucets	1.000	Job	2.000	0.00
Paddle faucets with spout	1.000	Ea.	0.800	126.00
Totals			2.800	126.00

Total per set including general contractor's overhead and profit: $392

Wrap pipe with insulation

Description	Quantity	Unit	Work Hours	Material
1" wall flexible closed cell foam insulation	4.000	L.F.	0.380	9.40
Totals			0.380	9.40

Total per each including general contractor's overhead and profit: $39

Remove base cabinets, add additional bracing

Description	Quantity	Unit	Work Hours	Material
Labor minimum to remove base cabinets	1.000	Job	2.000	0.00
Blocking	0.013	M.B.F.	0.631	7.30
Bracing cover (plywood)	32.000	S.F.	1.152	46.40
Add. labor for fitting plywood bracing cover	1.000	Job	4.000	0.00
Totals			7.783	53.70

Total per each including general contractor's overhead and profit: $420

Copyright R. S. Means Company, Inc., 1994

52. Modify Existing Sink *(continued)*

Remove apron below plastic laminate counter

Description	Quantity	Unit	Work Hours	Material
Labor minimum to remove apron	1.000	Job	2.000	0.00
Totals			2.000	0.00

Total per each including general contractor's overhead and profit — **$91**

Remove apron below synthetic stone counter

Description	Quantity	Unit	Work Hours	Material
Labor minimum to remove apron	1.000	Job	2.000	0.00
Totals			2.000	0.00

Total per each including general contractor's overhead and profit — **$91**

Remove apron below granite counter

Description	Quantity	Unit	Work Hours	Material
Labor minimum to remove apron	1.000	Job	2.000	0.00
Totals			2.000	0.00

Total per each including general contractor's overhead and profit — **$91**

Copyright R. S. Means Company, Inc, 1994

Notes

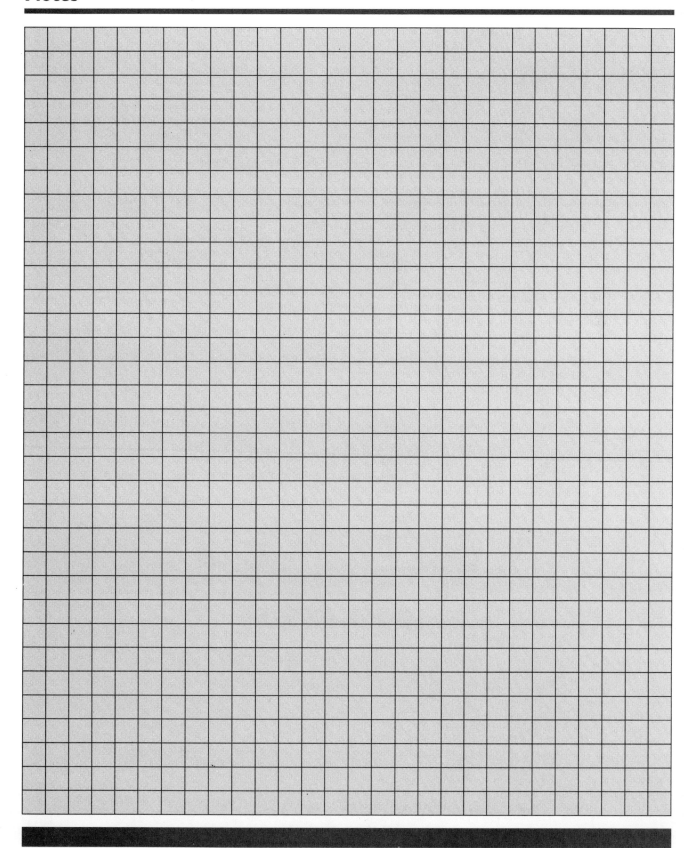

53
Install or Modify Grab Bars

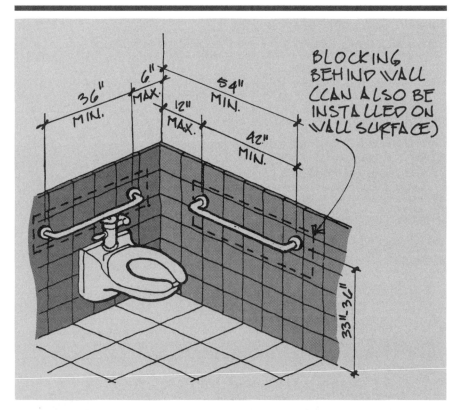

Grab bars are essential in enabling many people to use toilets, tubs, and showers. They are also one of the simplest access modifications to install. The types of available grab bars have increased, so colored, non-institutional grab bars are available to fit into an existing decor, and different configurations are available to assist people who might need bars in addition to those required in the ADA Standards.

ADAAG References

4.26 Handrails, Grab Bars, and Tub and Shower Seats
4.16 Water Closets
4.17 Toilet Stalls
4.20 Bathtubs
4.21 Shower Stalls

Where Applicable

All accessible toilets, tubs, and showers.

Design Requirements

- 1-1/4" to 1-1/2" diameter, 1-1/2" from wall.
- Capable of resisting 250 lbs. of force (as specified in 4.26.3).
- For toilet stalls, 33" to 36" a.f.f., 36" long on rear wall, 12" from corner; 42" long on side wall, 12" from corner. In 36" wide stalls, two 42" long grab bars on either side. *(See illustrations in ADAAG Sections 4.17, 4.20, and 4.21 for exact configurations.)*
- In roll-in showers:
  a. in 36" × 36" stalls, grab bars on side wall and wall opposite seat.
  b. in 30" × 60" stall, on both side wall and rear wall.
- In tubs:
  a. For 60" tub with in-tub seat: two grab bars on rear wall, 24" long, top bar 33" to 36" a.f.f., bottom 9" above rim of tub. One 24" grab bar on side wall; one 12" grab bar on side wall opposite controls.
  b. For 60" tub with in-tub seat: two grab bars on rear wall, 48" long for tub with end seat, top bar 33" to 36" a.f.f., bottom 9" above rim of

tub. One 24" long grab bar on side wall with controls.

Design Suggestions

A grab bar doesn't have to be battleship gray. Textured grab bars provide a better gripping surface than smooth bars. If studs or blocking are insufficient or difficult to locate, it might be possible to attach a painted wood 1 x 6 on the face of the wall to the studs, and attach the grab bar to that. In tight spaces, a fold-down grab bar allows flexibility in use, but only in addition to the required fixed grab bars. Grab bars are vital for safety, however, and it should be determined that any grab bar or installation method meets the strength requirements in ADAAG 4.26.3, cited above.

Key Items

Grab bars, fasteners, either blocking in wall or anchors, and possibly, finish materials.

Level of Difficulty

Moderate. May require finish work.

Estimates

Install grab bar, gypsum/metal stud wall

Description	Quantity	Unit	Work Hours	Material
Cutout demolition of partition	1.000	Ea.	0.333	0.00
2" x 4" blocking	0.005	M.B.F.	0.306	2.92
Miscellaneous materials for gypsum board repair	1.000	Job	0.000	25.00
Labor minimum to repair and paint gypsum board	1.000	Job	2.000	0.00
Grab bars	1.000	Ea.	0.400	35.00
Totals			3.039	62.92

Total per each including general contractor's overhead and profit **$267**

Install grab bar, gypsum/metal stud wall with ceramic tile

Description	Quantity	Unit	Work Hours	Material
Cutout demolition of partition	1.000	Ea.	0.333	0.00
2" x 4" blocking	0.005	M.B.F.	0.306	2.92
Misc. materials for gypsum board and ceramic tile repair	1.000	Job	0.000	100.00
Labor minimum to repair gypsum board and ceramic tile	1.000	Job	4.923	0.00
Grab bars	1.000	Ea.	0.400	35.00
Totals			5.962	137.92

Total per each including general contractor's overhead and profit **$531**

Copyright R. S. Means Company, Inc, 1994

54 Install or Modify Toilet Room Dispensers

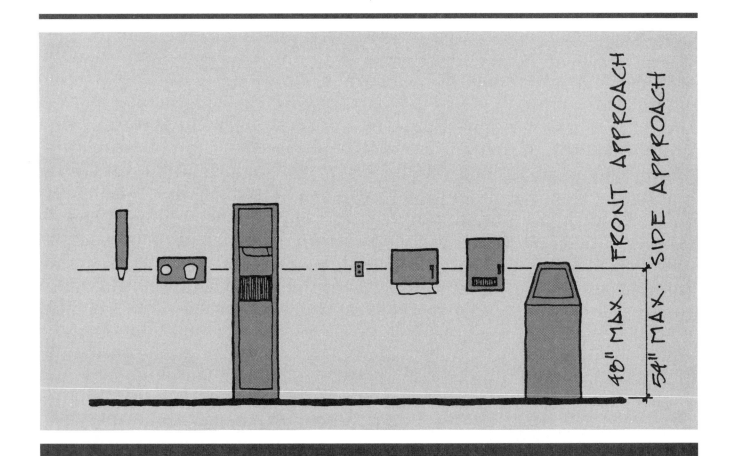

The most common problem with toilet room dispensers is simply that they are out of reach for people with disabilities. This makes relocating them within ADA compliant reach ranges a simple access modification. Some dispensers have controls that are difficult to operate for people with limited fine motor control. Replacing these is also an easy modification that could make a rest room more accessible and usable.

ADAAG Reference
4.27 Controls and Operating Mechanisms

Where Applicable
At least one of each type of dispenser must be accessible in accessible public toilet rooms.

Design Requirements
- Dispensers to be on an accessible route.
- Dispensers to be within reach range: 48" for front reach, 54" for side reach.
- Controls must be accessible (operable with a closed fist).

- 30" × 48" clear floor space in front of dispensers.

Design Suggestions
Specify dispensers with easy-to-operate controls. Locate all dispensers as close as possible to accessible fixtures (but still within easy reach) and at least 18" from an inside corner.

Key Items
Dispensers, fasteners, anchors or blocking and finish materials if necessary.

Level of Difficulty
Low to moderate. May involve finish or tile work.

Estimate

Lower dispenser, screwed to wall (gypsum board on metal studs)

Description	Quantity	Unit	Work Hours	Material
Cutout demolition of partition	1.000	Ea.	0.333	0.00
2" × 4" blocking	0.005	M.B.F.	0.306	2.92
Miscellaneous materials for gypsum board repair	1.000	Job	0.000	25.00
Labor minimum to repair and paint gypsum board	1.000	Job	2.000	0.00
Labor minimum to remove and re-install dispenser	1.000	Job	1.600	0.00
Totals			4.239	27.92

Total per each including general contractor's overhead and profit: $277

Copyright R. S. Means Company, Inc, 1994

55
Children's Accessible Bathroom Fixtures

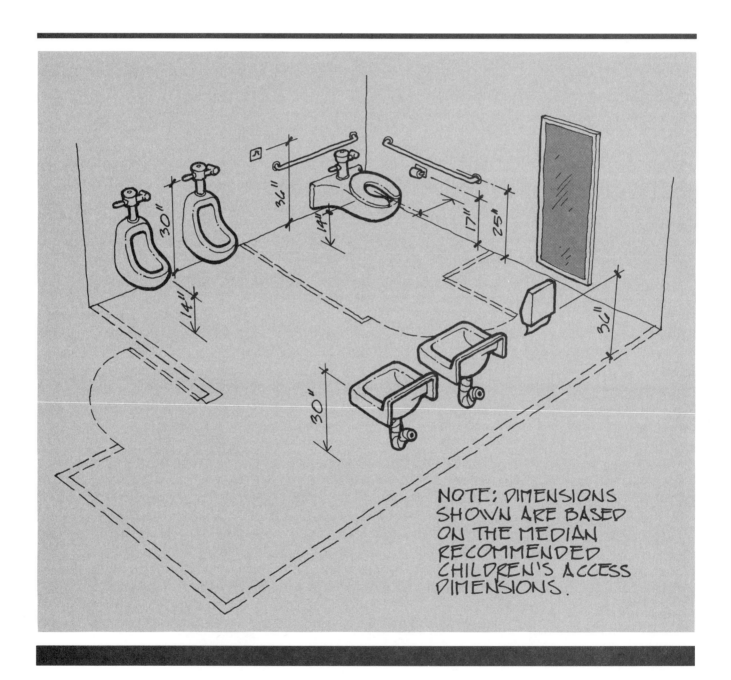

It states on page one of the ADAAG that "The specifications in these guidelines are based upon adult dimensions and anthropometrics." Designing for children with disabilities can require using dimensions not included in the ADA Standards. Since ADAAG allows for alternative technologies to "provide substantially equivalent or greater access to and usability of the facility," installation of children's bathroom fixtures can be compliant with ADA. This project is included to assist in designing for children's accessibility, an area of growing concern and research.

ADAAG Reference
4.1.3 (11) New Construction

Where Applicable
Children's bathrooms are subject to the same ADA scoping and technical requirements as adult bathrooms. Specific children's accessibility design standards, cited below, will be most useful in settings that have regular use by younger children, as in day care centers and elementary schools.

Design Requirements
There are no specific ADA Standards for children's fixtures. ADAAG 2.1 states, "The specifications and provisions in these guidelines are based upon adult dimensions and specifications." Accessible children's fixtures and dimensions can be used under ADAAG 2.2 Equivalent Facilitation, which states, "Departures from particular technical and scoping requirements of this guideline by the use of other designs and technologies are permitted where the alternative designs and technologies used will provide substantially equivalent or greater access to and usability of the facility." All public and common use toilet rooms need to be located on an accessible route, with accessible entrances and maneuvering spaces inside.

Design Suggestions
When designing bathrooms for children's access, it is important to allow for greater maneuvering space than usual, since many children who use wheelchairs are not yet able to control them as well as adults. Local codes should be consulted to determine whether accessible features are needed in addition to standard features.

Privacy is an important consideration; some children who use wheelchairs need assistance in the bathroom after the age which other children are able to use the bathroom unassisted, and visual privacy should be provided with full-height partitions where possible. If a school is being modified for an individual child, it is vital to get input on the student's needs, but the rest room should also allow for future use by other children with different needs. In situations where rest rooms are not divided by age, the third-to-fourth grade dimensions (such as 15" a.f.f. for toilet seat height) provide a usable compromise.

To give some guidance in designing for children's accessibility, the "Recommendations for Accessibility Standards for Children's Environments" are being reviewed by the Department of Justice for inclusion in the ADA Standards. These were done as *Task Five of the Accessibility Standards for Children's Environments* study done for the Architectural and Transportation Barriers Compliance Board by the Center for Accessible Housing, Raleigh, North Carolina, issued in November 1991.

Key Items
Same as standard plumbing installations: fixtures, water supply, drain pipes, faucets, insulation, counters, mirrors, dispensers.

Level of Difficulty
Moderate to high.

Following are the proposed standards given for children's bathrooms. Although at this point they are only recommendations, they represent some of the most comprehensive studies to date of children's access needs. Except where noted, all dimensions are for children using wheelchairs. Any bathroom access dimensions not cited are the same for children's access as for adult access.

4.2.1	Wheelchair passage clear width:	44"
4.2.5	Forward and side reach: 36" high, 20" low	
4.16.3	Water closets (toilets), to the top of the seat	
	Pre-kindergarten	11-1/2" to 12-1/2"
	Kindergarten to third grade	12" to 15"
	Fourth grade and older	15" to 17"

Copyright R. S. Means Company, Inc., 1994

55. Children's Accessible Bathroom Fixtures (continued)

4.16.4 Grab bars, side wall location: (Heights are given for side grab bars only, since standard location of the toilet tank prevents the installation of a rear bar at children's heights, and the rear bar should be installed at the standard height or 33"–36"; if there is no tank, the rear bar should be installed at the same height as the side bar).

Pre-kindergarten	18" to 20" to the top of the side bar;
Kindergarten to third grade	20" to 25"
Fourth grade and older	25" to 27"

4.16.4b Toilet centerline location to wall:

Pre-kindergarten	11"
Kindergarten to third grade	11" to 15"
Fourth grade and older	15" to 18"

4.16.5 Toilet flush valve location: 20" to 30" a.f.f., depending on the age of the children.

4.16.6 Toilet paper dispensers:

Pre-kindergarten	14"
Kindergarten to third grade	14" to 17"
Fourth grade and older	17" to 19"

4.17.3 Toilet stall dimensions: no allowance for a 36" wide stall, which is considered too narrow for a child to use.

4.17.4 Toilet partition toe clearance: 9" clear for pre-kindergarten, 12" for kindergarten and up.

4.18 Urinals: Stall-type, or with a rim at 14" a.f.f. maximum.

4.18.4 Urinal flush valves: 30" a.f.f. maximum.

4.19.2 Lavatories: rim or counter—30" a.f.f. maximum; knee space below—27" a.f.f. minimum. With two or more lavatories provided for pre-kindergarten, one shall be 22" a.f.f. maximum.

4.19.3 Lavatory clear floor space: 36" × 55" clear for a front approach. If faucets are mounted on the side of the rim or counter, 32" clear space shall be provided on the faucet side.

4.19.6 Mirrors: bottom reflecting surface no higher than 30".

4.21.3 Shower stall seat: 12" to 17" a.f.f. maximum.

4.21.5 Shower stall controls: 36" a.f.f. maximum.

Estimate

Install children's accessible bathroom fixtures (plumbing available)

Description	Quantity	Unit	Work Hours	Material
Wall hung one piece water closet	1.000	Ea.	5.300	700.00
Rough in supply, waste and vent	1.000	Ea.	5.634	113.00
Wall hung urinal	2.000	Ea.	10.666	760.00
Rough in supply, waste and vent	2.000	Ea.	11.308	169.00
16" x 14" porcelain enamel on cast iron lavatory	2.000	Ea.	4.000	620.00
Paddle faucets with spout	2.000	Ea.	1.600	252.00
Rough in supply, waste and vent	2.000	Ea.	19.278	256.00
Grab bars	2.000	Ea.	0.800	70.00
Totals			58.586	2940.00

Total for 2 fixtures as described including general contractor's overhead and profit $8,131

Copyright R. S. Means Company, Inc, 1994

56
Install Roll-In Shower

A shower that allows a person using a wheelchair to roll in is one of the most versatile access modifications, since it permits use both by people who use wheelchairs and those who don't. For people who shower while in their wheelchair, it is vital. Prefab fiberglass roll-in showers are now standard and make installation relatively simple.

ADAAG Reference
4.21 Shower Stalls

Where Applicable
Accessible public bathing facilities.

Design Requirements
- On accessible route.
- No lip at entry.
- 30" × 60" clear with grab bars with 36" × 60" clear floor space in front, 33" to 36" a.f.f., 1-1/4" to 1-1/2" in diameter, 1-1/2" from wall, on side and rear walls.
- Shower head adjustable and hand held with flexible hose at least 60" long on rear wall, 27" maximum from corner.
- Accessible controls, 48" a.f.f. maximum.

Design Suggestions
Recessing the floor pan below floor level allows for smooth roll-in. This can be very difficult as a retrofit, but is easier to do in new construction, since it requires strengthening the existing floor structure. Shower curtains should reach to the floor to prevent spillage.

Key Items
Either prefab shower stall or custom-framed stall and floor pan, water supply and drainage, ventilation, wall/ceiling finish materials.

Level of Difficulty
High. Requires plumbing, framing, finish work.

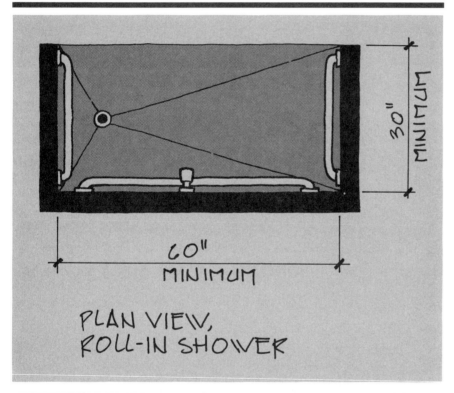

PLAN VIEW, ROLL-IN SHOWER

Estimates

Install prefabricated roll-in shower (plumbing available)

Description	Quantity	Unit	Work Hours	Material
Remove metal stud/gypsum board part. for shower opening	35.000	S.F.	1.610	0.00
Remove vinyl flooring	20.000	S.F.	0.320	0.00
Metal stud partition	77.000	S.F.	1.463	22.33
1/2" gypsum board, taped and finished	110.000	S.F.	1.870	22.00
Insulation	110.000	S.F.	0.660	18.70
Vinyl tile flooring	11.000	S.F.	0.176	15.40
Vinyl base, 6" high	12.000	L.F.	0.300	9.48
Painting	110.000	S.F.	1.980	18.70
Bar-mounted hand-held shower head	1.000	Ea.	0.400	137.00
Fiberglass shower, corner seat, and grab bars	1.000	Ea.	8.000	570.00
Rough in supply, waste and vent	1.000	Ea.	7.805	76.50
Totals			24.584	890.11

Total per each including general contractor's overhead and profit: $2,853

Install custom roll-in shower

Description	Quantity	Unit	Work Hours	Material
Remove metal stud/gypsum board part. for shower opening	35.000	S.F.	1.610	0.00
Remove vinyl flooring	20.000	S.F.	0.320	0.00
Wood stud partition	11.000	L.F.	1.760	41.58
Insulation	110.000	S.F.	0.660	18.70
Water-resistant 5/8" gypsum board	110.000	S.F.	0.880	30.80
5/8" gypsum board, taped and finished	110.000	S.F.	1.870	26.40
Vinyl tile flooring	11.000	S.F.	0.176	15.40
Vinyl base, 6" high	12.000	L.F.	0.300	9.48
Painting	110.000	S.F.	1.980	18.70
Copper shower pan	18.000	S.F.	1.440	45.00
Gypsum board ceiling	15.000	S.F.	0.315	3.60
Ceramic tile floor (pitched to drain)	15.000	S.F.	1.995	53.25
Ceramic tile walls, thin-set 4-1/4" x 4-1/4"	110.000	S.F.	9.240	209.00
Ceramic bath accessories	2.000	Ea.	0.390	18.10
Tub grab bar	1.000	Ea.	0.571	77.00
Grab bar vertical arms	2.000	Ea.	1.334	137.00
Bar-mounted hand-held shower head	1.000	Ea.	0.400	137.00
Rough in supply, waste and vent	1.000	Ea.	7.805	76.50
Totals			33.046	917.51

Total per each including general contractor's overhead and profit: $3,260

Copyright R. S. Means Company, Inc, 1994

57
Modify Existing Shower

Many existing showers have the approach and space to be accessible, but need modifications to comply with the ADA Standards. Even if it is too small, some showers have a non-plumbing wall which can be relocated to increase the shower's size. Creating a shower with grab bars, accessible controls and a seat can make an existing shower accessible to a person with a mobility or balance impairment.

ADAAG Reference
4.21 Shower Stalls

Where Applicable
Accessible public bathing facilities.

Design Requirements
- On accessible route.
- 1/2" maximum lip.
- Either 36" × 36" shower with seat and grab bars with 36" × 48" clear floor space in front (long dimension parallel to shower entrance), or 30" × 60" shower with grab bars with 36" × 60" clear floor space in front.
- Seat 17" to 19" above floor of stall.
- Grab bars 33" to 36" a.f.f., 1-1/4" to 1-1/2" in diameter, 1-1/2" from wall; in 36" × 36" stalls, grab bars on side wall and wall opposite seat; in 30" × 60" stall, on side and rear walls (see Figs. 35, 36, and 37 in ADAAG 4.21 for exact configurations).
- Shower head on pole with flexible hose at least 60" long.
- Controls operable with a closed fist, 48" a.f.f. maximum. In 36" × 36" stalls, controls mounted on wall opposite seat.

Design Suggestions

For 30" × 60" showers, include a shower seat to allow for people who shower in a wheelchair and who transfer to a seat (all 36" × 36" showers require a seat, since they are not roll-in showers). If removing the lip at the shower door leaves a threshold higher than 1/2", this requires a bevel. (Since the bevel can result in wet floors, escaping water should be controlled with either a long curtain or a shower door). Recessing the

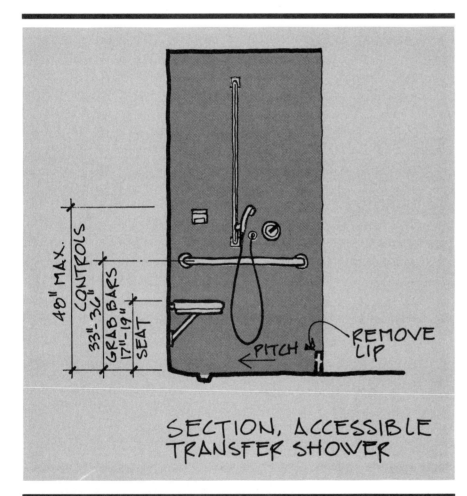

SECTION, ACCESSIBLE TRANSFER SHOWER

floor pan below floor level allows for a smooth roll-in. This is relatively easy to do in new construction, but can be difficult in existing buildings since it requires strengthening the existing floor structure.

Key Items
Grab bars, shower head on hose, fold-down seat, bevel at entrance, possibly new floor pan.

Level of Difficulty
Moderate to high. Requires plumbing, seat installation, finish work. Removing the lip requires extensive modification to an existing shower.

Estimates

Remove concrete lip at entry and patch flooring

Description	Quantity	Unit	Work Hours	Material
Labor minimum for demolition and tile installation	1.000	Job	4.923	0.00
Ceramic tile floor	5.000	S.F.	0.665	17.75
Totals			5.588	17.75

Total per linear foot including general contractor's overhead and profit	**$60**
Total per each 5 linear feet including general contractor's overhead and profit	**$300**

Replace existing floor pan with pan recessed below floor

Description	Quantity	Unit	Work Hours	Material
Selective demolition of floor (concrete)	4.000	C.F.	3.556	0.00
Remove thin-set ceramic tiles	18.000	S.F.	0.360	0.00
Remove shower pan	1.000	Job	2.000	0.00
Install new copper shower pan	18.000	S.F.	1.440	45.00
Ceramic tile floor	15.000	S.F.	1.995	53.25
Ceramic tile walls, thin-set 4-1/4" x 4-1/4"	9.000	S.F.	0.756	17.10
Shower drain	1.000	Ea.	2.000	65.00
Totals			12.107	180.35

Total per each including general contractor's overhead and profit	**$950**

Replace fixed shower head with shower head on hose

Description	Quantity	Unit	Work Hours	Material
Labor minimum to remove and reset	1.000	Job	2.000	0.00
Bar-mounted hand-held shower head	1.000	Ea.	0.400	137.00
Totals			2.400	137.00

Total per each including general contractor's overhead and profit	**$385**

Copyright R. S. Means Company, Inc., 1994

57. Modify Existing Shower (continued)

Add fold-down seat and grab bars to tiled shower wall

Description	Quantity	Unit	Work Hours	Material
Demolition of ceramic tile wall and gypsum board	3.000	Ea.	0.999	0.00
2" × 4" blocking	0.005	M.B.F.	0.306	2.92
Misc. materials for gypsum board and ceramic tile repair	1.000	Job	0.000	100.00
Labor minimum to repair gypsum board and ceramic tile	1.000	Job	4.923	0.00
Ceramic tile walls, thin-set 4-1/4" × 4-1/4" (material)	3.000	S.F.	0.252	5.70
Grab bars	1.000	Ea.	0.400	35.00
Fold-down seat	1.000	Ea.	1.000	250.00
Totals			7.880	393.62

Total per each including general contractor's overhead and profit $1,068

Copyright R. S. Means Company, Inc, 1994

Notes

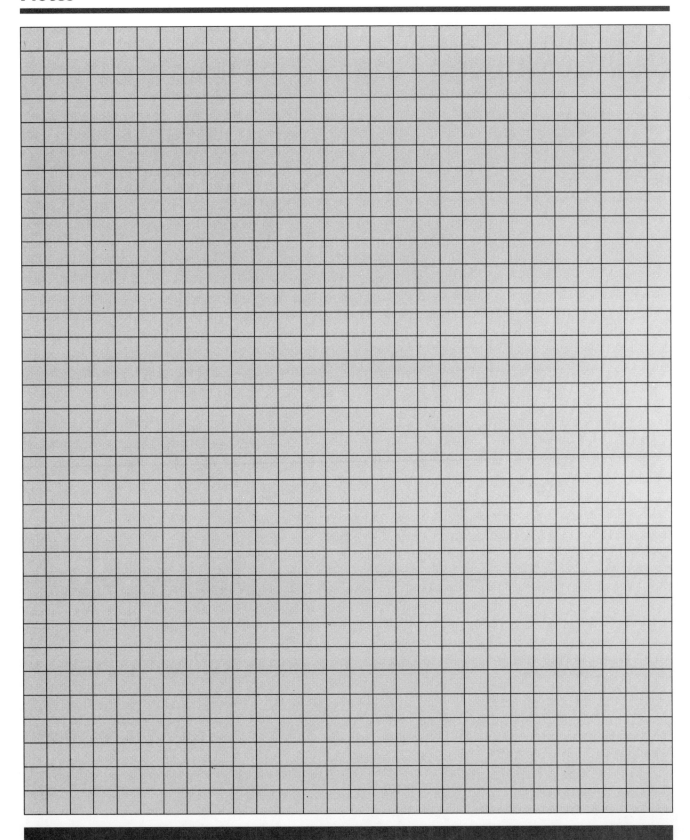

58
Replace Tub With Roll-In Shower

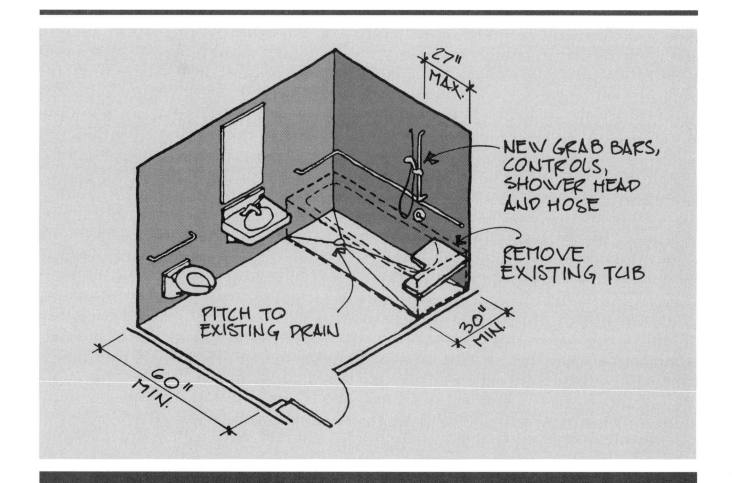

The floor space of a standard 30" × 60" tub is the same as what is required for a roll-in shower. In an existing structure with plumbing and drain locations in place, it is often possible to create an accessible roll-in shower by replacing an existing tub. In an accessible bathroom, this modification can create a shower usable by a wide range of people, including those who use wheelchairs.

ADAAG Reference
4.21 Shower Stalls

Where Applicable
Accessible bathing facilities are required to have at least one accessible tub or shower.

Design Requirements
- On accessible route.
- No lip at entry.

- 1/2" maximum lip.
- 30" × 60" shower with grab bars with 36" × 60" clear floor space in front.
- Folding seat 17" to 19" above floor of stall.
- Grab bars 33" to 36" a.f.f., 1-1/4" to 1-1/2" in diameter, 1-1/2" from wall; on both end and rear walls, but not behind seat.
- Shower head usable as either hand-held or fixed accessible shower head on flexible hose at least 60" long.
- Accessible controls, 48" a.f.f. maximum.

Design Suggestions

Include a fold-down seat to allow for both people who transfer to a seat and those who shower in a wheelchair. Recessing the floor pan below floor level allows for smooth roll-in. This is relatively easy to do in new construction, but can be difficult in existing buildings since it requires strengthening the existing floor structure. The shower curtain should reach all the way to the floor to prevent spillage.

Key Items

Either prefab shower stall or custom-framed stall and floor pan, water supply and drainage, and wall/ceiling finish materials.

Level of Difficulty

High. Requires plumbing, finish, and possibly framing work.

Estimates

Replace tub with accessible fiberglass shower

Description	Quantity	Unit	Work Hours	Material
Remove bathtub	1.000	Ea.	2.000	0.00
Labor minimum to disconnect plumbing	1.000	Job	4.000	0.00
Remove partition finishes	66.000	S.F.	0.594	0.00
Fiberglass shower, corner seat, and grab bars	1.000	Ea.	8.000	570.00
Bar-mounted hand-held shower head	1.000	Ea.	0.400	137.00
Rough in supply, waste and vent	1.000	Ea.	7.805	76.50
Totals			22.799	783.50

Total per each including general contractor's overhead and profit: $2,642

Replace tub with custom roll-in shower

Description	Quantity	Unit	Work Hours	Material
Remove bathtub	1.000	Ea.	2.000	0.00
Labor minimum to disconnect plumbing	1.000	Job	4.000	0.00
Remove partition finishes	66.000	S.F.	0.594	0.00
Water-resistant 5/8" gypsum board	110.000	S.F.	0.880	30.80
Copper shower pan	18.000	S.F.	1.440	45.00
Ceramic tile floor (pitched to drain)	15.000	S.F.	1.995	53.25
Ceramic tile walls, thin-set 4-1/4" × 4-1/4"	110.000	S.F.	9.240	209.00
Ceramic bath accessories	2.000	Ea.	0.390	18.10
Tub grab bar	1.000	Ea.	0.571	77.00
Grab bar vertical arms	2.000	Ea.	1.334	137.00
Bar-mounted hand-held shower head	1.000	Ea.	0.400	137.00
Rough in supply, waste and vent	1.000	Ea.	7.805	76.50
Totals			30.649	783.65

Total per each including general contractor's overhead and profit: $2,967

Copyright R. S. Means Company, Inc, 1994

59 Modify Existing Tub

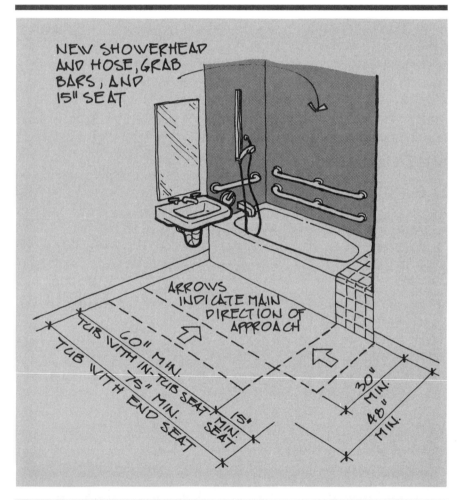

Many people who use wheelchairs do transfer to a seat in a tub, but a bathtub does not necessarily have to be part of a fully accessible bathroom to be modified for increased accessibility. Accessible grab bars, controls, and an in-tub seat can be added relatively simply to assist people with mobility, balance, and grasping impairments.

ADAAG Reference
4.20 Bathtubs

Where Applicable
Accessible bathing facilities are required to have at least one accessible tub or shower.

Design Requirements
- On accessible route.
- Clear floor space as least as wide as the tub and 30" to 48" deep, depending on the approach.
- Either an in-tub seat or end seat; end seat 15" wide minimum, full depth of tub.
- Grab bars:
  a. For 60" tub with in-tub seat: two grab bars on rear wall, 24" minimum long, top bar 33" to 36" a.f.f., bottom 9" above rim of tub. One 24" long minimum grab bar on wall at foot; one 12" minimum grab bar on wall at head, opposite controls.
  b. For 60" tub with end seat: two grab bars on rear wall, 48" minimum long, top bar 33" to 36" a.f.f., bottom 9" above rim of tub. One 24" minimum long grab bar on foot wall with controls.
- Controls on foot wall, below grab bar.
- Shower head on a 60" minimum hose, usable as a fixed or hand-held shower.
- No tub enclosure obstructing controls or transfer to a seat, or mounted on a rim.

Design Suggestions
In-tub seats prevent a shower curtain from closing fully; for an in-tub seat, two curtains would minimize spillage. Some sliding door tracks permit transfer to a seat.

Key Items
Shower head on a hose, tub seat, grab bars.

Level of Difficulty
Low to moderate. Replacing showerhead and faucets might require a plumber; grab bar installation may require finish (tile) work. Extending tub enclosure to include end seat requires additional framing.

Estimates

Extend tub enclosure

Description	Quantity	Unit	Work Hours	Material
Remove wood stud/gypsum board partition	36.000	S.F.	1.656	0.00
Remove partition finishes	12.000	S.F.	0.108	0.00
Wood stud partition	3.000	L.F.	0.480	11.34
Water-resistant 5/8" gypsum board	36.000	S.F.	0.288	10.08
Seat framing	0.030	M.B.F.	1.412	16.35
Seat sheathing	18.000	S.F.	0.198	6.66
Ceramic tile covering for seat	9.000	S.F.	1.197	31.95
Ceramic tile walls, thin-set 4-1/4" × 4-1/4"	45.000	S.F.	3.780	85.50
Totals			9.119	161.88

Total per each including general contractor's overhead and profit: $713

Add in-tub seat

Description	Quantity	Unit	Work Hours	Material
Portable in-tub or in-shower seating	1.000	Ea.	0.000	110.00
Totals			0.000	110.00

Total per each including general contractor's overhead and profit: $188

Add grab bars

Description	Quantity	Unit	Work Hours	Material
Cutout demolition of partition	1.000	Ea.	0.333	0.00
2" × 4" blocking	0.005	M.B.F.	0.306	2.92
Misc. materials for gypsum board and ceramic tile repair	1.000	Job	0.000	100.00
Labor minimum to repair gypsum board and ceramic tile	1.000	Job	4.923	0.00
Grab bars	1.000	Ea.	0.400	35.00
Totals			5.962	137.92

Total per each including general contractor's overhead and profit: $531

Copyright R. S. Means Company, Inc., 1994

59. Modify Existing Tub (continued)

Replace fixed shower head with shower head on hose

Description	Quantity	Unit	Work Hours	Material
Labor minimum to remove and install shower head	1.000	Job	2.000	0.00
Bar-mounted hand-held shower head	1.000	Ea.	0.400	137.00
Totals			2.400	137.00

Total per each including general contractor's overhead and profit: $385

Relocate controls, tiled wall

Description	Quantity	Unit	Work Hours	Material
Cutout demolition of partition	1.000	Ea.	0.333	0.00
Labor minimum to remove & replace mixing valve	1.000	Job	2.000	0.00
Misc. materials for gypsum board and ceramic tile repair	1.000	Job	0.000	100.00
Labor minimum to repair gypsum board and ceramic tile	1.000	Job	4.923	0.00
Totals			7.256	100.00

Total per each including general contractor's overhead and profit: $554

Copyright R. S. Means Company, Inc, 1994

Notes

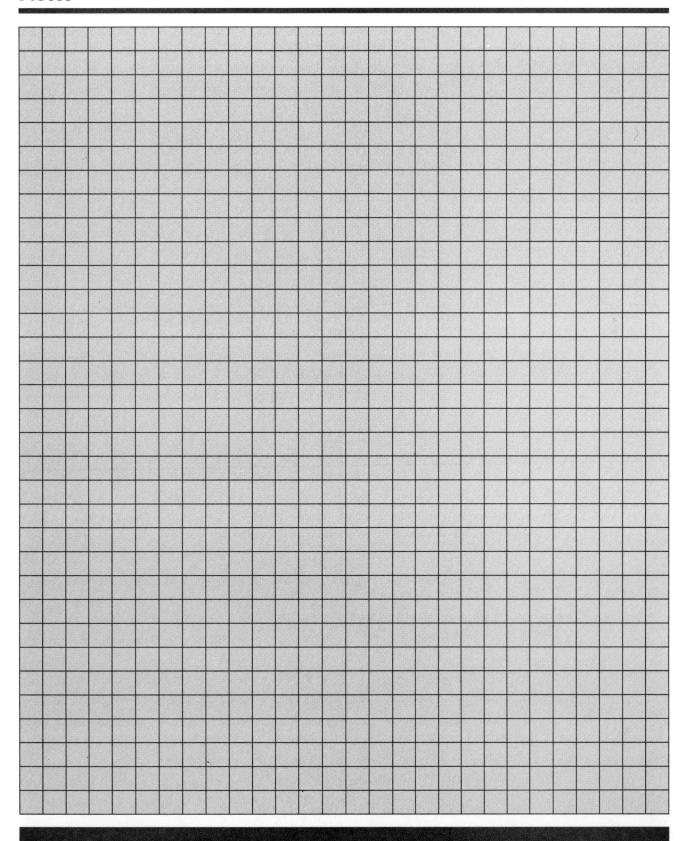

60 Create Accessible Gang Showers

It is impossible to predict the precise access needs of all people who use a shower; some people shower in a wheelchair, while some transfer to a seat; others with mobility impairments who do not use a wheelchair can walk to a shower, but still need a seat and/or grab bars for assistance.

Accommodating as many needs as possible is desirable, although only either a roll-in shower or transfer shower is needed for ADA compliance. Since gang showers are often located in facilities open to the public, they are important to modify for access in order to make them usable by as wide a range of people as possible. Such showers often have compliant maneuvering space along the approach and in the shower and can be made accessible by creating an accessible entrance, adding grab bars and modifying the controls.

Where there are multiple stalls, it is often possible to modify the front lip and add the other required elements. Even where existing materials are difficult to modify (such as marble stall partitions or concrete shower pans) it is possible to add accessible features relatively simply.

ADAAG References
4.21 Shower Stalls
4.23 Bathrooms, Bathing Facilities, and Shower Rooms

Where Applicable
At least one shower or shower stall in public shower rooms.

Design Requirements
- On an accessible route. Opening into the shower at least 32" wide clear with no lip greater than 1/2" which is not beveled.
- Slip-resistant flooring.
- Controls on an accessible route, no higher than 48", operable with a closed fist, with clear floor space 30" × 48" in front.
- Either a shower spray unit with a hose at least 60" long which can be used as both a shower head and hand

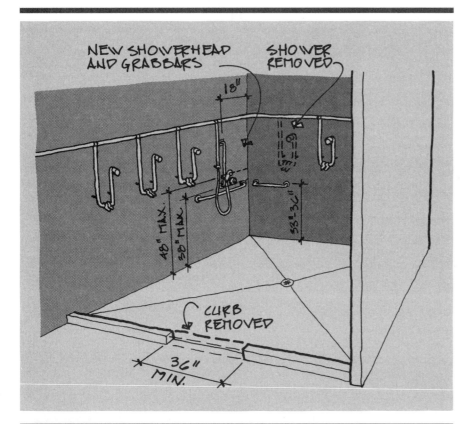

held, or, in unmonitored facilities, a fixed shower head mounted at 48" a.f.f.
- Fold-down shower seat at least 24" by 16", fixed to the wall on the seat's longer dimension, 17"–19" a.f.f.
- At least one grab bar, 33"–36" a.f.f., 1-1/2" diameter, 1-1/2" from wall, under the controls if the accessible shower is not located in a corner, with an additional grab bar on the side wall if it is.
- Shower stalls at least 36" × 36", fixed seat mounted at 17"–19" a.f.f., extended full depth of stall on wall opposite controls.

Design Suggestions

Gang showers may consist either of a row of shower heads against a wall, or a row of individual shower stalls (it is rare that there are collective tub areas). If there are no individual stalls, it would be preferable to locate the accessible shower in a corner to allow for grab bars on two walls. Very often there is a continuous lip a certain distance from the shower wall. A section will have to be removed to create an accessible route to the shower. An alternative is to ramp up to the lip and install a grating flush with the top of the lip. This can involve less demolition and maintain existing drainage, but can be expensive since the grating will have to be strong enough to support an adult using a wheelchair (a lip or some sort of protection will be needed on one side for safety).

A row of individual stalls is often more difficult to make accessible, since stalls are rarely large enough to allow a roll-in situation. If the stall is at least 36" × 36", a compliant transfer shower can be installed if there is an accessible route to the stall. This can involve the same solutions as creating access at the gang shower described above: removing a lip, or ramping up and installing a grating across the shower pan.

Key Items

Shower hose, seat, grab bar(s), accessible shower controls, and finish floor materials.

Level of Difficulty

Moderate to high.

Estimate

Modify gang showers

Description	Quantity	Unit	Work Hours	Material
Labor minimum for demolition and tile installation	1.000	Job	4.923	0.00
Ceramic tile floor	5.000	S.F.	0.665	17.75
Cutout demolition of partition	3.000	Ea.	0.999	0.00
2" × 4" blocking	0.005	M.B.F.	0.306	2.92
Remove and reset mixing valve, remove shower head	1.000	Ea.	2.000	0.00
Misc. materials for gypsum board and ceramic tile repair	1.000	Job	0.000	100.00
Labor minimum to repair gypsum board and ceramic tile	1.000	Job	4.923	0.00
Fold-down seat	1.000	Ea.	1.000	250.00
Grab bars	1.000	Ea.	0.400	35.00
Bar-mounted hand-held shower head	1.000	Ea.	0.400	137.00
Totals			15.616	542.67

Total per each including general contractor's overhead and profit: $1,731

Copyright R. S. Means Company, Inc, 1994

61 Create Accessible Counter

Just about every public facility has a service counter. Counters that are too high, even in accessible spaces, are difficult to use for people who use wheelchairs or have difficulty reaching over a certain height. Since the required height given in the ADA Standards is usable by most adults (and can actually be more convenient for transfering goods across a surface), creating an accessible counter or adding an accessible shelf will greatly increase the usability of a public space.

ADAAG Reference
7.2 Sales and Service Counters, Teller Windows, Information Counters

Where Applicable
Counters where sales or distribution of goods or services take place.

Design Requirements
- On accessible route.
- At least one portion of the counter no more than 36" high and no less than 36" wide.
- In alterations, if it is technically infeasible to have a portion of the counter no more than 36" high and no less than 36" wide, an auxiliary counter of the same dimensions is allowed.
- Clear floor space in front of the accessible counter at least 30" × 48".

Design Suggestions
If it is not possible to lower a section of counter due to historic preservation requirements or because of existing conditions (if the counter is made of marble or granite, for instance) it might be possible to install a fold-down shelf on the face of the existing counter, or to install a pass-through window in the face of the counter. Some areas behind the counter are entered through a door on a public route, and it might be possible to replace this with a split door that can be used on an as-needed basis. Although there is no single height that will accommodate everyone, it is

recommended that the accessible counter be lower than 36" (32"-34") to facilitate use by the greatest number of people.

Key Items
Low counter, or fold-down counter if a low counter is technically infeasible.

Level of Difficulty
Varies. Depends on material and extent of renovation.

Estimates

Lower 36" wide section of counter

Description	Quantity	Unit	Work Hours	Material
Remove 36" wide section of counter	3.000	L.F.	0.399	0.00
Remove counter support	1.000	Job	2.000	0.00
End caps at counter cuts (pine trim)	4.000	L.F.	0.132	1.60
Counter supports	0.032	M.B.F.	1.506	17.44
Additional labor for end caps and counter support	1.000	Job	2.000	0.00
Totals			6.037	19.04

Total per each including general contractor's overhead and profit **$348**

Install fold-down counter to face of existing counter

Description	Quantity	Unit	Work Hours	Material
Plastic laminate counter	3.000	S.F.	0.600	21.00
Continuous steel hinge	3.000	L.F.	0.750	12.00
Totals			1.350	33.00

Total per linear foot including general contractor's overhead and profit **$43**

Total per each 3 linear feet including general contractor's overhead and profit **$129**

Copyright R. S. Means Company, Inc, 1994

62
Create Accessible Aisles

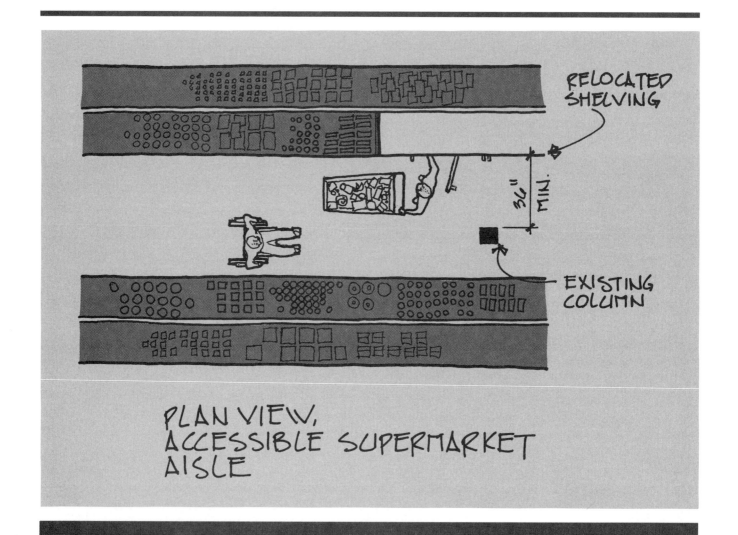

PLAN VIEW, ACCESSIBLE SUPERMARKET AISLE

Creating an accessible entrance is the first step in creating an accessible space; the next is ensuring that people can get around in the facility. Providing a 36" minimum width route of travel makes it possible for people using wheelchairs to get to the goods or services. It also makes it easier for others, since any point narrower than 36" is difficult for many people to use (including those carrying packages). All that is often required is moving or modifying shelving or display cases, with some possible replacement of the flooring surface.

200

ADAAG References
4.2 Space Allowance and Reach Ranges
4.3 Accessible Route
5. Restaurants and Cafeterias (5.3 Access Aisles, 5.4 Dining Areas)
7. Business and Mercantile (7.3 Check-out Aisles)
8. Libraries (8.5 Stacks)

Where Applicable
All aisles on accessible routes where goods or services are offered to the public.

Design Requirements
- 36" clear minimum route of travel to all goods and services.
- Route compliant with slope, protruding objects, and slip-resistant surface requirements.

Design Suggestions
There is no height restriction on shelving, but some method must be provided for obtaining goods out of reach range. One solution is to display goods such as food vertically instead of horizontally, having the same goods available at different heights. Where informational material is offered for distribution, such as maps, schedules, or forms, it should be within reach range (48" a.f.f. for front reach, 54"a.f.f. for side reach). Some method of conveying the information must be available to all users, and service assistance is acceptable.

Key Items
Existing shelving and information displays.

Level of Difficulty
Low. Relocating information displays can involve finish work.

Estimates

Make aisles accessible (patching of flooring after moving of F.F.E. type items by owner)

Description	Quantity	Unit	Work Hours	Material
Heavy duty vinyl sheet goods	1.000	S.F.	0.040	4.30
Totals			0.040	4.30

Total per square foot including general contractor's overhead and profit: $9

Make aisles accessible (moving of unfixed F.F.E. type items)

Description	Quantity	Unit	Work Hours	Material
Labor minimum	1.000	Job	2.000	0.00
Totals			2.000	0.00

Total per each including general contractor's overhead and profit: $91

Lower wall-mounted information rack

Description	Quantity	Unit	Work Hours	Material
Labor minimum for lowering signage	1.000	Job	2.000	0.00
Miscellaneous materials for gypsum board repair	1.000	Job	0.000	25.00
Labor minimum to repair and paint gypsum board	1.000	Job	2.000	0.00
Totals			4.000	25.00

Total per each including general contractor's overhead and profit: $259

Copyright R. S. Means Company, Inc, 1994

63
Modify Dining Area

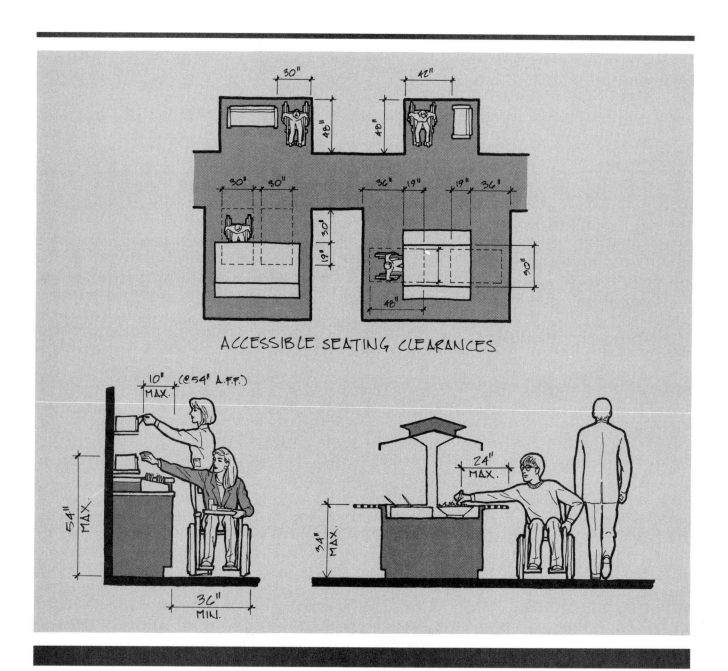

ACCESSIBLE SEATING CLEARANCES

Many dining areas are designed to maximize seating capacity, which can minimize access. In some dining areas or restaurants which are on an accessible route, it is often possible to create accessible seating and accessible services without major modifications to the space by relocating or removing some seating, and widening aisles. This can make the space easier to use for all customers, as well as the cleaning and wait staff.

ADAAG References

5. Restaurants and Cafeterias
4.32 Fixed or Built-in Seating and Tables

Where Applicable

Public and common-use dining areas.

Design Requirements

- 36" clear interior route of travel; floor surfaces slip-resistant and stable.
- Minimum of 5% fixed seating accessible (in all sections: smoking, nonsmoking, and range of quality).
- At least one 60" long section of counter where food is eaten 34" a.f.f. maximum (or accessible table service in the same area).
- All levels of dining areas required to be accessible in new construction of building with elevators (exceptions: new construction buildings without elevators, buildings with mezzanine that contain less than 33% of the total seating, and where an accessible area will be provided that offers the same services and decor. The same applies to construction alterations, except that there is no specific seating percentage.
- Tray slides 34" a.f.f. maximum.
- Self-service devices within reach range (48" a.f.f. for front reach, 54" a.f.f. for side reach), except when reaching over a shelf.
- Raised areas accessible.
- Space for vending machines accessible.

Design Suggestions

Ensure that movable tables have correct height and adequate space below (30" × 48" clear floor space, 19" under table, knee space 27" clear a.f.f. minimum, table height 28" to 34", 32" preferred.) The preferred food service line width is 42" minimum clear. Provide a continuous lowered shelf from the tray and utensil area to the check-out to avoid having to carry the tray from one to the other.

Key Items

Existing dispensers/slides to be relocated, new movable seating.

Level of Difficulty

Low to moderate. Removing fixed seating may require finish work to existing floors and/or walls. Could require schematic planning and design drawings.

Estimates

Lower tray slide (assume slide is mounted with brackets every 2 feet)

Description	Quantity	Unit	Work Hours	Material
Remove existing brackets and slide	1.000	L.F.	0.125	0.00
Layout and drill new mounting holes (4 per bracket)	2.000	Ea.	0.500	0.00
Re-install brackets and slide	1.000	L.F.	0.125	0.00
Totals			0.750	0.00

Total per linear foot including general contractor's overhead and profit: $80

Copyright R. S. Means Company, Inc., 1994

63. Modify Dining Area (continued)

Widen food service line (assume pipe rail line delineator, uprights 4' O.C.)

Description	Quantity	Unit	Work Hours	Material
Remove existing pipe rail (bolted to floor)	1.000	L.F.	0.125	0.00
Layout & drilling of bolt holes in conc. floor (per 1" of depth)	2.000	Ea.	0.320	0.24
Expansion anchors and shields (1/2" diameter)	1.000	Ea.	0.107	2.55
Re-install pipe rail	1.000	L.F.	0.200	0.00
Vinyl tiles 12" x 12" (1 tile at former upright locations)	0.250	S.F.	0.004	0.16
Totals			0.756	2.95

Total per linear foot including general contractor's overhead and profit: $50

Replace fixed seating with movable tables and chairs

Description	Quantity	Unit	Work Hours	Material
Remove four-seat, 24" x 48" table	1.000	Ea.	0.572	0.00
Patching concrete floor	3.000	S.F.	0.240	10.80
Vinyl tile	3.000	S.F.	0.048	8.25
Tile layer minimum	1.000	Job	2.000	0.00
Totals			2.860	19.05

Furniture materials and labor prices provided by supplier

Total per each including general contractor's overhead and profit: $174

Lower section of bar (wood)

Description	Quantity	Unit	Work Hours	Material
Remove 60" section of bar/counter	5.000	L.F.	0.665	0.00
Remove supporting framework for bar/counter	1.000	Job	2.000	0.00
Miscellaneous millwork demolition	5.000	L.F.	1.000	0.00
Install new framework for bar/counter	0.032	M.B.F.	1.506	17.44
Install new wood bar/counter	5.000	L.F.	1.430	150.00
Additional labor for framework/millwork	1.000	Job	2.000	0.00
Totals			8.601	167.44

Total per linear foot including general contractor's overhead and profit: $146

Total per each 5 foot section including general contractor's overhead and profit: $729

Copyright R. S. Means Company, Inc., 1994

Install ramp up to raised (6") area, 4' wide (carpet on wood, no rails)

Description	Quantity	Unit	Work Hours	Material
Ramp framing	0.040	M.B.F.	0.438	23.00
Ramp floor deck	24.000	S.F.	0.312	13.44
Additional labor required for ramp framing	1.000	Job	2.000	0.00
Additional carpet labor, minimum	1.000	Job	2.667	0.00
Carpet (minimum material quantity)	8.000	S.Y.	1.120	148.00
Totals			6.537	184.44

Total per square foot including general contractor's overhead and profit	$28
Total per each including general contractor's overhead and profit	$673

Copyright R. S. Means Company, Inc, 1994

64
Create Accessible Dressing Rooms

A standard dressing or fitting room is inaccessible to someone who uses a wheelchair. Increasing the size of the door and room and adding a compliant bench makes it possible for people with mobility impairments to try on clothes, as well as making it easier for large, frail or older people, and for parents with children.

ADAAG References
4.1.3 (21) New Construction
4.35 Dressing and Fitting Rooms

Where Applicable
At least one fitting room in public accommodations which provide them.

Design Requirements
- On an accessible route.
- Door 32" clear opening, hardware operable with a closed fist.
- Clear floor space allowing a person in a wheelchair to make a 180° turn if the room is entered through a swinging or sliding door (door cannot turn into maneuvering space).
- Sufficient maneuvering space in fitting room entered through a curtained opening at least 32" wide.
- Bench at least 24" × 48", with a slip-resistant surface fixed to the wall along the longer dimension, with clear floor space along the bench to allow transfer from a wheelchair. In wet locations such as swimming pools or showers, water shall not accumulate on the bench surface.
- Any mirror shall be full-length, at least 18" wide by 54" high, located to allow a view to a person sitting on the bench as well as standing.
- Any coat hooks mounted within reach range (54" a.f.f. maximum).
- Signage with International Symbol of Accessibility adjacent to door latch, 60" high.

Design Suggestions
An accessible fitting room should allow as much maneuvering space as possible. If the floor plan permits, the turning space inside should be a clear 60" circle that does not overlap with the space underneath the bench. The main

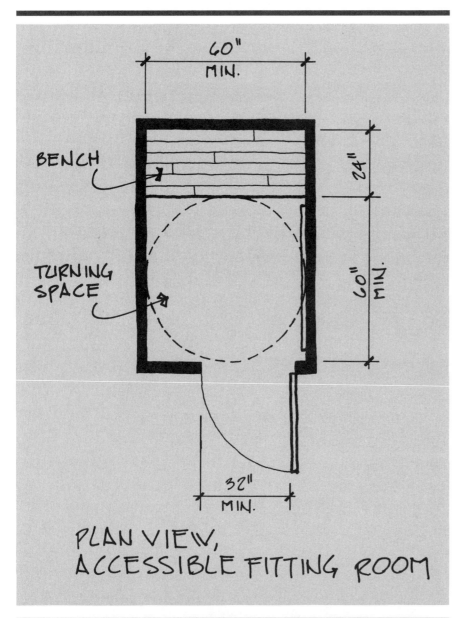

PLAN VIEW, ACCESSIBLE FITTING ROOM

danger of having a large changing room is that it will be used for storage. Care should be taken to ensure that this does not happen. Many people (such as parents with children) without physical limitations will find the larger changing room useful.

Key Items
Wallboard and stud wall, wood bench, door with accessible hardware, full-length mirror, coat hooks.

Level of Difficulty
Moderate. Basic carpentry and finish work.

Estimate

Install dressing rooms

Description	Quantity	Unit	Work Hours	Material
Metal stud partition (3-5/8" wide, 16" O.C.)	168.000	S.F.	3.192	48.72
1/2" gypsum board, taped and finished	336.000	S.F.	5.712	67.20
Painting	336.000	S.F.	6.048	57.12
Interior door frame	18.000	L.F.	0.774	57.60
Interior wood door, birch face, 3'-0" × 6'-8"	1.000	Ea.	1.143	65.00
Hinges	1.500	Pr.	0.000	69.00
Lever-handled lockset	1.000	Ea.	0.000	79.78
Bench, 24" × 4'	4.000	L.F.	1.144	120.00
Mirror	12.000	S.F.	1.200	68.40
Coat hook	1.000	Ea.	0.222	12.70
Totals			19.435	645.52

Total per square foot including general contractor's overhead and profit	**$434**
Total per each including general contractor's overhead and profit	**$2,170**

Copyright R. S. Means Company, Inc, 1994

65
Create Accessible Theater Seating

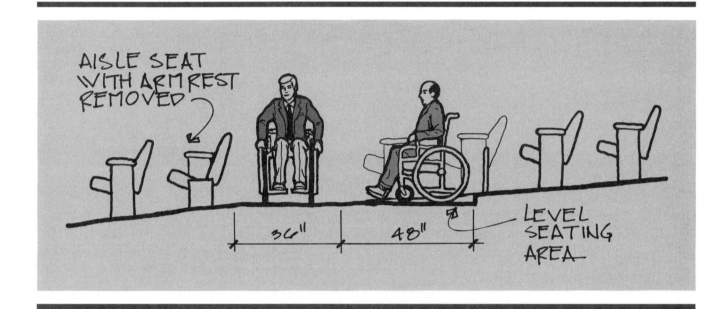

Theater design is complicated. Many factors are taken into account: sight lines, acoustics, maximizing seating, and fire egress. Accessibility, however, has not traditionally been a factor. Integrating accessible seating into existing theaters (or stadiums, etc.) is often a challenge, since they are complex facilities often built of materials and in structures that are difficult to modify. Nevertheless, it is critical to make these public facilities accessible and compliant. If a theater has an accessible route of travel, it is often possible to create accessible seating and, depending on the availability of accessible fire egress, to distribute the accessible seating throughout the space.

ADAAG References
4.1.3 (19) (a) Accessible Buildings, New Construction, Assembly Areas
4.33 Assembly Areas

Where Applicable
All public and common-use assembly areas with fixed seating.

Design Requirements
- Accessible route to the accessible seating.
- Comparable sight lines to inaccessible seating.
- Comparable choice of admission prices to inaccessible seating.
- Wheelchair-accessible seating integrated with the general seating plan; with auditoriums of over 300 seats, accessible seating in more than one location. (Exception for bleachers and balconies and other areas with sight lines that require slopes greater than 5%, where accessible seating can be grouped. Readily removable seats allowed to occupy wheelchair seating spaces when not occupied by wheelchair users).
- At least one fixed companion seat next to each accessible space.
- Two adjacent wheelchair spaces minimum in each seating area, 66" × 48" for front approach, 66" × 60" for side approach.
- Level area for wheelchair seating.
- 1% of total seats to be aisle seats without armrests or with folding armrests.

Design Suggestions

Locate accessible seats adjoining an accessible route of fire egress. If there is accessible egress from more than one part of the theater, disperse the accessible seating throughout the auditorium as much as possible. If a theater has different price ranges for seating areas, ADA requires that either accessible seats be located in each area or that policies be modified to offer the same price range for the accessible seats as for general seating (if accessible seats are located only in the most expensive section, wheelchair users cannot be charged only the most expensive price).

Key Items

Varies depending on location. Could require alteration(s) to accessible route: lifts, doors, flooring; leveling of sloped areas to create accessible seating spaces.

Level of Difficulty

Varies. Low for seating removal, if accessible seating area(s) are level and on an accessible route. High for creating on accessible route, and for leveling seating areas on existing sloping surface. Could require schematic planning and design drawings, compliant with local safety and building codes.

Estimate

Create accessible theater seating

Description	Quantity	Unit	Work Hours	Material
Remove existing seating	4.000	Ea.	0.920	0.00
Concrete formwork	20.000	SFCA	3.560	37.40
Cast in place concrete, finishing included	25.000	S.F.	0.650	48.25
Layout & drilling of anchor holes	4.000	Ea.	0.212	0.16
Plastic shields and screws	4.000	Ea.	0.000	0.12
Signage	1.000	Ea.	0.457	8.75
Totals			5.799	94.68

Total for 2 accessible locations including general contractor's overhead and profit: $474

Copyright R. S. Means Company, Inc, 1994

66
Install Assistive Listening Systems

Assistive listening systems are amplification systems designed to reduce background noise and room reverberations and to pick up sound at the source, amplify it, and direct it to the ear of the listener. Certain assistive listening devices can be used with hearing aids.

ADAAG References
4.1.3 Accessible Buildings, New Construction (19) (b) Assembly Areas
4.33 Assembly Areas

Where Applicable
All assembly areas with more than a 50-person capacity or with audio-amplification systems, or those that have fixed seating.

Design Requirements
- Permanently installed assistive listening system, with the number of devices equal to a minimum of 4% of the total number of seats, and not less than two.
- Signage indicating availability of services.
- Where devices are used with particular seats, those seats must be located within 50' of the stage.

Design Suggestions
It is important to provide the most appropriate assistive listening system for the size and type of theater. There are several types: infra-red, magnetic induction (or audio) loop radio frequency.
- Infrared: uses invisible infrared light beams to carry information from a transmitter connected to a PA system or microphone to a receiver worn by the listener. The light emitter picks up sound from the microphone and changes it into infrared light, which is directed towards the listener's receiver. It cannot be used in direct sunlight, but can be used in all other public (or private) spaces, including theaters. Infrared systems mostly benefit people with mild to moderate hearing loss. They can be useful in multi-theater complexes where sound distraction from other theaters can be a problem.
- Audio loop: consists of a microphone, amplifier, and wire loop that is placed around the listening area of the room, and a receiver worn by the listener. Sound is transmitted by a magnetic field created within the loop, and the listener picks up the amplified sound either with T switches on their hearing aids or with special receivers. Audio loops can be used indoors and outdoors, in large or small rooms, but people must sit within the "loop." They are helpful to people with mild to profound hearing loss. It is necessary to have different systems set at different frequencies since FM systems transmit through walls into adjacent rooms.
- FM system: a wireless amplification system consisting of a microphone, transmitter, and receiver resembling a small pocket radio worn around the neck. The transmitter sends the sound to the listener's receiver using FM radio signals. FM systems apply in any sized space and benefit people with mild to profound hearing loss.

There are several factors involved (size, shape, and acoustics of the space, whether or not it is indoors, the type of sound being projected, and the level of background noise) in the selection of audio systems. It is strongly recommended that manufacturer's representatives and local agencies or societies for people who are deaf or hard of hearing be consulted prior to the installation of any system.

Key Items
Wiring, outlets, assistive listening equipment.

Level of Difficulty
Varies, depending on system. Infrared or audio loop systems can require electricians and skilled technicians for installation; FM system requires equipment and outlets.

Estimates

Provide magnetic induction loop assistive listening system

Description	Quantity	Unit	Work Hours	Material
100 W amp., 200' loop, 20' lead, 1 receiver	1.000	Ea.	0.000	1755.00
Totals			0.000	1755.00

Total per each including general contractor's overhead and profit **$2,495**

Copyright R. S. Means Company, Inc., 1994

66. Install Assistive Listening Systems *(continued)*

Provide infrared assistive listening system (existing P.A. system is necessary)

Description	Quantity	Unit	Work Hours	Material
Basic system (transmitter, emitter, receiver)	1.000	Ea.	0.000	4900.00
Totals			0.000	4900.00

Total per each including general contractor's overhead and profit: $6,966

Provide radio frequency assistive listening system (existing P.A. system is necessary)

Description	Quantity	Unit	Work Hours	Material
Basic system (transmitter, 3 receivers)	1.000	Ea.	0.000	1200.00
Totals			0.000	1200.00

Total per each including general contractor's overhead and profit: $1,706

Install electrical outlet for assistive listening system

Description	Quantity	Unit	Work Hours	Material
Cutout demolition of partition	1.000	Ea.	0.333	0.00
Conductor	0.100	C.L.F.	0.296	1.55
Install junction box	1.000	Ea.	0.400	4.65
Install outlet	1.000	Ea.	0.296	5.25
Install cover plate	1.000	Ea.	0.100	2.30
Miscellaneous materials for gypsum board repair	1.000	Job	0.000	25.00
Labor minimum to repair and paint gypsum board	1.000	Job	2.000	0.00
Totals			3.425	38.75

Total per each including general contractor's overhead and profit: $252

Install signage for assistive listening system

Description	Quantity	Unit	Work Hours	Material
Signage	1.000	Ea.	0.457	8.75
Layout & drilling of anchor holes	4.000	Ea.	0.212	0.16
Plastic shields and screws	4.000	Ea.	0.000	0.12
Totals			0.669	9.03

Total per each including general contractor's overhead and profit: $58

Copyright R. S. Means Company, Inc, 1994

Notes

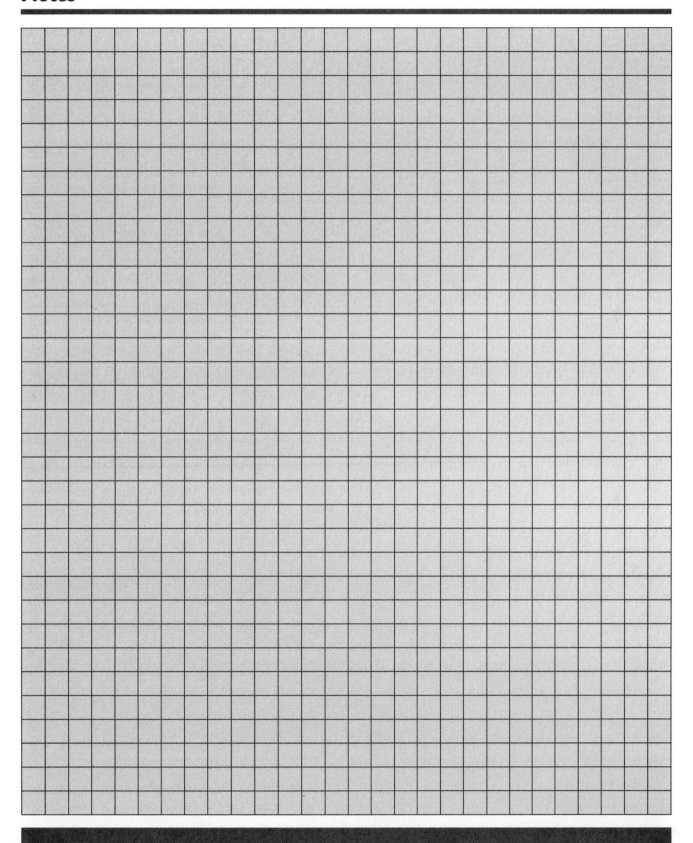

67
Install Detectable Warnings

Detectable warnings serve as a tactile indication of potentially dangerous locations, such as railway tracks for people who cannot use visual warnings alone. Research found that raised bumps (called "truncated domes" in the ADA Standards) can be felt by a cane and by a person's foot. It was also found that such domes made snow removal almost impossible. For that reason, this section of ADAAG has been reserved as of November, 1993 until further research is completed in July, 1996. Until that time, detectable warnings are required only at train platforms. However, detectable warnings have proven extremely useful in helping people with visual impairments to identify hazards on a route of travel (as well as reinforcing such locations for sighted people). Such materials should be considered where no other tactile clues exist to warn people of hazardous situations.

ADAAG References
4.29 Detectable Warnings
10.3.1(8) Fixed Facilities and Stations

Where Applicable
Edges of all train platforms.

Note: In November 1993, the Federal Access Board, jointly with the Department of Justice and the Department of Transportation, issued a proposed rule suspending the requirement for truncated dome detectable warnings at all areas except at train platforms. This is due to many comments from designers and facility managers regarding the inherent difficulty of snow removal and the potential buildup of ice between the domes, although many advocates for people with vision impairments feel that the regulation should remain. Until the Department of Justice and the Access Board issue a final rule modifying the ADA standards in the Federal Register, however, detectable warnings as specified or as equivalent facilitation are required for ADA compliance at all locations cited above.

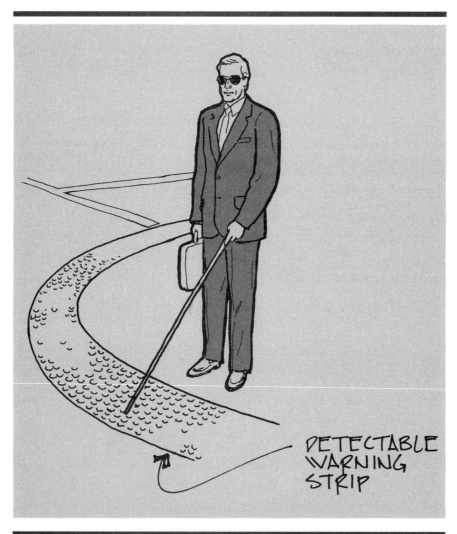
DETECTABLE WARNING STRIP

Design Requirements

- Visual contrast with surrounding paving.
- Domes detectable by cane and foot: .9" across, .2" high, beveled sides, 2.35" between centerlines.
- Slip-resistant material.
- 36" wide strip at vehicular crossings, 24" at train platforms.
- Detectable warnings on running slope portion of all curb ramps.

Design Suggestions

Detectable warning materials are currently being developed. Several types are now available which manufacturers claim comply with current regulations:

- Plastic tile, usually glued to the paving surface. While inexpensive and relatively easy to install, these do not weather well and are not very resistant to foot traffic. Edges should be beveled where exposed. Not recommended for climates with high snow/rainfall.
- Metal tile, usually screwed to the paving surface. Somewhat more expensive than plastic, but more durable. Fasteners can come loose over time, especially when subjected to freeze/thaw. Edges should be beveled where exposed.
- Ceramic tile, set with grout. Durable and weather-resistant, but thicker than plastic or metal. They can present a tripping hazard if installed on top of existing paving, and are best set in a recess flush with paving. This method can be expensive in existing situations.
- Rolled-on: a layer of plastic adhered to the existing surface, with domes set on top of the plastic, covered by another layer of plastic inset with rubber pieces for slip-resistance. Made with recycled materials, high impact and weather-resistant.

In addition, most products have the tops of domes textured for slip-resistance. Check manufacturer's specifications to see that this detail is included. Due to the difficulties cited above, installing detectable warnings on existing curb cuts might be technically infeasible unless a new curb cut is installed.

Key Items

Detectable warning material and fasteners or adhering material.

Level of Difficulty

Moderate. Setting tile flush with concrete paving may involve a paving contractor.

Estimates

Adhere 36" wide plastic strip detectable warning to concrete

Description	Quantity	Unit	Work Hours	Material
Embossed plastic tile	3.000	S.F.	0.048	2.46
Adhesive	0.012	Gal.	0.000	0.15
Totals			0.048	2.46

Total per linear foot including general contractor's overhead and profit: $7

Fasten 36" wide metal tile detectable warning to concrete

Description	Quantity	Unit	Work Hours	Material
Stainless steel metal tiles, mechanically fastened	3.000	S.F.	0.093	46.05
Totals			0.093	46.05

Total per linear foot including general contractor's overhead and profit: $84

Copyright R. S. Means Company, Inc., 1994

67. Install Detectable Warnings (continued)

Install 36" wide ceramic tile detectable warning into new construction

Description	Quantity	Unit	Work Hours	Material
Porcelain-type 2" × 2" ceramic tile	3.000	S.F.	0.252	11.07
Add for detectable finish on ceramic tile	3.000	S.F.	0.000	1.41
Totals			0.252	12.48

Total per linear foot including general contractor's overhead and profit — **$33**

Install 36" wide ceramic tile detectable warning into existing floor or slab

Description	Quantity	Unit	Work Hours	Material
Saw cutting per inch of depth	8.000	L.F.	0.232	2.72
Cutout demolition of slab	3.000	S.F.	1.599	0.00
Concrete walkway (set below adjacent walkway)	3.000	S.F.	0.120	2.70
Porcelain-type 2" × 2" ceramic tile	3.000	S.F.	0.252	11.07
Add for detectable finish on ceramic tile	3.000	S.F.	0.000	1.41
Totals			2.203	17.90

Total per linear foot including general contractor's overhead and profit — **$144**

Install 36" wide rubber strip detectable warning

Description	Quantity	Unit	Work Hours	Material
Raised detectable strip	3.000	S.F	0.060	12.90
Adhesive	0.012	Gal.	0.000	0.15
Totals			0.060	13.05

Total per linear foot including general contractor's overhead and profit — **$25**

Copyright R. S. Means Company, Inc, 1994

Notes

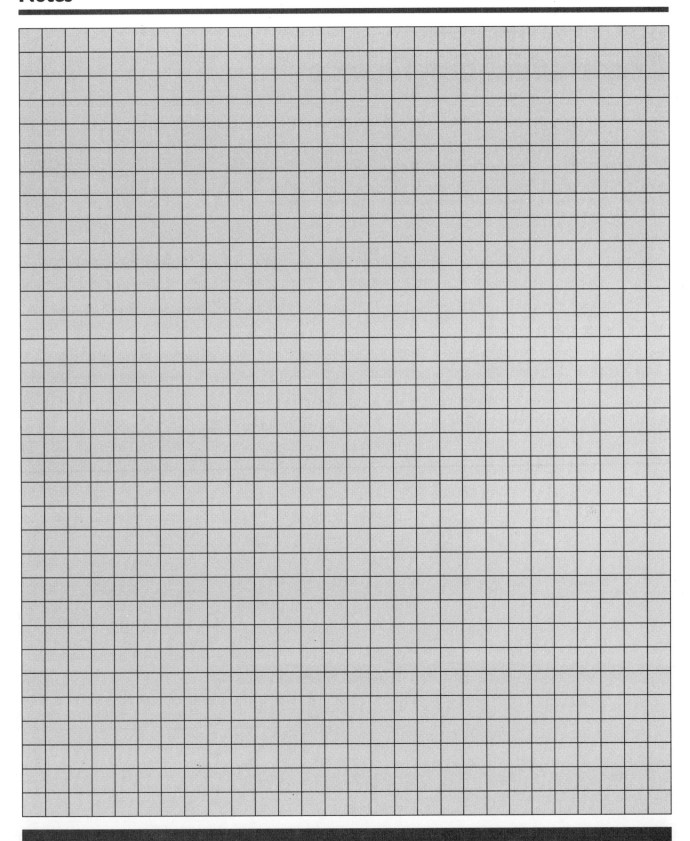

68 Install Emergency Communication Device

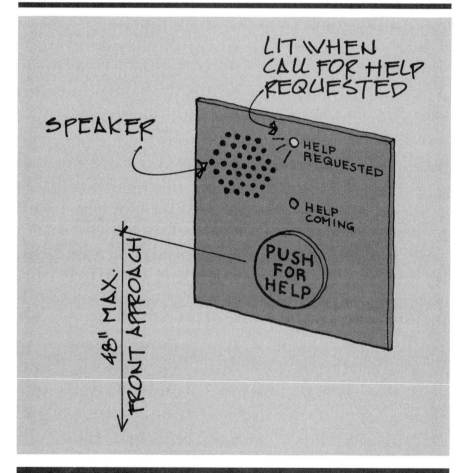

Enabling all people to summon assistance in an emergency situation is one of the most critical elements of making a facility accessible. People with mobility impairments may not be able to leave the scene of a fire, and those with visual or hearing impairments may not know where to go to call for assistance when needed. It is possible to enable all people to call for assistance, and a range of physical limitations must be addressed in new construction and renovations where emergency communications systems are being installed to achieve compliance with ADA.

ADAAG References
4.2 Space Allowances and Reach Ranges
4.27 Controls and Operating Mechanisms

Where Applicable
Wherever emergency communication devices are installed for use by facility occupants.

Design Requirements
- Within reach range: 48" for front reach, 54" for side reach.
- Accessible controls, operable with one hand without tight grasping, pinching, or twisting of the wrist with a force no greater than 5 pounds per foot.
- Voice and non-voice communication included (such as push-button emergency assistance call).

Design Suggestions
ADAAG specifically address voice and non-voice communication only for elevators, but the Department of Justice Final Rules for both Title II and Title III cover this situation generally. We recommend this modification wherever emergency call boxes are used based on regulations in Title II and Title III. Title II Section 35.160(a) states, "A public entity shall take appropriate steps to ensure that communications with applicants, participants, and members of the public with disabilities are as effective as communications

with others." The DOJ Title III Final Rule, Section 36.303(4)(a) Auxiliary Aids and Services, states, "*General.* A public accommodation shall take those steps that may be necessary to ensure that no individual with a disability is excluded, denied services, segregated or otherwise treated differently than other individuals because of the absence of auxiliary aids and services." Section 36.303(4)(c) states, "*Effective Communication.* A public accommodation shall furnish appropriate auxiliary aids and services where necessary to ensure effective communication with individuals with disabilities."

This means that wherever emergency communication is installed, it must be usable not only by people with mobility impairments, but also by those who may not be able to communicate by voice. Voice communication is often preferred as a deterrent against false alarms, but this must not prevent someone who cannot communicate verbally from obtaining assistance in an emergency situation. (Non-voice communication also enables people not fluent in English to summon help.) It is possible to install an emergency communication box that summons help just by pushing a button or pulling a lever (like a fire alarm), or that uses a TDD.

Emergency communications boxes should be located in clearly visible areas, with compliant signage indicating their use. Any controls should be large and easily operable. Some call buttons are small and meant to be pushed with one finger, but this can be difficult for individuals with limited fine motor control. If the voice box is separate from the call buttons, it should be located between 48"–54" a.f.f. to allow access to people using a wheelchair. The color of the box should contrast with the wall finish to make it easier for people with low vision to see.

Key Items
Communications device, electrical wiring to emergency power source.

Level of Difficulty
Moderate.

Estimate

Install emergency communications device

Description	Quantity	Unit	Work Hours	Material
Master panel, four stations, amp., battery back-up	1.000	Ea.	4.000	2500.00
Conductor	5.000	C.L.F.	17.390	145.00
Cutout demolition of partition	5.000	Ea.	1.665	0.00
Miscellaneous materials for gypsum board repair	1.000	Job	0.000	50.00
Labor minimum to repair and paint gypsum board	2.000	Job	4.000	0.00
Totals			27.055	2695.00

Total per each including general contractor's overhead and profit: $6,197

Copyright R. S. Means Company, Inc, 1994

69
Modify Kitchenette

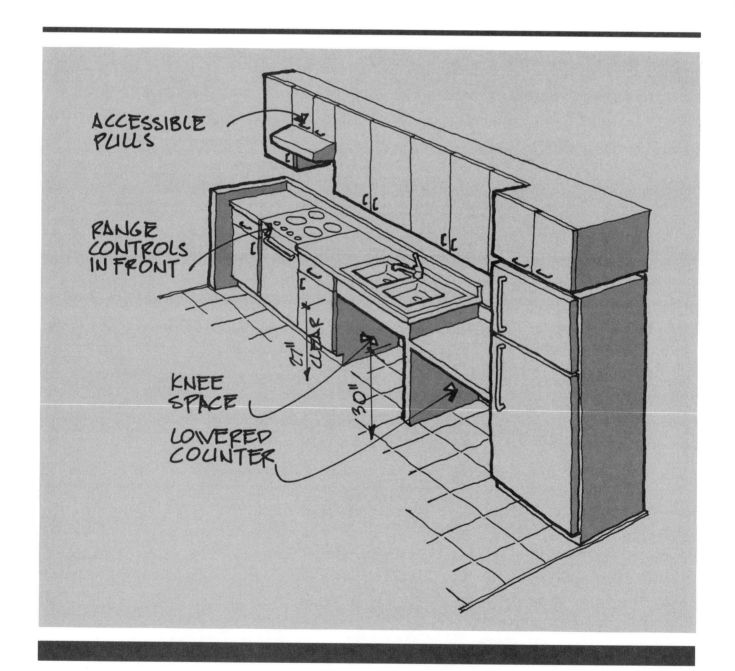

Common kitchens in public spaces covered under ADA are not subject to the same requirements as residential accessible kitchens covered by other access regulations, but they still need some degree of access. As with any other space, a kitchen located on an accessible route can have an entrance modified to create access; even kitchen spaces not located on an accessible route can be altered to include accessible sink faucets, work counters, drawer and cabinet pulls, and outlets with relatively little modification to the space.

ADAAG References

9.2 Requirements for Accessible Units (9.2.2 (7) Kitchens, Kitchenettes, or Wet Bars.)
4.2 Space Allowance and Reach Ranges
4.3 Accessible Route
4.27 Controls and Operating Mechanisms

Where Applicable

All common-use kitchens.

Design Requirements

Kitchen guidelines are covered under 4.1 Minimum Requirements as common areas under Title II and Title III. The requirements below apply to accessible lodging facilities and dormitories, and are recommended for all accessible common-use kitchens.

- On an accessible route, with an accessible entrance.
- Clear floor space (30" × 48") for frontal or parallel approach to all features.
- Countertops and sinks 34" a.f.f. maximum.
- Minimum 50% shelf space and refrigerator/freezer space within reach range.
- Accessible controls and handles (acceptable if operable with a closed fist).
- Slip-resistant surface on accessible route.

Design Suggestions

L-shaped or U-shaped kitchens work best, since they allow objects to slide without having to be picked up. A possible renovation is to join two separate counter sections to create a single counter. If the cabinet floor under the sink is not installed and the finish floor continues for the full depth, opening the doors can provide knee space. Base cabinets shouldn't structurally support the sink so they can be removed for knee space if necessary. A hose at the sink is recommended even at sinks with a faucet located near the front. Loop-type handles on drawers and cabinets allow for ease of use (cabinets with routed holds are difficult to use for people with low fine motor control, and also do not comply with ADAAG). Pull-out drawers and lazy Susans prevent the need for reaching to the back of storage spaces.

Some appliances are easier to use by a wide range of people. Side-by-side refrigerator/freezers allow a range of storage space on both sides. Stove and range controls should be located in front, and staggered burners prevent having to reach over a hot surface from a seated position. It is important to make the kitchen usable for people with low vision: lighting levels should be high at work stations, and placed so as not to cast shadows on the work space. Light-colored finishes and matte surfaces help people with low vision.

Key Items

Varies: drawer hardware, sink hardware, cabinets, slide drawers, and new appliances, possibly relocated to create maneuvering space.

Level of Difficulty

Varies. Low for cabinet and storage modifications; moderate for sink and stove alterations; high for kitchen reconfiguration or total rehabs.

69. Modify Kitchenette (continued)

Estimate

Modify kitchens

Description	Quantity	Unit	Work Hours	Material
Remove cabinet hardware (labor minimum)	1.000	Job	2.000	0.00
Remove drawer hardware (labor minimum)	1.000	Job	2.000	0.00
Remove & replace faucet with lever handle faucet (labor only)	1.000	Job	4.000	0.00
Kitchen faucet, gooseneck spout, paddle handles	1.000	Ea.	0.800	126.00
Add for spray hose	1.000	Ea.	0.333	10.00
Install D pull drawer/cabinet handles	13.000	Ea.	0.650	11.70
Install new drawer guides	2.000	Pr.	0.334	11.00
Under-cabinet task lighting	4.000	Ea.	2.000	204.00
Conductor	0.300	C.L.F.	1.043	8.70
Outlet boxes	4.000	Ea.	1.392	6.00
Switch	1.000	Ea.	0.200	3.65
Miscellaneous materials for gypsum board repair	1.000	Job	0.000	25.00
Labor minimum to repair and paint gypsum board	1.000	Job	2.000	0.00
Totals			16.752	406.05

Total per each including general contractor's overhead and profit $1,663

Copyright R. S. Means Company, Inc, 1994

Notes

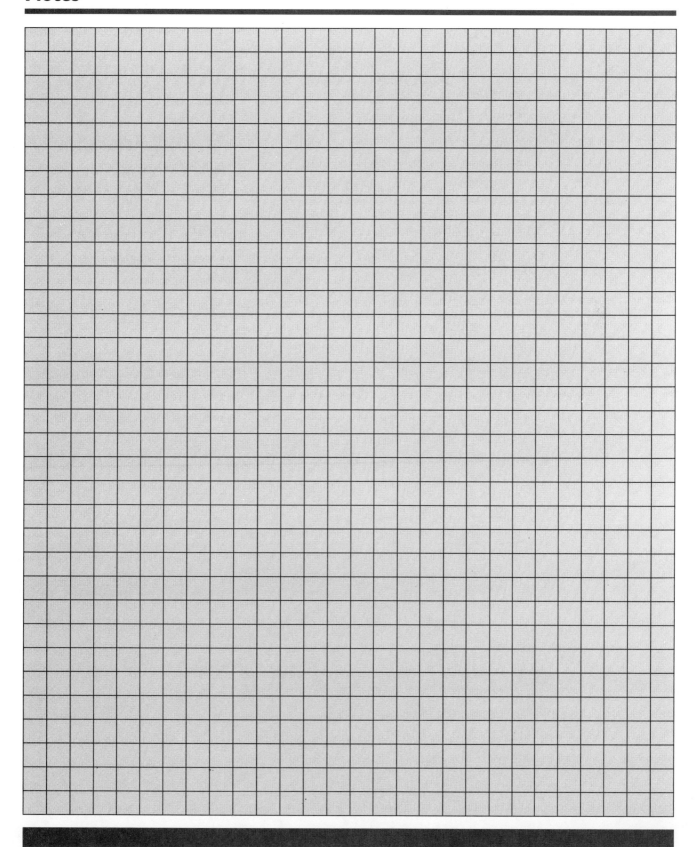

70
Modify Closet

Most closets can be modified with some basic carpentry work, in many cases without altering the doorway. Simple modifications like lowering a shelf and pole or installing them on adjustable brackets are nevertheless vital in making common storage spaces accessible.

ADAAG Reference
4.25 Storage

Where Applicable
At least on closets or storage area (if provided) in accessible spaces.

Design Requirements
- 32" minimum door clear opening width.
- Shelves, poles, and hooks within reach range: 48" for front reach, 54" for side reach.
- Accessible door hardware.
- Clear floor space in front, 30" × 48" minimum.

Design Suggestions
Adjustable shelves and poles, set at 12" heights, allow for flexibility in use. If the closet is deep, consider adding hooks to side walls or inside of doors for easy reach.

Key Items
Shelves, brackets, poles, coat hooks, closet door hardware.

Level of Difficulty
Low.

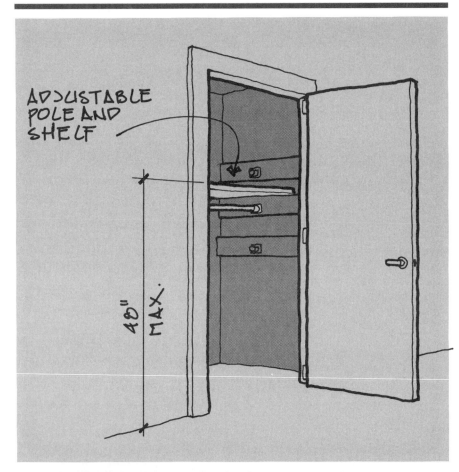

Estimates

Install accessible wood shelf and pole

Description	Quantity	Unit	Work Hours	Material
Shelf supports	7.000	L.F.	0.231	2.80
Closet pole	3.000	L.F.	0.120	2.55
Shelving	6.000	L.F.	0.684	12.30
Additional labor required for work within closet area	1.000	Job	2.000	0.00
Totals			3.035	17.65

Total per each including general contractor's overhead and profit	**$207**

Install adjustable wood shelf and pole

Description	Quantity	Unit	Work Hours	Material
Adjustable closet rod and shelf, 12" wide, 3' long	1.000	Ea.	0.400	8.05
Additional labor required for work within closet area	1.000	Job	2.000	0.00
Totals			2.400	8.05

Total per each including general contractor's overhead and profit	**$155**

Lower 2 coat hooks

Description	Quantity	Unit	Work Hours	Material
Labor minimum to lower 2 coat hooks	1.000	Job	2.000	0.00
Totals			2.000	0.00

Total per each including general contractor's overhead and profit	**$59**
Total per 2 coat hooks including general contractor's overhead and profit	**$117**

Copyright R. S. Means Company, Inc, 1994

71
Install Accessible Play Area Pathways

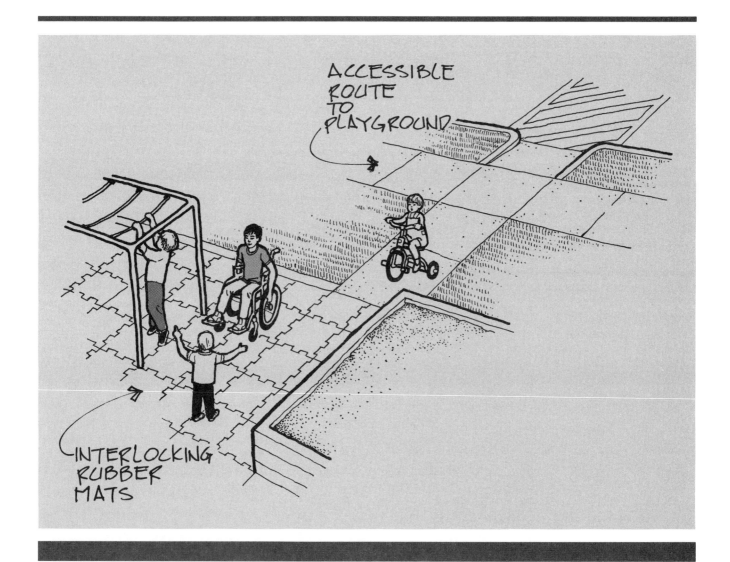

Play areas are covered under both Title II and Title III, but they do not as yet have any technical requirements in the ADA Standards. It is vital that children with disabilities have access to as much of a play area as possible. At a minimum, an accessible route up to and through the facility is needed to comply with ADA. Combining the ADA requirements for an accessible route of travel (usually a hard, firm surface) and requirements for child fall safety zones (usually a soft surface) is a challenge. It is possible to comply with both, however, and the particulars of play area pathways are included here in order to facilitate maximum levels of compliance, access and safety.

ADAAG References
2.2 Equivalent Facilitation
4.1.2 Accessible Sites and Exterior Facilities, New Construction
4.1.3 Accessible Buildings, New Construction
4.2 Space Allowance and Reach Ranges
4.3 Accessible Route
4.5 Ground and Floor Surfaces

Where Applicable
All public or common use play areas.

Design Requirements
- Accessible route to the play area.
- An accessible route or surface to all play areas and equipment.

Design Suggestions/ Discussion
Play areas are covered within the scoping sections of ADA, either under the program access requirements of Title II (State and Local Government) or as public accommodations under Title III (either category 9, places of recreation, or category 10, places of education). Accessible children's play equipment can be used under ADAAG 2.2 Equivalent Facilitation (allowing alternative designs and technologies which provide equivalent or greater accessibility to a facility than required in ADAAG), but there are no specific technical requirements in the ADA Accessibility Guidelines for playgrounds or play equipment for children who use wheelchairs and other mobility aids or who have visual, hearing or cognitive impairments. This makes defining or designing an accessible playground extremely difficult. The strongest recommendation is to discuss proposed playground designs and renovations with disability organizations, users, and parents to determine the optimal design for individual and general use. The suggestions that follow provide some guidance on designing an accessible and compliant play area.

General requirements: Public spaces and playgrounds are within the scope of ADA, and therefore must conform to basic construction requirements as to accessible routes and reach ranges, but a fully accessible playground is the subject of much debate and study. It is very difficult to say if a play area complies with ADA where not all play structures are fully accessible. Generally, play areas need to be located on an accessible route, with accessible routes to the play structures, with the play equipment as accessible as possible.

Accessible route to the play area: It can never be assumed that an accessible route to a play area (or anywhere else) is unnecessary "because it would never be needed." As a facility covered under Title II and Title III of ADA, play areas need an accessible route from the street or the main facility (many playgrounds are located in the middle of a field, which can be inaccessible especially in inclement weather). The route should be stable, firm, and slip-resistant as required, and should, if possible, be the same as the general route of travel.

Accessible route into the play area: An accessible route should be installed to access all equipment, even if the play structures themselves are inaccessible. It is problematic as to whether ADA requires this (4.1.2 requires an accessible route to all "accessible elements"), but the intent of ADA clearly drives the play area to be as accessible as possible. ADA allows exceptions in new construction only where it is "structurally impracticable" (see definition in Glossary), and it is usually structurally practicable to install an accessible route on a level surface.

The playground route and surface must not only conform to access guidelines, which require an accessible route to be "stable, firm, and slip-resistant," but also to child safety guidelines, which require a safe fall area around all equipment. Some materials seem to comply with both: interlocking rubber matting (which is expensive), plastic matting which allows grass to grow between the holes, poured-in-place rubber surfacing, and wood fibers ("fibar") which works well on a relatively level area. Some traditional playground materials, such as sand, are clearly inappropriate for accessible routes unless there are rubber or plastic mats along the route to and around all play equipment.

Play Equipment
This is the most difficult area of accessible playground design. ADAAG has no requirements for play equipment design. As with providing access to any type of space, wheelchair access is often the first issue considered, but it is only one type of access that needs to be provided. ADA generally does not provide technical requirements for pieces of equipment, but the lack of specific requirements may not relieve the controlling authority of its obligation to provide access.

Many challenging types of designs can accommodate children with a wide range of abilities, such as climbing structures, play structures that allow transfer from a wheelchair, sand tables and water tables which allow children using wheelchairs to roll up to and under them, and others. Consider installing play equipment that is usable by people with differing abilities. Input from children with disabilities, their parents, teachers, and others (who have experience in accessible playground design) is vital in designing an accessible play area. (Also consult *Play for All Guidelines: Planning, Design and Management of Outdoor Play Settings for All Children*, MIG Communications, 1802 Fifth Street, Berkeley, CA, 94710 Tel. 510–845–0953.)

Key Items
Accessible route surface material.

Level of Difficulty
Moderate to high. Installing prefab play equipment is fairly easy, but an accessible play area can require architectural and landscape planning, and landscape contractors.

71. Install Accessible Play Area Pathways (continued)

Estimates

Install accessible route to play area, 36" wide asphalt pathway

Description	Quantity	Unit	Work Hours	Material
Excavating	0.100	C.Y.	0.018	0.00
Asphalt sidewalk base	3.000	S.F.	0.030	0.36
2-1/2" asphalt sidewalk	0.340	S.Y.	0.025	1.22
Install topsoil, 4" deep	0.230	S.Y.	0.002	0.43
Install sod	0.002	M.S.F.	0.010	0.31
Totals			0.085	2.32

Total per linear foot including general contractor's overhead and profit: $9

Install accessible route in play area, rubber mats, interlocking over concrete

Description	Quantity	Unit	Work Hours	Material
Excavating	0.030	C.Y.	0.005	0.00
Concrete sidewalk base	1.000	S.F.	0.010	0.12
4" sidewalk, broom finish	1.000	S.F.	0.040	0.90
Install topsoil, 4" deep	0.230	S.Y.	0.002	0.43
Install sod	0.002	M.S.F.	0.010	0.31
Interlocking rubber mats	1.000	S.F.	0.038	2.20
Totals			0.105	3.96

Total per square foot including general contractor's overhead and profit: $12

Install accessible route in play area, woven rubber mats on existing sand

Description	Quantity	Unit	Work Hours	Material
Woven rubber mats	1.000	S.F.	0.052	11.70
Totals			0.052	11.70

Total per square foot including general contractor's overhead and profit: $22

Install wood mulch on existing sand

Description	Quantity	Unit	Work Hours	Material
Wood mulch in place	1.000	S.Y.	0.080	1.52
Totals			0.080	1.52

Total per square yard of 3-inch-deep mulch including general contractor's overhead and profit: $6

Copyright R. S. Means Company, Inc., 1994

Provide woven plastic mats on existing grass

Description	Quantity	Unit	Work Hours	Material
Woven plastic mats	1.000	S.F.	0.052	12.80
Totals			0.052	12.80

Total per square foot including general contractor's overhead and profit — **$24**

Copyright R. S. Means Company, Inc, 1994

72
Create Beach Access

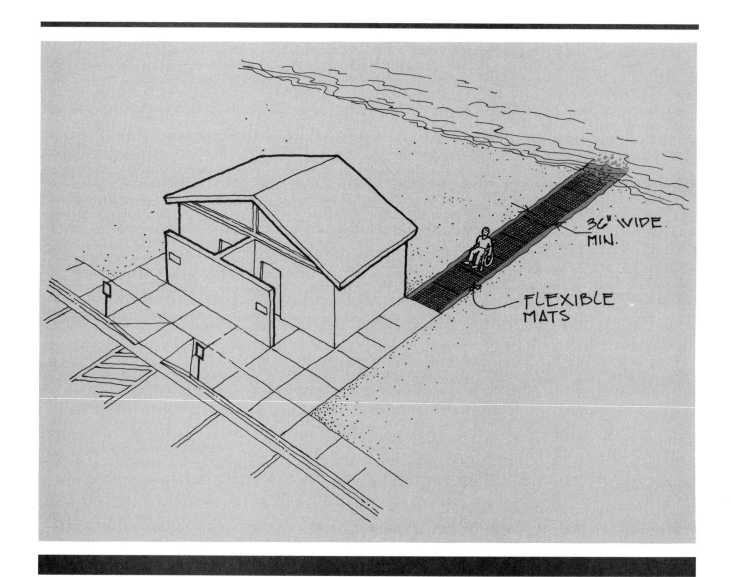

Sand is one of the most difficult ground surfaces to make accessible, and many of the traditional materials used for creating an accessible route of travel conflict with environmental regulations. Beaches are covered under Titles II and III of the ADA, however, either as programs offered by state or local governments or as a public accommodation. Beaches are used by people with a wide range of abilities, and there are several design solutions which can help create access to the water, despite the lack of technical guidelines in the ADA Standards.

ADAAG References
2.2 Equivalent Facilitation
4.3 Accessible Route
4.5 Ground and Floor Surfaces

Where Applicable
Public beaches.

Design Requirements
- Sufficient number of all common facilities accessible (bathhouses, showers, changing rooms, concession stands, telephones, etc.)
- Accessible route of travel to the beach: free of steps, minimum 36" wide, no protruding objects, surface stable, firm and slip-resistant; compliant with all ramp requirements if slope is between 1:20 and 1:12.

Design Suggestions
Most beaches are owned and operated by a municipal entity, which requires compliance with the program access sections of Title II (if state or locally run) or similar legislation if federally operated. Those beaches located at public accommodations (at resorts, etc.) need to comply with Title III of ADA. As a minimum, to comply with the ADA standards, the beach should be on an accessible route and at least one type of any built structures serving the beach (such as changing rooms, showers, concession stands) needs to be accessible and located on an accessible route.

However, it is not clear to what extent the beach itself has to be accessible: whether only an accessible route up to the beach is required, or to the water line (and if so, what tide line), if the sand needs to be re-graded to create a 1:20 maximum slope, or if more than one accessible path of travel is necessary. People *do* use wheelchairs to enter the water at both fresh and saltwater beaches. It is recommended that as much of the beach be accessible as possible, with the path(s) of travel complying with as many of the requirements in ADAAG (width, surface, slope) as possible without violating environmental concerns.

Beach access is an area of developing technology without a large body of experience; most sources for beach access products draw more on a range of local experiments than hard research. Some types of beach access involve placing movable rubber or plastic mats down over the sand. These mats are often 36" wide, but 48" is recommended since rolling off the edge is easy when the mat is placed over an unstable surface like sand. Some localities extend one length down to the high tide line; others to the low tide water line; others form a "T" to allow access to different sections of the beach. The mats can be set daily in place during a given time period, say from May to September (or whenever the beach is open), but some beaches have them available on an as-needed basis. They may need to be rolled out and put away on a daily basis.

It is possible to install permanent pathways, but these are subject to limitations placed by environmental concerns and ordinances. Concrete or wood pathways can inhibit natural movement of sand, and even wood walkways placed over the sand can block sunlight and adversely affect the growth of grasses needed to stabilize sand dunes. Generally, mats placed in areas where the public is allowed to walk will not disturb the beach any more than typical foot traffic, but local environmental legislation and advisers should be consulted.

There are beach chairs for individual users which can roll on sand, but these are not readily available. They create the greatest degree of mobility, and it would be possible to have one or more available for use on an as-needed basis. For maximum accessibility, they should be used in conjunction with some sort of mat system or accessible pathway to ensure that people who prefer to use (or must use) their own wheelchairs have access to the beach.

Key Items
Flexible mats, wood boardwalks, or concrete pathways.

Level of Difficulty
Low to Moderate.

Estimates

Provide plastic mats for beach access (3' wide, 40' long)

Description	Quantity	Unit	Work Hours	Material
Woven plastic mats	120.000	S.F.	6.240	1536.00
Totals			6.240	1536.00

Total per each including general contractor's overhead and profit: **$2,888**

Copyright R. S. Means Company, Inc., 1994

72. Create Beach Access (continued)

Provide plastic mats for beach access

Description	Quantity	Unit	Work Hours	Material
Woven plastic mats	32.000	S.F.	1.664	409.60
Totals			1.664	409.60

Total per 4' × 8' section including general contractor's overhead and profit — **$770**

Provide wood duckboards for beach access

Description	Quantity	Unit	Work Hours	Material
Hardwood strips on rubber backing (3 foot wide)	3.000	S.F.	0.156	32.10
Totals			0.156	32.10

Total per linear foot including general contractor's overhead and profit — **$62**

Install concrete pathway to high water line, 36" wide

Description	Quantity	Unit	Work Hours	Material
Excavating	0.123	C.Y.	0.022	0.00
Concrete sidewalk base	2.000	S.F.	0.020	0.24
4" sidewalk, broom finish	3.000	S.F.	0.120	2.70
Totals			0.162	2.94

Total per linear foot including general contractor's overhead and profit — **$14**

Copyright R. S. Means Company, Inc, 1994

Notes

73
Install Swimming Pool Access

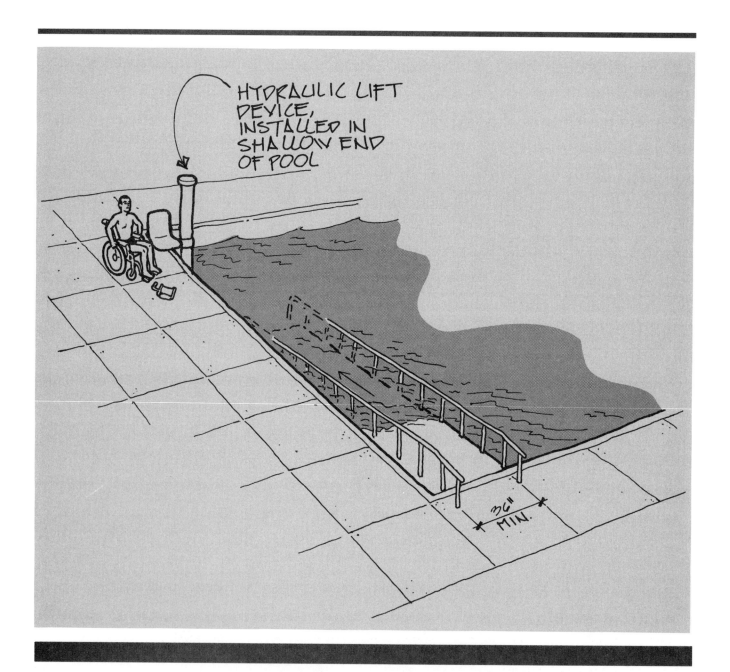

Swimming pool access solutions have been evolving for many years, and in some cases, minimal retrofit is necessary. A wide range of lift, transfer wall, and ramp solutions are available to help a facility with a swimming pool comply with ADA's requirements under Title II, and as a public accommodation under Title II.

ADAAG References

2.2 Equivalent Facilitation
4.1.3 Accessible Buildings, New Construction

Where Applicable

Public swimming pools.

Design Requirements

On an accessible route.

Design Suggestions

Swimming pools located in facilities which are covered under Title III Public Accommodations of ADA category 9 (places of recreation) or 12 (places of exercise and recreation) would be considered common areas and a primary function area, and publicly supported pools are covered under Title II as part of programs offered by state and local governments. Under both Titles, to comply with the ADA Standards they have to be on an accessible route, but ADAAG gives no further technical guidance. Swimming pools are relatively easy to make accessible for people with mobility impairments. Several methods which have been in use for some time are outlined under "Design Suggestions."

- Extend the pool to create an accessible area, including a ramp or a roll-in area. Many pool surroundings have the area to expand, but this solution can be expensive.
- Create a ramp into the pool leading to the shallow end. The ramp should comply with all standard ramp dimensions. This, too, can be expensive, even without the excavation required.
- Build a transfer section of wall 18" wide and 18" high around the edge of the pool to be used by someone transferring from their wheelchair into the pool. This requires the water level to be higher than the accessible route of travel around the pool.
- Install a pool transfer device. These lifts transfer a person from the side of the pool into the water by two means of operation: (1) a seat mounted on a vertical pole operated hydraulically and swinging out from the side of the pool into the water, and (2) a sling mounted on a vertical in which the person sits, and is lowered into the water. All devices require assistance for use.
- Handrails on both sides of stairs leading into the pool help people who have some walking ability.
- For people with visual impairments, it is recommended that a 24" wide detectable warning strip be installed around the edge of the pool if there is no continuous wall or lip. (ADA specifically requires detectable warning strips only at reflecting pools.) This is a useful safety feature, and the usual drawback of snow removal (trying to shovel over the bumps in the materials) is not a factor.

Key Items

If pool configuration is not altered, involves installation of a concrete or prefab ramp, and/or low wall, and/or transfer device.

Level of Difficulty

Moderate. Involves concrete and/or concrete block work.

Estimate

Install transfer device for swimming pool access

Description	Quantity	Unit	Work Hours	Material
Saw cutting, concrete per inch of depth	32.000	L.F.	0.544	8.00
Remove concrete deck	0.450	S.Y.	0.071	0.00
Hand excavation	0.500	C.Y.	1.000	0.00
Spread footing	0.500	C.Y.	1.507	37.00
Anchor bolts	4.000	Ea.	0.212	7.20
Lifting device	1.000	Ea.	0.000	1825.00
Labor to install lifting device	1.000	Job	8.000	0.00
Totals			11.334	1877.20

Total per each including general contractor's overhead and profit	**$3,620**

Copyright R. S. Means Company, Inc, 1994

74
Create Accessible Trails

Creating access is often thought of only in terms of modifying buildings. However, outdoor facilities are used by all people and are covered both under Title II of ADA as programs offered by state or local governments, and under Title III as a public accommodation. Creating or modifying certain trails to comply with ADA's requirements for accessible routes can be relatively simple, and still comply with environmental regulations.

ADAAG References
4.3 Accessible Route
4.4 Protruding Objects
4.5 Ground and Floor Surfaces

Where Applicable
Public trails.

Design Requirements
- Pathway stable, firm, slip-resistant.
- 36" minimum clear width.
- No protruding objects.
- Maximum slope 5%, cross-slope 2% for drainage.
- All public buildings used by the public (ranger stations, rest rooms, etc.) accessible and on an accessible route.

Design Suggestions
As with beach access, public pathways are covered under Title II as public programs or facilities, or under Title III as public common areas. They are, however, given no strict design guidance in ADAAG. At this time, the Access Board has a working group on recreation, developing recommendations for ADAAG. The preliminary proposals suggest that there will be a range of standards, depending on the type of setting. In designing accessible trails, there can be conflict between accessibility and environmental guidelines, if ordinances prohibit removal of plants and/or installation of any paving materials in wooded areas (even gravel, which is not usually fully accessible as a pathway material). Also, it is not clear to what extent trails need to be re-graded to comply with the maximum slope of 1:20 allowed in ADAAG as an accessible path of travel. Any facilities serving the public, such as ranger stations, first aid stations, and concession stands on an accessible route need to be accessible.

The accessible trails should be representative of all trails in that recreational area: If a nature park has trails in wooded areas and others in open grassland, it is recommended that at least part of the trails in both areas be made accessible to make as much of the program accessible, rather than creating access to only one part of the area. If not disruptive environmentally, the pathways should be graded at a compliant slope (1:20 or less); if this is not possible, signage should be installed at the head of the trail and along all parts steeper than 1:20 indicating that the slope could be difficult to negotiate. Where the slope is steeper than 1:20, handrails should be installed where possible. As always, consultation with people with disabilities is necessary both as part of a good faith effort and to gain data for modifications.

Key Item
Landscaping equipment.

Level of Difficulty
Moderate. Basic landscaping.

Estimate

Create accessible trails (no heavy trees)

Description	Quantity	Unit	Work Hours	Material
Clear trail	0.070	Acre	3.360	0.00
Stump rem. (assume ten 6" dia. stumps per 100 L.F.)	10.000	Ea.	4.000	0.00
Fine grading by hand	11.120	S.Y.	0.378	0.00
Gravel and cinders over stone base	11.120	S.Y.	3.503	41.70
Compaction	10.000	C.Y.	0.400	0.00
Totals			11.641	41.70

Total per square foot including general contractor's overhead and profit: **$3**

Total per 100 linear feet of 3 foot wide trail including general contractor's overhead and profit: **$1,044**

Copyright R. S. Means Company, Inc, 1994

75 Create Accessible Sleeping Rooms

When inaccessible sleeping rooms are located on an accessible route, it is almost always possible to have at least some modifications made to increase accessibility. Widened doorways, lever handles, visual and audible alarms, drawer pulls, closet poles, grab bars, and other modifications are relatively simple changes that can make facilities in the hospitality industry even more accessible and hospitable.

ADAAG References
9.1.2 Accessible Units, Sleeping Rooms, and Suites
9.1.3 Sleeping Accommodations for Persons with Hearing Impairments

Where Applicable
All transient lodging buildings except where there are five rooms or less for rent or hire, and the building is the owner's residence.

Design Requirements
In new construction, at least one accessible sleeping room or suite is required. Additional accessible rooms, accessible rooms with roll-in showers, and sleeping accommodations for persons with hearing impairments are required as listed in ADAAG 9.2.3, dispersed among the different classes of sleeping accommodations. (The number varies between 2% and 7%, with the percentage diminishing with a greater number of rooms.) Existing lodging facilities must remove barriers where it is readily achievable to do so as public accommodations.

Requirements for Accessible Units, Sleeping Rooms, and Suites
- On an accessible route.
- An accessible entrance.
- 36" clear width on both sides of a bed or between two beds.
- Accessible route (36" clear minimum, with no protruding objects) connecting all accessible

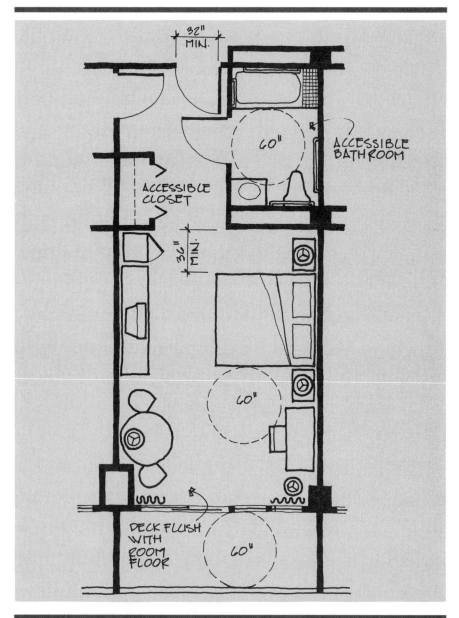

spaces and elements (including phones) within the unit.

- At least one of each type of cabinet, dresser, or storage area on an accessible route, hardware operable with a closed fist, and storage within reach range (48" high for front approach).
- All controls operable with a closed fist and within reach range (48" high for front approach, 54" for side approach).
- Living area, dining area, at least one sleeping area, and any patios, terraces, and balconies on an accessible route and accessible. If a threshold or change in level is necessary for any patios, terraces, and balconies for water and/or wind protection, some sort of equivalent facilitation (such as ramping to the balcony) is necessary.
- At least one accessible full bathroom (toilet, sink, bathtub or shower), or at least an accessible half bath where only half baths are provided.
- Accessible carports, garages, and parking.
- Accessible kitchens, kitchenettes, and wet bars, including sufficient clear floor space for a front or side approach at all cabinets, sinks, and appliances, with counters and sinks mounted at 34" high maximum, 50% of shelf space in cabinets and refrigerator/freezers within reach range, sufficient floor space to allow all doors to be accessible and usable, and controls and pulls operable with a closed fist.

Requirements for Sleeping Accommodations for Persons with Hearing Impairments

- Visual fire alarms.
- Telephone call alert devices and visitor alert devices, not connected to visual alarms.
- Volume controls on permanently installed telephones.
- An accessible electrical outlet within 4' of a telephone connection to allow use of a text telephone.
- If all the above are not installed, equivalent facilitation must be provided through the installation of accessible electrical outlets and phone jacks to allow use of portable visual alarms and communication devices, provided by the operator of the facility.

Design Suggestions

In lodging rooms, readily achievable barrier removal might include changing door knobs to lever handles, installing a peephole at an accessible height, adding paddle faucets and grab bars in the bathroom, or a door signal with a light to let a guest with a hearing impairment know when someone is at the door.

If carpeting is installed, it should be unbacked, with 1/2" pile maximum. The 36" clear width on both sides of a bed and to all accessible features is, as always, a minimum. Extra storage space for a wheelchair is very useful (otherwise a wheelchair user is forced to leave the wheelchair in the route of travel). Corner guards and door kickplates help protect walls from damage. Crank or lever hardware on a window is easier to operate than locks, which require pinching, and blinds on a continuous pull chain (or on a machine-operated push button) are easier to use than blinds on a standard string. Clear floor space is necessary to reach window drape or blind controls.

It can be a readily achievable modification to remove a bed to create additional maneuvering space. Accessible rooms sometimes have one single-person bed instead of a double to create additional maneuvering space. Double beds are a standard feature in lodging rooms for many reasons, and are strongly recommended for all accessible rooms. Attention should be paid to detail; reading lamps should be within reach range of the bed (swing-arm lamps are useful), with easily operated controls (touch-switches are ideal, but chains are acceptable). Just as important as the width of the route through the room is the arrangement of the furniture; features in the room should not block environmental controls (thermostats, air conditioning vents, etc.).

Key Items

Standard room partition materials, finishes, and features installed to compliant dimensions; accessible door and cabinet hardware; either visual alarms, phone alerts, and door alerts or outlets for the same; accessible bathroom features.

Level of Difficulty

Varies. Carpentry, wiring, finish work and possibly plumbing involved.

75. Create Accessible Sleeping Rooms (continued)

Estimates

Modify bedroom area of hotel room

Description	Quantity	Unit	Work Hours	Material
Remove interior door	1.000	Ea.	0.400	0.00
Remove door frame	2.000	Ea.	1.142	0.00
Remove metal stud/gypsum board partition	28.000	S.F.	1.288	0.00
Re-frame door opening	96.000	S.F.	4.416	67.20
Interior door frame	2.000	Ea.	2.000	131.00
Hollow metal flush door, 3'-0" x 6'-8"	2.000	Ea.	1.882	294.00
Hinges	3.000	Pr.	0.000	138.00
Lever-handled lockset	1.000	Ea.	0.000	79.78
Threshold	1.000	Ea.	0.400	27.50
Paint door	2.000	Ea.	1.882	9.92
Painting	136.000	S.F.	2.448	23.12
Remove closet hardware (labor minimum)	1.000	Job	2.000	0.00
Install D-pull closet handles	4.000	Ea.	0.200	3.60
Adjustable closet rod and shelf, 12" wide, 3' long	1.000	Ea.	0.400	8.05
Cutout demolition of partition	1.000	Ea.	0.333	0.00
Conductor	0.100	C.L.F.	0.296	1.55
Install junction box	1.000	Ea.	0.400	4.65
20-amp rocker switch	1.000	Ea.	0.296	13.25
Install plate	1.000	Ea.	0.100	1.85
Cutout demolition of partition	1.000	Ea.	0.333	0.00
Conductor	0.100	C.L.F.	0.296	1.55
Install junction box	1.000	Ea.	0.400	4.65
Install outlet	1.000	Ea.	0.296	5.25
Install plate	1.000	Ea.	0.100	1.85
Misc. materials for gypsum board painting and repair	1.000	Job	0.000	75.00
Repair gypsum board	1.000	Job	2.000	0.00
Paint gypsum board—minimum	1.000	Job	2.000	0.00
Fire alarm horn	1.000	Ea.	1.194	33.00
#18 fire alarm conductor	0.100	C.L.F.	0.100	6.40
Fire alarm light	1.000	Ea.	1.509	96.00
#18 fire alarm conductor	0.100	C.L.F.	0.100	6.40
Telephone company labor minimum	1.000	Job	4.000	0.00
Totals			32.211	1033.57

Total per each including general contractor's overhead and profit: $3,728

Copyright R. S. Means Company, Inc., 1994

Modify bathroom in hotel room

Description	Quantity	Unit	Work Hours	Material
Remove interior door	1.000	Ea.	0.400	0.00
Remove door frame	2.000	Ea.	1.142	0.00
Remove metal stud/gypsum board partition	28.000	S.F.	1.288	0.00
Re-frame door opening	96.000	S.F.	4.416	67.20
Interior door frame	2.000	Ea.	2.000	131.00
Hollow metal flush door, 3'-0" × 6'-8"	2.000	Ea.	1.882	294.00
Hinges	3.000	Pr.	0.000	138.00
Lever-handled lockset	1.000	Ea.	0.000	79.78
Paint door	2.000	Ea.	1.882	9.92
Painting	136.000	S.F.	2.448	23.12
Remove bathtub	1.000	Ea.	2.000	0.00
Remove partition finishes	66.000	S.F.	0.594	0.00
Water-resistant 5/8" gypsum board	110.000	S.F.	0.880	30.80
Copper shower pan	18.000	S.F.	1.440	45.00
Ceramic tile floor (pitched to drain)	15.000	S.F.	1.995	53.25
Ceramic tile walls, thin-set 4-1/4" × 4-1/4"	110.000	S.F.	9.240	209.00
Ceramic bath accessories	2.000	Ea.	0.390	18.10
Tub grab bar	2.000	Ea.	1.142	154.00
Grab bar vertical arms	4.000	Ea.	2.668	274.00
Bar-mounted hand-held shower head	1.000	Ea.	0.400	137.00
Rough in supply, waste and vent for shower	1.000	Ea.	7.805	76.50
Rough in supply, waste and vent for sink	1.000	Ea.	9.639	128.00
Wall-hung porcelain enamel lavatory (22" × 19")	1.000	Ea.	2.000	315.00
Faucet, handles and drain	1.000	Ea.	0.800	126.00
Mirror with stainless steel shelf	1.000	Ea.	0.400	67.50
Labor minimum for toilet relocation	1.000	Job	4.000	0.00
Rough in supply, waste and vent	1.000	Ea.	5.634	113.00
Misc. materials for gypsum board and ceramic tile repair	1.000	Job	0.000	100.00
Labor minimum to repair gypsum board and ceramic tile	1.000	Job	4.923	0.00
Totals			71.408	2590.17

Total per each including general contractor's overhead and profit: $8,231

Copyright R. S. Means Company, Inc, 1994

Part Three
Case Studies

The case studies are included as examples of the access design and construction process. Every case is different, but all projects—no matter how small—involve a survey to identify existing barriers, an examination of design options, a design process, and construction planning. Deciding what might be a satisfactory solution is where much of the expense, and the headaches, of access modifications come into play. To say the least, the design process is not always smooth, especially if it is not possible to make a facility or renovation fully compliant with ADA Standards. The headaches can be reduced if users are included in the design process as early as possible. There is always an interplay between need and cost, but with more information in the design program, a more successful balance may be achieved. Each case study is divided to show what was existing, what was needed, what was required, and what was done. Again, cost estimates are a key component in the design process.

Existing Bathroom Rehab

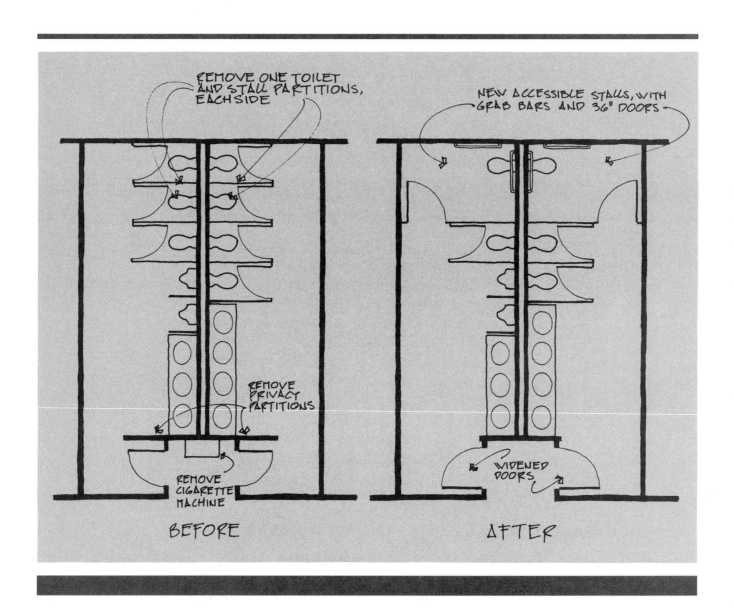

Facility:
Restaurant with accessible entrance on the first floor of a 1920's-era Art Deco office building.

Space Being Modified:
Men's and Women's Rest Rooms.

Level of Accessibility of Rest Rooms:
Inaccessible, due to narrow entrance, lack of maneuvering space, no accessible toilet stalls, sinks, or dispensers.

Reason for Modification:
The restaurant is undergoing renovation due to a change of ownership, with an estimated construction cost of $85,000. The rest rooms are in good condition, and the owners were only planning to make minimal modifications for reasons unrelated to ADA. The rest rooms area contains a primary function and the route to this area, including rest rooms, needs to be modified up to 20% of the total cost of the renovation. Since the cost estimates for bringing the rest rooms into compliance with ADA are less than 20% of the total cost of the renovation (in this case about 7.6%), bringing the rest rooms into compliance is triggered by ADA.

Existing Conditions:
Inaccessible entrances: Narrow doorways (30" with 28" clear opening), knob hardware, 3/4" marble thresholds, 30" wide entry vestibules created by vision screens, tiled floors.

Men's Room: Three 36" toilet stalls with enameled metal partitions, toilets centered within stall, seats 15" a.f.f., two urinals with rims 22" a.f.f., four sinks set in a plastic laminate countertop at 35" a.f.f. with 10" apron, ball faucets, mirror 44" a.f.f., paper towels 60" a.f.f., coat hooks on toilet doors 68" a.f.f.

Women's Room: Four 36" toilet stalls, toilets centered within stall, seats 15" a.f.f., four sinks set in a plastic laminate countertop at 35" a.f.f. with 10" apron, ball faucets, mirror 42" a.f.f., paper towels 60" a.f.f., coat hooks on toilet doors 68" a.f.f., sanitary napkin dispenser 60" a.f.f.

Design Options:
1. Leave the existing rest rooms and add two new single-use accessible rest rooms. This was rejected by the owner since it would take up too much space in the restaurant, and there is no room on the site for exterior expansion.
2. Leave the existing rest rooms and add one new single unisex rest room. This was rejected by the plumbing inspector. This solution is allowed by ADA, but is prohibited by the local plumbing code which mandates separate accessible rest rooms for men and women.
3. Modify both rest rooms for compliance. This was approved and accepted by the owners after approval by the plumbing inspector, which was required since one fixture would be lost in each rest room.

Preferred Solution:
Modify both restrooms. Estimated cost: $6,500 total.

Plans prepared by architect, reviewed by:
Building inspector, plumbing inspector, owner.

Modifications:
Entrances: Remove cigarette machine at entrance to create more space, widen rest room doors to 36", reverse door swings to create sufficient clearance on the latch side of the door, replace with a door that has lever handles, add raised character/Braille signage to the door jamb to the right of the door, remove existing thresholds.

Entry vestibules: Remove vision panels to allow more maneuvering space.

Toilet stalls: Remove one toilet in each rest room, combine two stalls to create one accessible stall with a wider door (other existing toilet partitions in good enough condition to remain) and add grab bars. The route of travel is narrow (34") but is allowed to remain as is, since widening the path of travel would require decreasing the other stalls, making them unusable. Existing toilet allowed to remain, despite low seat height, but flush valve reversed to face open side. Urinals in Men's Room to remain as is, since the renovations will create an accessible toilet.

Sinks: Sink heights allowed to remain; 10" aprons replaced with 3" aprons to allow knee space below. All pipes wrapped and faucets replaced with blade handles. Mirrors lowered to 38" a.f.f.

Dispensers: All dispensers lowered to 48" a.f.f.

Historic Entry Modification

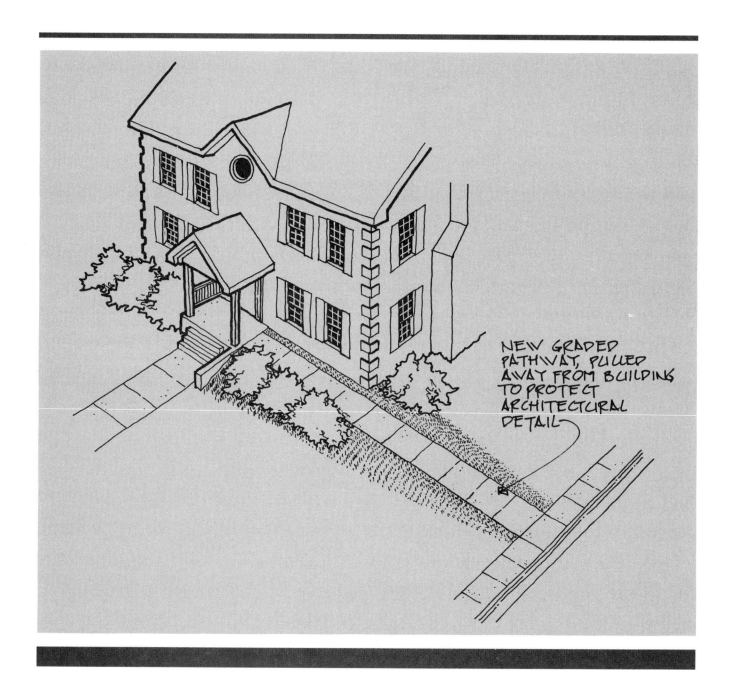

Existing Conditions:
Two-story library with basement, town-owned, 1840s Federalist design, brick and limestone construction, on local and state historic registers.

Level of Accessibility:
Building set back 50' from sidewalk. Two entrances, front and back; front entrance up a set of four steps, 6" risers, to extended entrance porch; rear entrance on grade, but opens onto stairwell.

Reason for Modification:
Title II compliance. A non-structural program access not possible since the town has no other library and no means of providing alternative forms of program accessibility.

Design Options:
1. Stair lift on front stairs. Rejected by historic commission.
2. Stair lift on rear stair. Accepted by historic commission, rejected by fire marshall; designated fire egress stair cannot be blocked with lift.
3. Exterior elevator. Rejected by historic commission, and by town board since cost would require floating a bond issue (estimated cost, two-stop elevator with brick shaft, $90,000).
4. Ramp at front entrance. Acceptable, if design conforms to existing style. Requires concrete ramp with limestone facing, and metal rails attached to painted fluted posts. Estimated cost: $19,000.
5. Graded pathway, rough-finish stone pavers, up to entrance at slope of 1:20, cover portion of existing steps. Acceptable, with approval from historic commission. Estimated cost: $14,000, with landscaping.

Plans reviewed by:
Library manager, town architect, local historic commission, and town disability rights committee.

Preferred Solution:
Graded pathway, changed to broom-finish concrete. It is the only affordable solution that satisfies the historic commission. Concern is raised about necessary snow removal and drainage, since the pathway is flush with the entrance level. This issue is addressed by the landscape contractor.

Part Four
Unit Costs

This section of the book is provided as a back-up cost resource. It enables you to modify the project estimates in the main part of the book to your particular conditions and allows experienced estimators to develop additional modification project estimates. For example, if your project involves not only installing a new pathway, but also removing an existing pathway, curb or fence, you can find a cost for the additional demolition work in this Unit Cost Section. To find that information, you would look under Division 2 (020—Site Work), Section 554 (Site Demolition), and choose the appropriate item.

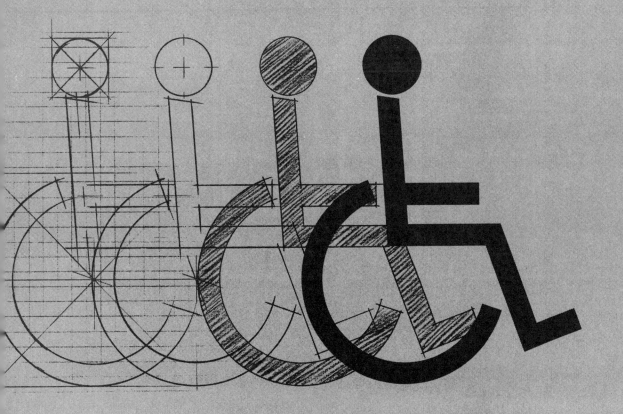

How To Use the Unit Costs

The Unit Cost Section is organized according to the 16-Division Construction Specifications Institute (CSI) MasterFormat, the most commonly accepted system for classifying construction information in the U.S. and Canada.

For a listing of these divisions and an outline of their subdivisions, see the Table of Contents of this Unit Cost Section, following this introduction.

The following is a detailed explanation of a sample entry in the Unit Cost Section. Next to each bold number is a definition of the item being described, with the appropriate component of the sample entry following in parenthesis.

1 Division Number/Title (025/Paving & Surfacing)

Use the Unit Cost Section Table of Contents to locate specific items. The sections are classified according to the CSI MasterFormat.

2 Line Number (025 128 0020)

Each unit price line item has been assigned a unique 10-digit code based on the 5-digit CSI MasterFormat classification.

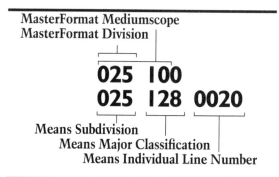

025	Paving and Surfacing					
025 100	**Walk/Rd/Parkng Paving**	WORK-HOURS	UNIT	BARE COSTS MAT.	TOTAL INCL. O&P	
0010	SIDEWALKS, DRIVEWAYS, & PATIOS No base					128
0020	Asphaltic concrete, 2" thick	.067	S.Y.	2.93	8.45	
0100	2-1/2" thick	.073	"	3.59	9.90	
0300	Concrete, 3000 psi, cast in place with 6 x 6 - W1.4 x W1.4 mesh,					
0310	broomed finish, no base, 4" thick	.040	S.F.	.90	3.61	
1700	Redwood, prefabricated, 4' x 4' sections	.05		3.30	.55	
1750	Plywood planks, 1" thick, on sleepers	.06		.45	3.10	
9000	Minimum labor/equipment charge	8	Job		15	

025 150	**Unit Pavers**					
0010	ASPHALT BLOCKS, 6"x12"x1-1/4", w/bed & neopr. adhesive	.119	S.F.	2.50	10.45	154
0100	3" thick	.123		3.50	12.35	
0300	Hexagonal tile, 8" wide, 1-1/4" thick	.119		2.50	10.45	
0400	2" thick	.123		3.50	12.35	
0500	Square, 8" x 8", 1-1/4" thick	.119		2.50	10.45	
0600	2" thick	.123		3.50	12.35	
9000	Minimum labor/equipment charge	4	Job		232	

3 Description (Asphaltic Concrete, 2" Thick)

Each line item is described in detail. Sub-items and additional sizes are indented beneath the appropriate line items. The first line or two after the main item (in boldface) may contain descriptive information that pertains to all line items beneath this boldface listing.

4 Work-Hours (.067)

The "Work-Hours" figure represents the number of hours required to install one unit of work. To find out the number of work-hours required for your particular task, multiply the quantity of item times the number of work-hours shown. For example:

Quantity	×	Productivity Rate	=	Duration
1,000 S.Y.	×	.067 S.Y.	=	67 Work-Hours

5 Unit (S.Y.)

The abbreviated designation indicates the unit of measure upon which the price and productivity are based (S.Y. = Square Yard). See "Abbreviations" at the back of the book for a complete list.

6 Bare Costs: Bare Mat. (Bare Material Cost) (2.93)

This figure for the unit material cost for the line item is the "bare" material cost with no markups included. *Costs shown reflect national average material prices for January 1994 and include delivery to the job site. No sales taxes are included.*

7 Total Costs Including O & P (8.45)

The figure in this column represents the total cost of the work and materials described in the line item. This figure is the sum of the bare material cost plus 20% (compensation for the contractor's risk in the purchase and handling of materials), plus the marked-up costs for labor and equipment as determined by R.S. Means Company. (Labor is marked up by an average of 66.8%* to cover labor burdens, such as Workers' Compensation and employee benefits, and the subcontractor's overhead and profit. Equipment is marked up 10% to cover the contractor's risk related to this item.) This sum is then increased by 15% to cover general conditions items such as permits, trash removal fees, and site supervision, 7.5% for the general contractor coordinating the subcontractors, and 15% for the general contractor's overhead and profit.

* Each trade as tracked by the R.S. Means Company has a different markup. The markups range from a high of 108.7% to a low of 55.4%, due to variations in Workers' Compensation rates for different trades.

Note: The line items in the Unit Cost Section were selected for their appropriateness to ADA modification projects. R. S. Means Company, Inc. publishes 23 annual cost data books covering a wide range of construction disciplines.

Unit Costs
Table of Contents

Div. No.		Page
	Site Work	
020	Subsurface Investigation and Demolition	254
022	Earthwork	264
025	Paving and Surfacing	265
	Concrete	
033	Cast-In-Place Concrete	266
	Masonry	
042	Unit Masonry	268
045	Masonry Restoration, Cleaning and Refractories	269
	Metals	
050	Metal Materials, Coatings and Fastenings	269
055	Metal Fabrications	271
	Wood and Plastics	
061	Rough Carpentry	273
062	Finish Carpentry	276
064	Architectural Woodwork	278
	Thermal & Moisture Protection	
071	Waterproofing and Dampproofing	280
073	Shingles and Roofing Tiles	280
079	Joint Sealers	281
	Doors and Windows	
081	Metal Doors and Frames	282
082	Wood and Plastic Doors	283
084	Entrances and Storefronts	286
087	Hardware	286
	Finishes	
092	Lath, Plaster and Gypsum Board	291
093	Tile	293
095	Acoustical Treatment and Wood Flooring	295
096	Flooring and Carpet	295
099	Painting and Wall Coverings	298
	Specialties	
101	Visual Display Boards, Compartments and Cubicles	301
104	Identifying and Pedestrian Control Devices	303
105	Lockers, Protective Covers and Postal Specialties	303
107	Telephone Specialties	304
108	Toilet and Bath Accessories and Scales	304
	Conveying Systems	
142	Elevators	305
	Mechanical	
151	Pipe and Fittings	305
152	Plumbing Fixtures	310
153	Plumbing Appliances	313
	Electrical	
160	Raceways	313
161	Conductors and Grounding	316
162	Boxes and Wiring Devices	318
166	Lighting	318
168	Special Systems	319

020 | Subsurface Investigation and Demolition

020 550 | Site Demolition

			WORK-HOURS	UNIT	BARE COSTS MAT.	TOTAL INCL. O&P	
554	0010	SITE DEMOLITION No hauling, abandon catch basin or manhole	3.429	Ea.		210	554
	0020	Remove existing catch basin or manhole	6			365	
	0030	Catch basin or manhole frames and covers stored	1.846			113	
	0040	Remove and reset	3.429			210	
	0100	Roadside delineators, remove only	.183			12.80	
	0110	Remove and reset	.320			22.50	
	0400	Minimum labor/equipment charge	6	Job		365	
	0600	Fencing, barbed wire, 3 strand	.037	L.F.		1.69	
	0650	5 strand	.057			2.60	
	0700	Chain link, remove only, 8' to 10' high	.054			3.30	
	0800	Guide rail, corrugated steel, remove only	.188			8.55	
	0850	Remove and reset	.457			21	
	0860	Guide posts, remove only	.267	Ea.		22.50	
	0870	Remove and reset	.480	"		40.50	
	0890	Minimum labor/equipment charge	4	Job		182	
	0900	Hydrants, fire, remove only	3.404	Ea.		216	
	0950	Remove and reset	11.429	"		725	
	0990	Minimum labor/equipment charge	8	Job		505	
	1000	Masonry walls, block or tile, solid, remove	.036	C.F.		2.60	
	1100	Cavity wall	.029			2.13	
	1200	Brick, solid	.071			5.20	
	1300	With block back-up	.057			4.14	
	1400	Stone, with mortar	.071			5.20	
	1500	Dry set	.043			3.11	
	1600	Median barrier, precast concrete, remove and store	.112	L.F.		11.15	
	1610	Remove and reset	.123	"		12.30	
	1650	Minimum labor/equipment charge	2	Job		114	
	1710	Pavement removal, bituminous, 3" thick	.058	S.Y.		5.70	
	1750	4" to 6" thick	.095			9.40	
	1800	Bituminous driveways	.059			5.80	
	1900	Concrete to 6" thick, mesh reinforced	.157			15.45	
	2000	Rod reinforced	.200			19.70	
	2100	Concrete 7" to 24" thick, plain	3.053	C.Y.		300	
	2200	Reinforced	4.211	"		415	
	2250	Minimum labor/equipment charge	6.667	Job		655	
	2300	With hand held air equipment, bituminous	.025	S.F.		1.34	
	2320	Concrete to 6" thick, no reinforcing	.040			2.10	
	2340	Mesh reinforced	.058			3.04	
	2360	Rod reinforced	.063			3.28	
	2390	Minimum labor/equipment charge	6.667	Job		655	
	2400	Curbs, concrete, plain	.074	L.F.		4.52	
	2500	Reinforced	.109			6.65	
	2600	Granite curbs	.068			4.14	
	2700	Bituminous curbs	.029			1.76	
	2790	Minimum labor/equipment charge	4	Job		245	
	2900	Pipe removal, concrete, no excavation, 12" diameter	.137	L.F.		8.40	
	2930	15" diameter	.160			9.75	
	2960	24" diameter	.200			12.25	
	3000	36" diameter	.267			16.35	
	3200	Steel, welded connections, 4" diameter	.150			9.15	
	3300	10" diameter	.300			18.35	
	3390	Minimum labor/equipment charge	8	Job		490	
	3500	Railroad track removal, ties and track	.436	L.F.		23.50	
	3600	Ballast	.096	C.Y.		5.20	
	3700	Remove and re-install ties & track using new bolts & spikes	.960	L.F.		52	
	3800	Turnouts using new bolts and spikes	48	Ea.		2,600	

020 | Subsurface Investigation and Demolition

020 550 | Site Demolition

			WORK-HOURS	UNIT	BARE COSTS MAT.	TOTAL INCL. O&P	
554	3890	Minimum labor/equipment charge	9.600	Job		520	554
	4000	Sidewalk removal, bituminous, 2-1/2" thick	.074	S.Y.		4.52	
	4050	Brick, set in mortar	.130			7.95	
	4100	Concrete, plain, 4"	.150			9.15	
	4200	Mesh reinforced	.160	↓		9.75	
	4290	Minimum labor/equipment charge	4	Job		210	

020 600 | Building Demolition

608	0010	DISPOSAL ONLY Urban buildings with salvage value allowed					608
	0020	Including loading and 5 mile haul to dump					
	0200	Steel frame	.112	C.Y.		11.15	
	0300	Concrete frame	.132			13.10	
	0400	Masonry construction	.108			10.75	
	0500	Wood frame	.194	↓		19.40	
612	0010	DUMP CHARGES Typical urban city, fees only					612
	0100	Building construction materials		C.Y.		49.70	
	0200	Demolition lumber, trees, brush				52.54	
	0300	Rubbish only		↓		42.60	
	0500	Reclamation station, usual charge		Ton		92.30	
620	0010	RUBBISH HANDLING The following are to be added to the					620
	0020	selective demolition prices					
	0400	Chute, circular, prefabricated steel, 18" diameter	.600	L.F.	10	45.50	
	0440	30" diameter	.800	"	18.75	69.50	
	0600	Dumpster, (debris box container), 5 C.Y., rent per week		Ea.		248.50	
	0700	10 C.Y. capacity				312.40	
	0800	30 C.Y. capacity				426	
	0840	40 C.Y. capacity		↓		497	
	1000	Dust partition, 6 mil polyethylene, 4' x 8' panels, 1" x 3" frame	.008	S.F.	.20	.80	
	1080	2" x 4" frame	.008	"	.29	.95	
	2000	Load, haul to chute & dumping into chute, 50' haul	.667	C.Y.		30.50	
	2040	100' haul	.970			44	
	2080	Over 100' haul, add per 100 L.F.	.451			20.50	
	2120	In elevators, per 10 floors, add	.114			5.20	
	3000	Loading & trucking, including 2 mile haul, chute loaded	1			66	
	3040	Hand loading truck, 50' haul	.744			34	
	3080	Machine loading truck	.300			18.35	
	3120	Wheeled 50' and ramp dump loaded	.667			30.50	
	5000	Haul, per mile, up to 8 C.Y. truck	.007			.85	
	5100	Over 8 C.Y. truck	.005	↓		.64	

020 700 | Selective Demolition

702	0010	CEILING DEMOLITION					702
	0200	Drywall, on wood frame	.020	S.F.		.91	
	0220	On metal frame	.021			.95	
	0240	On suspension system, including system	.022			1.01	
	1000	Plaster, lime and horse hair, on wood lath, incl. lath	.023			1.04	
	1020	On metal lath	.028			1.28	
	1100	Gypsum, on gypsum lath	.022			1.01	
	1120	On metal lath	.032			1.45	
	1200	Suspended ceiling, mineral fiber, 2'x2' or 2'x4'	.011			.48	
	1250	On suspension system, incl. system	.013			.61	
	1500	Tile, wood fiber, 12" x 12", glued	.018			.81	
	1540	Stapled	.011			.48	
	1580	On suspension system, incl. system	.021			.95	
	2000	Wood, tongue and groove, 1" x 4"	.016	↓		.73	

020 | Subsurface Investigation and Demolition

020 700 | Selective Demolition

			WORK-HOURS	UNIT	BARE COSTS MAT.	TOTAL INCL. O&P	
702	2040	1" x 8"	.015	S.F.		.67	702
	2400	Plywood or wood fiberboard, 4' x 8' sheets	.013	↓		.61	
	9000	Minimum labor/equipment charge	4	Job		182	
704	0010	**CUTOUT DEMOLITION** Conc., elev. slab, light reinf., under 6 C.F.	.615	C.F.		32	704
	0050	Light reinforcing, over 6 C.F.	.533	"		28	
	0200	Slab on grade to 6" thick, not reinforced, under 8 S.F.	.471	S.F.		24.50	
	0250	Not reinforced, over 8 S.F.	.229	"		11.95	
	0600	Walls, not reinforced, under 6 C.F.	.667	C.F.		35	
	0650	Not reinforced, over 6 C.F.	.615			32	
	1000	Concrete, elevated slab, bar reinforced, under 6 C.F.	.889			46	
	1050	Bar reinforced, over 6 C.F.	.800	↓		42	
	1200	Slab on grade to 6" thick, bar reinforced, under 8 S.F.	.533	S.F.		28	
	1250	Bar reinforced, over 8 S.F.	.381	"		19.90	
	1400	Walls, bar reinforced, under 6 C.F.	.800	C.F.		42	
	1450	Bar reinforced, over 6 C.F.	.727	"		38.50	
	2000	Brick, to 4 S.F. opening, not including toothing					
	2040	4" thick	1.333	Ea.		69.50	
	2060	8" thick	2.222			116	
	2080	12" thick	4			209	
	2400	Concrete block, to 4 S.F. opening, 2" thick	1.143			59.50	
	2420	4" thick	1.333			69.50	
	2440	8" thick	1.481			77.50	
	2460	12" thick	1.667			87.50	
	2600	Gypsum block, to 4 S.F. opening, 2" thick	.500			26	
	2620	4" thick	.571			30	
	2640	8" thick	.727			38.50	
	2800	Terra cotta, to 4 S.F. opening, 4" thick	.571			30	
	2840	8" thick	.615			32	
	2880	12" thick	.800	↓		42	
	4000	For toothing masonry, see Division 045-290					
	4010						
	6000	Walls, interior, not including re-framing,					
	6010	openings to 5 S.F.					
	6100	Drywall to 5/8" thick	.333	Ea.		19	
	6200	Paneling to 3/4" thick	.400			23	
	6300	Plaster, on gypsum lath	.400			23	
	6340	On wire lath	.571	↓		32.50	
	7000	Wood frame, not including re-framing, openings to 5 S.F.					
	7200	Floors, sheathing and flooring to 2" thick	1.600	Ea.		91	
	7310	Roofs, sheathing to 1" thick, not including roofing	1.333			75.50	
	7410	Walls, sheathing to 1" thick, not including siding	1.143	↓		64.50	
	8500	Minimum labor/equipment charge	2	Job		114	
706	0010	**DOOR DEMOLITION**					706
	0200	Doors, exterior, 1-3/4" thick, single, 3' x 7' high	.500	Ea.		23	
	0220	Double, 6' x 7' high	.667			30.50	
	0500	Interior, 1-3/8" thick, single, 3' x 7' high	.400			18.20	
	0520	Double, 6' x 7' high	.500			23	
	0700	Bi-folding, 3' x 6'-8" high	.400			18.20	
	0720	6' x 6'-8" high	.444			20	
	0900	Bi-passing, 3' x 6'-8" high	.500			23	
	0940	6' x 6'-8" high	.571			26	
	1500	Remove and reset, minimum	1			57	
	1520	Maximum	1.333			76	
	2000	Frames, including trim, metal	1	↓		57	

020 | Subsurface Investigation and Demolition

020 700 | Selective Demolition

			WORK-HOURS	UNIT	BARE COSTS MAT.	TOTAL INCL. O&P	
706	2200	Wood	.571	Ea.		32.50	706
	2201	Alternate pricing method	.040	L.F.		2.27	
	2950	Minimum labor/equipment charge	2	Job		91	
	3000	Special doors, counter doors	2.667	Ea.		156	
	3100	Double acting	1.600			94	
	3200	Floor door (trap type)	2			117	
	3300	Glass, sliding, including frames	1.333			78	
	3400	Overhead, commercial, 12' x 12' high	4			235	
	3440	20' x 16' high	5.333			315	
	3500	Residential, 9' x 7' high	2			117	
	3540	16' x 7' high	2.286			134	
	3600	Remove and reset, minimum	4			235	
	3620	Maximum	6.400			375	
	3700	Roll-up grille	3.200			188	
	3800	Revolving door	8			470	
	3900	Swing door	5.333	↓		315	
	9000	Minimum labor/equipment charge	2	Job		114	
708	0010	**ELECTRICAL DEMOLITION**					708
	0020	Conduit to 15' high, including fittings & hangers					
	0100	Rigid galvanized steel, 1/2" to 1" diameter	.033	L.F.		2	
	0120	1-1/4" to 2"	.040			2.43	
	0140	2" to 4"	.053			3.21	
	0200	Electric metallic tubing (EMT) 1/2" to 1"	.020			1.24	
	0220	1-1/4" to 1-1/2"	.025			1.49	
	0240	2" to 3"	.034	↓		2.06	
	0400	Wiremold raceway, including fittings & hangers					
	0420	No. 3000	.032	L.F.		1.95	
	0440	No. 4000	.037	"		2.25	
	0500	Channels, steel, including fittings & hangers					
	0520	3/4" x 1-1/2"	.026	L.F.		1.58	
	0540	1-1/2" x 1-1/2"	.030	"		1.81	
	0600	Copper bus duct, indoor, 3 ph, incl. removal of hangers					
	0610	hangers & supports					
	0620	225 amp	.119	L.F.		7.25	
	0640	400 amp	.151			9.15	
	0660	600 amp	.186			11.30	
	0720	3000 amp	.800	↓		48.50	
	0800	Plug-in switches, 600V 3 ph, incl. disconnecting					
	0820	wire, pipe terminations, 30 amp	.516	Ea.		31.50	
	0840	60 amp	.576			35	
	0850	100 amp	.769			47	
	0860	200 amp	1.290			78	
	0940	1200 amp	8			485	
	0960	1600 amp	9.412	↓		570	
	1010	Safety switches, 250 or 600V, incl. disconnection					
	1050	of wire & pipe terminations					
	1100	30 amp	.650	Ea.		40	
	1120	60 amp	.909			55.50	
	1140	100 amp	1.096			67	
	1160	200 amp	1.600	↓		97.50	
	1210	Panel boards, incl. removal of all breakers,					
	1220	pipe terminations & wire connections					
	1230	3 wire, 120/240V, 100A, to 20 circuits	3.077	Ea.		188	
	1240	200 amps, to 42 circuits	6.154			375	
	1260	4 wire, 120/208V, 125A, to 20 circuits	3.333	↓		203	

020 | Subsurface Investigation and Demolition

020 700 | Selective Demolition

			WORK-HOURS	UNIT	BARE COSTS MAT.	TOTAL INCL. O&P	
708	1270	200 amps, to 42 circuits	6.667	Ea.		405	708
	1300	Transformer, dry type, 1 ph, incl. removal of					
	1320	supports, wire & pipe terminations					
	1340	1 KVA	1.039	Ea.		63.50	
	1360	5 KVA	1.702			104	
	1420	75 KVA	6.400	▼		390	
	1440	3 Phase to 600V, primary					
	1460	3 KVA	2.078	Ea.		127	
	1480	15 KVA	3.810			232	
	1500	30 KVA	4.598			280	
	1530	112.5 KVA	6.897			420	
	1560	500 KVA	13.793			840	
	1570	750 KVA	17.778	▼		1,075	
	1600	Pull boxes & cabinets, sheet metal, incl. removal					
	1620	of supports and pipe terminations					
	1640	6" x 6" x 4"	.257	Ea.		15.65	
	1660	12" x 12" x 4"	.343			21	
	1720	Junction boxes, 4" sq. & oct.	.100			6.10	
	1740	Handy box	.075			4.55	
	1760	Switch box	.075			4.55	
	1780	Receptacle & switch plates	.031	▼		1.89	
	1800	Wire, THW-THWN-THHN, removed from					
	1810	in place conduit, to 15' high					
	1830	#14	.123	C.L.F.		7.45	
	1840	#12	.145			8.80	
	1850	#10	.176			10.65	
	1880	#4	.302			18.35	
	1890	#3	.320			19.50	
	1910	1/0	.482			29	
	1920	2/0	.548			33.50	
	1930	3/0	.640			39	
	1980	400 MCM	.941			57	
	1990	500 MCM	.988	▼		59.50	
	2000	Interior fluorescent fixtures, incl. supports					
	2010	& whips, to 15' high					
	2100	Recessed drop-in 2' x 2', 2 lamp	.457	Ea.		28	
	2120	2' x 4', 2 lamp	.485			29	
	2140	2' x 4', 4 lamp	.533			32.50	
	2160	4' x 4', 4 lamp	.800	▼		48.50	
	2180	Surface mount, acrylic lens & hinged frame					
	2200	1' x 4', 2 lamp	.364	Ea.		22	
	2220	2' x 2', 2 lamp	.364			22	
	2260	2' x 4', 4 lamp	.485			29	
	2280	4' x 4', 4 lamp	.696	▼		42	
	2300	Strip fixtures, surface mount					
	2320	4' long, 1 lamp	.302	Ea.		18.35	
	2340	4' long, 2 lamp	.320			19.50	
	2360	8' long, 1 lamp	.381			23	
	2380	8' long, 2 lamp	.400	▼		24.50	
	2400	Pendant mount, industrial, incl. removal					
	2410	of chain or rod hangers, to 15' high					
	2420	4' long, 2 lamp	.457	Ea.		28	
	2440	8' long, 2 lamp	.593	"		36.50	
	2460	Interior incandescent, surface, ceiling					
	2470	or wall mount, to 12' high					
	2480	Metal cylinder type, 75 Watt	.258	Ea.		15.70	

020 | Subsurface Investigation and Demolition

020 700 | Selective Demolition

			WORK-HOURS	UNIT	BARE COSTS MAT.	TOTAL INCL. O&P	
708	2500	150 Watt	.258	Ea.		15.70	708
	2520	Metal halide, high bay					
	2540	400 Watt	1.067	Ea.		64.50	
	2560	1000 Watt	1.333	↓		81	
	2580	150 Watt, low bay	.800	↓		48.50	
	2600	Exterior fixtures, incandescent, wall mount					
	2620	100 Watt	.320	Ea.		19.50	
	2640	Quartz, 500 Watt	.485	↓		29	
	2660	1500 Watt	.593	↓		36.50	
	2680	Wall pack, mercury vapor					
	2700	175 Watt	.640	Ea.		39	
	2720	250 Watt	.640	"		39	
	9000	Minimum labor/equipment charge	2	Job		122	
712	0010	**FLOORING DEMOLITION**					712
	0200	Brick with mortar	.027	S.F.		1.51	
	0400	Carpet, bonded, including surface scraping	.008			.37	
	0480	Tackless	.002			.09	
	0600	Composition, acrylic or epoxy	.040			2.27	
	0800	Resilient, sheet goods (linoleum)	.011			.53	
	0820	For gym floors	.018			.81	
	0900	Tile, 12" x 12"	.016			.73	
	2000	Tile, ceramic, thin set	.020			1.14	
	2020	Mud set	.023			1.29	
	2200	Marble, slate, thin set	.020			1.14	
	2220	Mud set	.023			1.29	
	2600	Terrazzo, thin set	.032			1.82	
	2620	Mud set	.036			2.03	
	2640	Cast in place	.046			2.60	
	3000	Wood, block, on end	.020			1.14	
	3200	Parquet	.018			1.01	
	3400	Strip flooring, interior, 2-1/4" x 25/32" thick	.025			1.41	
	3500	Exterior, porch flooring, 1" x 4"	.036			2.06	
	3800	Subfloor, tongue and groove, 1" x 6"	.025			1.41	
	3820	1" x 8"	.019			1.07	
	3840	1" x 10"	.015			.87	
	4000	Plywood, nailed	.013			.77	
	4100	Glued and nailed	.020	↓		1.14	
	9000	Minimum labor/equipment charge	2	Job		114	
714	0010	**FRAMING DEMOLITION**					714
	1020	Concrete, average reinforcing, beams, 8" x 10"	.333	L.F.		17.40	
	1040	10" x 12"	.364			19.05	
	1060	12" x 14"	.444			23	
	1200	Columns, 8" x 8"	.333			17.40	
	1240	10" x 10"	.333			17.40	
	1280	12" x 12"	.364			19.05	
	1320	14" x 14"	.400			21	
	1400	Girders, 14" x 16"	.727			38.50	
	1440	16" x 18"	1	↓		52	
	1600	Slabs, elevated, 6" thick	.067	S.F.		3.48	
	1640	8" thick	.089			4.63	
	1680	10" thick	.111			5.80	
	1900	Add for heavy reinforcement		↓		28.40%	
	1910	Minimum labor/equipment charge	8	Job		455	
	2000	Steel framing, beams, 4" x 6"	.112	L.F.		6.95	

020 | Subsurface Investigation and Demolition

020 700 | Selective Demolition

			WORK-HOURS	UNIT	BARE COSTS MAT.	TOTAL INCL. O&P	
714	2020	4" x 8"	.140	L.F.		8.65	714
	2080	8" x 12"	.224			13.85	
	2200	Columns, 6" x 6"	.140			8.65	
	2240	8" x 8"	.160			9.95	
	2280	10" x 10"	.175			10.90	
	2400	Girders, 10" x 12"	.249			15.45	
	2440	10" x 14"	.280			17.35	
	2480	10" x 16"	.339			21	
	2520	10" x 24"	.448	▼		28	
	2950	Minimum labor/equipment charge	8	Job		455	
	3000	Wood framing, beams, 6" x 8"	.145	L.F.		6.75	
	3040	6" x 10"	.182			8.45	
	3080	6" x 12"	.216			10	
	3120	8" x 12"	.286			13.30	
	3160	10" x 12"	.364			16.90	
	3400	Fascia boards, 1" x 6"	.016			.73	
	3440	1" x 8"	.018			.81	
	3480	1" x 10"	.020	▼		.91	
	3520	For trim boards, see division 020-720					
	3800	Headers over openings, 2 @ 2" x 6"	.073	L.F.		3.31	
	3840	2 @ 2" x 8"	.080			3.64	
	3880	2 @ 2" x 10"	.089			4.04	
	4200	Joists, 2" x 4"	.016			.73	
	4230	Joists, 2" x 6"	.016			.75	
	4240	2" x 8"	.017			.77	
	4250	2" x 10"	.018			.80	
	4280	2" x 12"	.018			.82	
	5400	Posts, 4" x 4"	.020			.91	
	5440	6" x 6"	.040			1.82	
	5480	8" x 8"	.053			2.43	
	5500	10" x 10"	.067			3.03	
	5800	Rafters, ordinary, 2" x 6"	.019			.85	
	5840	2" x 8"	.022			1.01	
	5900	Hip & valley, 2" x 6"	.032			1.45	
	5940	2" x 8"	.038	▼		1.73	
	6200	Stairs and stringers, minimum	.400	Riser		18.20	
	6240	Maximum	.615	"		28	
	6600	Studs, 2" x 4"	.008	L.F.		.37	
	6640	2" x 6"	.010	"		.45	
	9000	Minimum labor/equipment charge	2	Job		91	
	9500	See Div. 020-620 for rubbish handling					
720	0010	**MILLWORK AND TRIM DEMOLITION**					720
	1000	Cabinets, wood, base cabinets	.200	L.F.		9.10	
	1020	Wall cabinets	.200	"		9.10	
	1060	Remove and reset, base cabinets	.889	Ea.		50.50	
	1070	Wall cabinets	.800	"		45.50	
	1100	Steel, painted, base cabinets	.267	L.F.		12.15	
	1120	Wall cabinets	.267	"		12.15	
	1200	Casework, large area	.050	S.F.		2.27	
	1220	Selective	.080	"		3.64	
	1500	Counter top, minimum	.080	L.F.		3.64	
	1510	Maximum	.133			6.05	
	1550	Remove and reset, minimum	.320			18.25	
	1560	Maximum	.400	▼		23	
	2000	Paneling, 4' x 8' sheets, 1/4" thick	.008	S.F.		.37	

020 | Subsurface Investigation and Demolition

020 700 | Selective Demolition

			WORK-HOURS	UNIT	BARE COSTS MAT.	TOTAL INCL. O&P	
720	2100	Boards, 1" x 4"	.023	S.F.		1.04	720
	2120	1" x 6"	.021			.97	
	2140	1" x 8"	.020	▼		.91	
	3000	Trim, baseboard, to 6" wide	.013	L.F.		.61	
	3040	12" wide	.016			.73	
	3080	Remove and reset, minimum	.040			2.27	
	3090	Maximum	.053			3.04	
	3100	Ceiling trim	.016			.73	
	3120	Chair rail	.013			.61	
	3140	Railings with balusters	.067	▼		3.03	
	3160	Wainscoting	.023	S.F.		1.04	
	9000	Minimum labor/equipment charge	2	Job		91	
724	0010	**PLUMBING DEMOLITION**					724
	1020	Fixtures, including 10' piping					
	1100	Bath tubs, cast iron	2	Ea.		127	
	1120	Fiberglass	1.333			84.50	
	1140	Steel	1.600			102	
	1200	Lavatory, wall hung	.800			50.50	
	1220	Counter top	1			63.50	
	1300	Sink, steel or cast iron, single	1			63.50	
	1320	Double	1.143			72.50	
	1400	Water closet, floor mounted	1			63.50	
	1420	Wall mounted	1.143			72.50	
	1500	Urinal, floor mounted	2			127	
	1520	Wall mounted	1.143			72.50	
	1600	Water fountains, free standing	1			63.50	
	1620	Recessed	1.333	▼		84.50	
	2000	Piping, metal, to 2" diameter	.040	L.F.		2.53	
	2050	To 4" diameter	.053			3.38	
	2100	To 8" diameter	.160			10.15	
	2150	To 16" diameter	.267	▼		16.90	
	2240	Toilet partitions, see division 020-732					
	2250	Water heater, 40 gal.	1.333	Ea.		84.50	
	6000	Remove and reset fixtures, minimum	1.333			84.50	
	6100	Maximum	2	▼		127	
	9000	Minimum labor/equipment charge	4	Job		253	
726	0010	**ROOFING AND SIDING DEMOLITION**					726
	1000	Deck, roof, concrete plank	.033	S.F.		2.06	
	1100	Gypsum plank	.014			.90	
	1150	Metal decking	.016			.98	
	1200	Wood, boards, tongue and groove, 2" x 6"	.017			.75	
	1220	2" x 10"	.015			.70	
	1280	Standard planks, 1" x 6"	.015			.67	
	1320	1" x 8"	.014			.63	
	1340	1" x 12"	.013	▼		.61	
	2000	Gutters, aluminum or wood, edge hung	.033	L.F.		1.52	
	2100	Built-in	.080	"		3.64	
	2500	Roof accessories, plumbing vent flashing	.571	Ea.		26	
	2600	Adjustable metal chimney flashing	.889	"		40.50	
	3000	Roofing, built-up, 5 ply roof, no gravel	.025	S.F.		1.17	
	3100	Gravel removal, minimum	.008			.37	
	3120	Maximum	.020			.92	
	3400	Roof insulation board	.010			.48	
	4000	Shingles, asphalt strip	.011			.53	
	4100	Slate	.016			.74	
	4300	Wood	.018	▼		.84	

020 | Subsurface Investigation and Demolition

020 700 | Selective Demolition

			WORK-HOURS	UNIT	BARE COSTS MAT.	TOTAL INCL. O&P	
726	4500	Skylight to 10 S.F.	1	Ea.		45.50	726
	5000	Siding, metal, horizontal	.018	S.F.		.82	
	5020	Vertical	.020			.91	
	5200	Wood, boards, vertical	.020			.91	
	5220	Clapboards, horizontal	.021			.95	
	5240	Shingles	.023			1.04	
	5260	Textured plywood	.011	↓		.50	
	9000	Minimum labor/equipment charge	4	Job		182	
728	0010	**SAW CUTTING** Asphalt over 1000 L.F., 3" deep	.021	L.F.	.21	1.93	728
	0020	Each additional inch of depth	.013		.05	1.04	
	0400	Concrete slabs, mesh reinforcing, per inch of depth	.017		.25	1.70	
	0420	Rod reinforcing, per inch of depth	.029		.34	2.79	
	0800	Concrete walls, plain, per inch of depth	.080		.23	4.55	
	0820	Rod reinforcing, per inch of depth	.133		.34	7.50	
	1200	Masonry walls, brick, per inch of depth	.055		.23	3.25	
	1220	Block walls, solid, per inch of depth	.066		.23	3.80	
	5000	Wood sheathing to 1" thick, on walls	.040			2.27	
	5020	On roof	.032	↓		1.82	
	9000	Minimum labor/equipment charge	4	Job		227	
732	0010	**WALLS AND PARTITIONS DEMOLITION**					732
	0100	Brick, 4" to 12" thick	.182	C.F.		9.55	
	0200	Concrete block, 4" thick	.040	S.F.		2.09	
	0280	8" thick	.049			2.57	
	1000	Drywall, nailed	.008			.37	
	1020	Glued and nailed	.009			.40	
	1500	Fiberboard, nailed	.009			.40	
	1520	Glued and nailed	.010			.45	
	2000	Movable walls, metal, 5' high	.027			1.21	
	2020	8' high	.020			.91	
	2200	Metal or wood studs, finish 2 sides, fiberboard	.046			2.18	
	2250	Lath and plaster	.092			4.35	
	2300	Plasterboard (drywall)	.046			2.18	
	2350	Plywood	.053			2.52	
	3000	Plaster, lime and horsehair, on wood lath	.020			.91	
	3020	On metal lath	.024			1.08	
	3400	Gypsum or perlite, on gypsum lath	.020			.88	
	3420	On metal lath	.027	↓		1.21	
	3800	Toilet partitions, slate or marble	1.600	Ea.		72.50	
	3820	Hollow metal	1	"		45.50	
	9000	Minimum labor/equipment charge	2	Job		91	
734	0010	**WINDOW DEMOLITION**					734
	0200	Aluminum, including trim, to 12 S.F.	.500	Ea.		26	
	0240	To 25 S.F.	.727			38.50	
	0280	To 50 S.F.	1.600			82.50	
	0320	Storm windows, to 12 S.F.	.296			15.45	
	0360	To 25 S.F.	.381			19.75	
	0400	To 50 S.F.	.500	↓		26	
	0600	Glass, minimum	.040	S.F.		1.82	
	0620	Maximum	.053	"		2.43	
	1000	Steel, including trim, to 12 S.F.	.615	Ea.		32	
	1020	To 25 S.F.	.889			46	
	1040	To 50 S.F.	2	↓		104	

020 | Subsurface Investigation and Demolition

020 700 | Selective Demolition

			WORK-HOURS	UNIT	BARE COSTS MAT.	TOTAL INCL. O&P	
734	2000	Wood, including trim, to 12 S.F.	.364	Ea.		16.55	734
	2020	To 25 S.F.	.444			20	
	2060	To 50 S.F.	.615			28	
	5020	Remove and reset window, minimum	1.333			76	
	5040	Average	2			114	
	5080	Maximum	4	↓		227	
	9000	Minimum labor/equipment charge	2	Job		91	

020 750 | Concrete Removal

754	0010	**FOOTINGS AND FOUNDATIONS DEMOLITION**					754
	0200	Floors, concrete slab on grade,					
	0240	4" thick, plain concrete	.080	S.F.		4.17	
	0280	Reinforced, wire mesh	.085			4.45	
	0300	Rods	.100			5.25	
	0400	6" thick, plain concrete	.107			5.55	
	0420	Reinforced, wire mesh	.118			6.15	
	0440	Rods	.133	↓		6.95	
	1000	Footings, concrete, 1' thick, 2' wide	.213	L.F.		15.55	
	1080	1'-6" thick, 2' wide	.256			18.70	
	1120	3' wide	.320			23.50	
	1140	2' thick, 3' wide	.366			26.50	
	1200	Average reinforcing, add				14.20%	
	1220	Heavy reinforcing, add		↓		28.40%	
	2000	Walls, block, 4" thick	.040	S.F.		2.27	
	2040	6" thick	.042			2.40	
	2080	8" thick	.044			2.53	
	2100	12" thick	.046			2.60	
	2200	For horizontal reinforcing, add				14.20%	
	2220	For vertical reinforcing, add				28.40%	
	2400	Concrete, plain concrete, 6" thick	.250			13	
	2420	8" thick	.286			14.95	
	2440	10" thick	.333			17.40	
	2500	12" thick	.400			21	
	2600	For average reinforcing, add				14.20%	
	2620	For heavy reinforcing, add		↓		28.40%	
	9000	Minimum labor/equipment charge	4	Job		227	
758	0010	**MASONRY DEMOLITION**					758
	1000	Chimney, 16" x 16", soft old mortar	.333	V.L.F.		19	
	1020	Hard mortar	.444			25	
	1080	20" x 20", soft old mortar	.667			38.50	
	1100	Hard mortar	.800			45.50	
	1140	20" x 32", soft old mortar	.800			45.50	
	1160	Hard mortar	1			57	
	1200	48" x 48", soft old mortar	1.600			91	
	1220	Hard mortar	2			114	
	2000	Columns, 8" x 8", soft old mortar	.167			9.55	
	2020	Hard mortar	.200			11.35	
	2060	16" x 16", soft old mortar	.500			28.50	
	2100	Hard mortar	.571			32.50	
	2140	24" x 24", soft old mortar	1			57	
	2160	Hard mortar	1.333			75.50	
	2200	36" x 36", soft old mortar	2			114	
	2220	Hard mortar	2.667	↓		152	
	3000	Copings, precast or masonry, to 8" wide					

020 | Subsurface Investigation and Demolition

020 750 | Concrete Removal

			WORK-HOURS	UNIT	BARE COSTS MAT.	TOTAL INCL. O&P	
758	3020	Soft old mortar	.044	L.F.		2.53	758
	3040	Hard mortar	.050	"		2.84	
	3100	To 12" wide					
	3120	Soft old mortar	.050	L.F.		2.84	
	3140	Hard mortar	.057	"		3.26	
	4000	Fireplace, brick, 30" x 24" opening					
	4020	Soft old mortar	4	Ea.		227	
	4040	Hard mortar	6.400			365	
	4100	Stone, soft old mortar	5.333			305	
	4120	Hard mortar	8			455	
	5000	Veneers, brick, soft old mortar	.057	S.F.		3.26	
	5020	Hard mortar	.064			3.64	
	5100	Granite and marble, 2" thick	.044			2.53	
	5120	4" thick	.047			2.69	
	5140	Stone, 4" thick	.044			2.53	
	5160	8" thick	.046			2.60	
	5400	Alternate pricing method, stone, 4" thick	.133	C.F.		7.60	
	5420	8" thick	.094	"		5.35	
	9000	Minimum labor/equipment charge	4	Job		227	

022 | Earthwork

022 200 | Excav./Backfill/Compact.

			WORK-HOURS	UNIT	BARE COSTS MAT.	TOTAL INCL. O&P	
204	0010	**BACKFILL** By hand, no compaction, light soil	.571	C.Y.		26	204
	0100	Heavy soil	.727			33.50	
	0300	Compaction in 6" layers, hand tamp, add to above	.388			17.70	
	0400	Roller compaction operator walking, add	.120			7.55	
	0500	Air tamp, add	.211			11	
	0600	Vibrating plate, add	.133			7.60	
	0800	Compaction in 12" layers, hand tamp, add to above	.235			10.75	
	1000	Air tamp, add	.140			7.30	
	1100	Vibrating plate, add	.089			5.05	
	1200	Dozer, backfilling, bulk, 75 H.P., 50' haul, no compaction	.011			.95	
	1300	Dozer backfilling, bulk, up to 300' haul, no compaction	.010			1.59	
	1400	Air tamped	.067			10.10	
	1900	Dozer backfilling, trench, up to 300' haul, no compaction	.013			2.12	
	2000	Air tamped	.068			10.30	
	2350	Spreading in 8" layers, small dozer	.011			1.81	
	2450	Compacting with vibrating plate, 8" lifts	.110			6.25	
212	0010	**BORROW** Buy and load at pit, haul 2 miles round trip					212
	0020	and spread, with 200 H.P. dozer, no compaction					
	0100	Bank run gravel	.047	C.Y.	3.75	12.90	
	0200	Common borrow	.047		3.50	12.45	
	0300	Crushed stone, 1-1/2"	.047		13.75	30	
	0320	3/4"	.047		13.95	30.50	
	0340	1/2"	.047		14.75	32	
	0360	3/8"	.047		15.25	32.50	
	0400	Sand, washed, concrete	.047		9.50	22.50	
	0500	Dead or bank sand	.047		3.50	12.45	

022 | Earthwork

022 200 | Excav./Backfill/Compact.

			WORK-HOURS	UNIT	BARE COSTS MAT.	TOTAL INCL. O&P	
212	0600	Select structural fill	.047	C.Y.	7.50	19.25	212
	0700	Screened loam	.047		15.50	33	
	0800	Topsoil, weed free	.047		12.75	28.50	
	0900	For 5 mile haul, add	.040	↓		4.95	
250	0010	**EXCAVATING, STRUCTURAL** Hand, pits to 6' deep, sandy soil	1	C.Y.		45.50	250
	0100	Heavy soil or clay	2			91	
	0300	Pits 6' to 12' deep, sandy soil	1.600			72.50	
	0500	Heavy soil or clay	2.667			122	
	0700	Pits 12' to 18' deep, sandy soil	2			91	
	0900	Heavy soil or clay	4			182	
	1100	Hand loading trucks from stock pile, sandy soil	.667			30.50	
	1300	Heavy soil or clay	1	↓		45.50	
	1500	For wet or muck hand excavation, add to above		%		71%	
	9000	Minimum labor/equipment charge	2	Job		91	
262	0010	**FILL** Spread dumped material, by dozer, no compaction	.012	C.Y.		1.91	262
	0100	By hand	.667	"		30.50	
	9000	Minimum labor/equipment charge	2	Job		91	
286	0010	**LOAM OR TOPSOIL** Remove and stockpile on site					286
	0700	Furnish and place, truck dumped @ $17.00 per C.Y., 4" deep	.009	S.Y.	1.87	4.06	
	0800	6" deep	.015	"	2.82	6.20	
	0810	Minimum labor/equipment charge	6	Ea.		560	
	0900	Fine grading and seeding, incl. lime, fertilizer & seed,					
	1000	With equipment	.048	S.Y.	.18	2.91	
	2000	Minimum labor/equipment charge	2	Job		91	

025 | Paving and Surfacing

025 100 | Walk/Rd/Parkng Paving

			WORK-HOURS	UNIT	BARE COSTS MAT.	TOTAL INCL. O&P	
128	0010	**SIDEWALKS, DRIVEWAYS, & PATIOS** No base					128
	0020	Asphaltic concrete, 2" thick	.067	S.Y.	2.93	8.45	
	0100	2-1/2" thick	.073	"	3.59	9.90	
	0300	Concrete, 3000 psi, cast in place with 6 x 6 - W1.4 x W1.4 mesh,					
	0310	broomed finish, no base, 4" thick	.040	S.F.	.90	3.61	
	1700	Redwood, prefabricated, 4' x 4' sections	.051		3.30	8.55	
	1750	Redwood planks, 1" thick, on sleepers	.067	↓	2.45	8.10	
	9000	Minimum labor/equipment charge	8	Job		415	

025 150 | Unit Pavers

			WORK-HOURS	UNIT	BARE COSTS MAT.	TOTAL INCL. O&P	
154	0010	**ASPHALT BLOCKS**, 6"x12"x1-1/4", w/bed & neopr. adhesive	.119	S.F.	2.50	10.45	154
	0100	3" thick	.123		3.50	12.35	
	0300	Hexagonal tile, 8" wide, 1-1/4" thick	.119		2.50	10.45	
	0400	2" thick	.123		3.50	12.35	
	0500	Square, 8" x 8", 1-1/4" thick	.119		2.50	10.45	
	0600	2" thick	.123	↓	3.50	12.35	
	9000	Minimum labor/equipment charge	4	Job		232	
158	0010	**BRICK PAVING** 4" x 8" x 1-1/2", without joints (4.5 brick/S.F.)	.145	S.F.	2.25	11.40	158
	0100	Grouted, 3/8" joint (3.9 brick/S.F.)	.178		2.15	12.90	

025 | Paving and Surfacing

025 150 | Unit Pavers

			WORK-HOURS	UNIT	BARE COSTS MAT.	TOTAL INCL. O&P	
158	0200	4" x 8" x 2-1/4", without joints (4.5 bricks/S.F.)	.145	S.F.	2.45	11.70	158
	0300	Grouted, 3/8" joint (3.9 brick/S.F.)	.178		2.25	13.10	
	1500	Brick on 4" thick sand bed laid flat, 4.5 per S.F.	.145		2.62	12.05	
	2000	Laid on edge, 7.2 per S.F.	.229		4.32	19.25	
	2500	For 4" thick concrete bed and joints, add	.027		.69	2.57	
	2800	For steam cleaning, add	.008	↓	.06	.59	
	9000	Minimum labor/equipment charge	4	Job		232	
166	0010	**STONE PAVERS**					166
	1300	Slate, natural cleft, irregular, 3/4" thick	.174	S.F.	1.50	11.60	
	1350	Random rectangular, gauged, 1/2" thick	.152		3.25	13.40	
	1400	Random rectangular, butt joint, gauged, 1/4" thick	.107		3.50	11.50	
	1450	For sand rubbed finish, add			2.30	3.92	
	1550	Granite blocks, 3-1/2" x 3-1/2" x 3-1/2"	.174		4.61	16.85	
	1600	4" to 12" long, 3" to 5" wide, 3" to 5" thick	.163	↓	3.84	15.05	

025 250 | Curbs

254	0010	**CURBS** Asphaltic, machine formed, 8" wide, 6" high, 40 L.F./ton	.032	L.F.	.65	2.69	254
	0100	8" wide, 8" high, 30 L.F. per ton	.036		.90	3.31	
	0150	Asphaltic berm, 12"W, 3'-6"H, 35 L.F./ton, before pavement	.046		.75	3.55	
	0200	12"W, 1-1/2"to 4" H, 60 L.F. per ton, laid with pavement	.038		.45	2.54	
	0300	Concrete, 6" x 18", wood forms, straight	.096		3.10	10.75	
	0400	6" x 18", radius	.240		3.26	19.30	
	0550	Precast, 6" x 18", straight	.080		5.25	14.20	
	0600	6" x 18", radius	.172		7.95	25	
	1000	Granite, split face, straight, 5" x 16"	.112		12.80	29	
	1100	6" x 18"	.124		16.80	36.50	
	1300	Radius curbing, 6" x 18", over 10' radius	.215	↓	20.50	49	
	1400	Corners, 2' radius	.700	Ea.	69	163	
	1600	Edging, 4-1/2" x 12", straight	.187	L.F.	6.40	23	
	1800	Curb inlets, (guttermouth) straight	1.366	Ea.	154	350	

025 800 | Pavement Marking

804	0010	**LINES ON PAV'T**, latex or chlorinated, white or yellow, 4" wide	.002	L.F.	.05	.25	804
	0200	6" wide	.004	"	.06	.38	
	0760	Arrows	.061	S.F.	1.24	6.40	
	0800	Parking stall, paint, white	.056	Stall	.75	4.73	
	1000	Street letters and numbers	.030	S.F.	.25	2.28	

033 | Cast-In-Place Concrete

033 100 | Structural Concrete

			WORK-HOURS	UNIT	BARE COSTS MAT.	TOTAL INCL. O&P	
130	0010	**CONCRETE IN PLACE** Including forms (4 uses), reinforcing					130
	0050	steel, including finishing unless otherwise indicated					
	0060	Lines without crews include forming, reinforcing, placing and finishing					
	0204	Base, granolithic, 1" x 5" high, straight	.137	L.F.	.13	7.85	
	0220	Cove	.171	"	.13	9.70	
	0300	Beams, 5 kip per L.F., 10' span	13.285	C.Y.	171	1,100	

033 | Cast-In-Place Concrete

033 100 | Structural Concrete

		WORK-HOURS	UNIT	BARE COSTS MAT.	TOTAL INCL. O&P	
130	0350	25' span		C.Y.	163	278	130
	0500	Chimney foundations, industrial, minimum	3.034		106	370	
	0510	Maximum	4.112		121	460	
	0700	Columns, square, 12" x 12", minimum reinforcing	18.916		186	1,400	
	0720	Average reinforcing	21.005		280	1,700	
	0740	Maximum reinforcing	23.240		355	2,000	
	0800	16" x 16", minimum reinforcing	13.507		157	1,050	
	0820	Average reinforcing	16.752		263	1,450	
	0840	Maximum reinforcing	20.010		370	1,875	
	3800	Footings, spread under 1 C.Y.	3.014		74	292	
	3850	Over 5 C.Y.	1.385		68	193	
	3900	Footings, strip, 18" x 9", plain	2.795		64.50	262	
	3950	36" x 12", reinforced	1.828		68	217	
	4000	Foundation mat, under 10 C.Y.	2.729		97.50	330	
	4050	Over 20 C.Y.	1.836		86	257	
	4200	Grade walls, 8" thick, 8' high	4.756		83.50	415	
	4250	14' high	7.714		111	640	
	4260	12" thick, 8' high	3.387		77	325	
	4270	14' high	5.264		88.50	460	
	4300	15" thick, 8' high	2.721		73.50	281	
	4350	12' high	4.101	▼	78	375	
	4751	Slab on grade, incl. troweled finish, not incl. forms					
	4760	or reinforcing, over 10,000 S.F., 4" thick slab	.020	S.F.	.64	2.27	
	4820	6" thick slab	.020		.94	2.76	
	4840	8" thick slab	.022		1.29	3.45	
	4900	12" thick slab	.026		1.93	4.75	
	4950	15" thick slab	.029	▼	2.43	5.75	
	5000	Slab on grade, incl. textured finish, not incl. forms					
	5001	or reinforcing, 4" thick slab	.022	S.F.	1.14	3.01	
	5010	6" thick	.024		1.44	3.62	
	5020	8" thick	.027	▼		1.29	
	5200	Lift slab in place above the foundation, incl. forms,					
	5210	reinforcing, concrete and columns, minimum	.098	S.F.	3.28	12	
	5250	Average	.122		3.53	13.95	
	5300	Maximum	.137	▼	3.82	15.45	
	5900	Pile caps, incl. forms and reinf., sq. or rect., under 5 C.Y.	2.099	C.Y.	73	241	
	5950	Over 10 C.Y.	1.511		70	203	
	6000	Triangular or hexagonal, under 5 C.Y.	2.147		65	228	
	6050	Over 10 C.Y.	1.291		69.50	193	
	6200	Retaining walls, gravity, 4' high see division 022-708	3.345		66.50	300	
	6250	10' high	1.807		58	197	
	6300	Cantilever, level backfill loading, 8' high	3.026		74	305	
	6350	16' high	2.294	▼	72	258	
	6800	Stairs, not including safety treads, free standing	.600	LF Nose	4.95	42	
	6850	Cast on ground	.400	"	3.47	28	
	7000	Stair landings, free standing	.253	S.F.	1.98	17.35	
	7050	Cast on ground	.105	"	1.12	7.70	
	9000	Minimum labor/equipment charge	16	Job		940	

042 | Unit Masonry

042 200 | Concrete Unit Masonry

		Description	WORK-HOURS	UNIT	BARE COSTS MAT.	TOTAL INCL. O&P	
216	0010	CONCRETE BLOCK, BACK-UP, Scaffolding not included					216
	0020	Sand aggregate, tooled joint 1 side					
	0050	Not-reinforced, 8" x 16", 2" thick, 2000 psi	.084	S.F.	.84	5.90	
	0200	Regular block, 4" thick	.091		.80	6.20	
	0300	6" thick	.095		.96	6.70	
	0350	8" thick	.100		1.15	7.25	
	0400	10" thick	.103		1.90	8.65	
	0450	12" thick	.130	↓	1.95	10.10	
	9000	Minimum labor/equipment charge	8	Job		415	
232	0010	CONCRETE BLOCK, PARTITIONS Scaffolding not included					232
	1000	Lightweight block, tooled joints, 2 sides, hollow					
	1100	Not reinforced, 8" x 16" x 4" thick	.091	S.F.	.90	6.35	
	1150	6" thick	.098		1.22	7.25	
	1200	8" thick	.104		1.58	8.20	
	1250	10" thick	.108		2.08	9.30	
	1300	12" thick	.137	↓	2.35	11.15	
	4000	Regular block, tooled joints, 2 sides, hollow					
	4100	Not reinforced, 8" x 16" x 4" thick	.093	S.F.	.80	6.30	
	4150	6" thick	.100		.98	7	
	4200	8" thick	.107		1.15	7.65	
	4250	10" thick	.111		1.90	9.15	
	4300	12" thick	.141	↓	1.95	10.65	
	9000	Minimum labor/equipment charge	8	Job		415	

042 300 | Reinforced Unit Masonry

		Description	WORK-HOURS	UNIT	BARE COSTS MAT.	TOTAL INCL. O&P	
310	0010	CONCRETE BLOCK, EXTERIOR Not including scaffolding					310
	0020	Reinforced, tooled joints 2 sides, styrofoam inserts					
	0100	Regular, 8" x 16" x 6" thick	.103	S.F.	1.80	8.50	
	0200	8" thick	.110		2	9.25	
	0250	10" thick	.113		2.89	10.95	
	0300	12" thick	.145	↓	2.98	12.65	
	9000	Minimum labor/equipment charge	8	Job		415	
320	0010	CONCRETE BLOCK FOUNDATION WALL Scaffolding not included					320
	0050	Sand aggregate, trowel cut joints, not reinf., parged 1/2" thick					
	0200	Regular, 8" x 16" x 6" thick	.089	S.F.	1.24	6.85	
	0250	8" thick	.093		1.47	7.45	
	0300	10" thick	.095		2.15	8.75	
	0350	12" thick	.122	↓	2.27	10.20	
	1000	Reinforced					
	1100	Regular, 8" x 16" block, 6" thick	.090	S.F.	1.34	7.05	
	1150	8" thick	.094		1.58	7.70	
	1200	10" thick	.096		2.27	9	
	1250	12" thick	.123	↓	2.39	10.50	
	9000	Minimum labor/equipment charge	8	Job		415	

042 550 | Masonry Veneer

		Description	WORK-HOURS	UNIT	BARE COSTS MAT.	TOTAL INCL. O&P	
554	0010	BRICK VENEER Scaffolding not included, truck load lots					554
	0015	Material costs incl. a 3% brick waste allowance					
	2000	Standard, sel. common, 4" x 2-2/3" x 8", (6.75/S.F.)	.174	S.F.	1.94	12.55	
	2020	Stnd, 4" x 2-2/3" x 8", running bond, red, (6.75/S.F.)	.182		2.18	13.40	
	2050	Full header every 6th course (7.88/S.F.)	.216		2.54	15.85	
	2100	English, full header every 2nd course (10.13/S.F.)	.286		3.27	21	
	2150	Flemish, alternate header every course (9.00/S.F.)	.267		2.91	19.10	
	2200	Flemish, alt. header every 6th course (7.13/S.F.)	.195	↓	2.30	14.35	

042 | Unit Masonry

042 550 | Masonry Veneer

			WORK-HOURS	UNIT	BARE COSTS MAT.	TOTAL INCL. O&P	
554	2250	Full headers throughout (13.50/S.F.)	.381	S.F.	4.36	27.50	554
	2300	Rowlock course (13.50/S.F.)	.400		4.36	28.50	
	2350	Rowlock stretcher (4.50/S.F.)	.129		1.45	9.30	
	2400	Soldier course (6.75/S.F.)	.200		2.18	14.40	
	2450	Sailor course (4.50/S.F.)	.138		1.45	9.80	
	2600	Running bond, buff or gray face (6.75/S.F.)	.182		2.77	14.40	
	2700	Glazed face brick, running bond	.190		7.30	22.50	
	2750	Full header every 6th course (7.88/S.F.)	.235		8.55	27	
	3000	Jumbo, 6" x 4" x 12" running bond (3.00/S.F.)	.092		3.19	10.35	
	3050	Norman, 4" x 2-2/3" x 12" running bond, (4.50/S.F.)	.125		2.75	11.35	
	3100	Norwegian, 4" x 3-1/5" x 12" (3.75/S.F.)	.107		2.56	10.05	
	3150	Economy, 4" x 4" x 8" (4.50/S.F.)	.129		2.52	11.15	
	3200	Engineer, 4" x 3-1/5" x 8" (5.63/S.F.)	.154		2.51	12.45	
	3250	Roman, 4" x 2" x 12" (6.00/S.F.)	.160		3.79	14.95	
	3300	SCR, 6" x 2-2/3" x 12" (4.50/S.F.)	.129		3.79	13.35	
	3350	Utility, 4" x 4" x 12" (3.00/S.F.)	.089		2.70	9.35	
	9000	Minimum labor/equipment charge	8	Job		415	

045 | Masonry Restoration, Cleaning and Refractories

045 200 | Masonry Restoration

			WORK-HOURS	UNIT	BARE COSTS MAT.	TOTAL INCL. O&P	
290	0010	**TOOTHING MASONRY**					290
	0500	Brickwork, soft old mortar	.200	V.L.F.		9.10	
	0520	Hard mortar	.267			12.15	
	0700	Blockwork, soft old mortar	.114			5.20	
	0720	Hard mortar	.160			7.25	
	9000	Minimum labor/equipment charge	2	Job		91	

050 | Metal Materials, Coatings and Fastenings

050 500 | Metal Fastening

			WORK-HOURS	UNIT	BARE COSTS MAT.	TOTAL INCL. O&P	
515	0010	**DRILLING** And layout for anchors, per					515
	0050	inch of depth, concrete or brick walls					
	0100	1/4" diameter	.107	Ea.	.08	6.20	
	0200	3/8" diameter	.127		.10	7.40	
	0300	1/2" diameter	.160		.12	9.35	
	0400	5/8" diameter	.167		.18	9.85	
	0500	3/4" diameter	.178		.22	10.55	
	0600	7/8" diameter	.186		.30	11.15	
	0700	1" diameter	.200		.38	12	
	0800	1-1/4" diameter	.211		.66	13.15	
	0900	1-1/2" diameter	.229		1.04	14.80	
	1100	Drilling & layout for drywall or plaster walls					

050 | Metal Materials, Coatings and Fastenings

050 500 | Metal Fastening

			WORK-HOURS	UNIT	BARE COSTS MAT.	TOTAL INCL. O&P	
515	1200	Holes, 1/4" diameter	.053	Ea.	.04	3.10	515
	1300	3/8" diameter	.057		.05	3.35	
	1400	1/2" diameter	.062		.06	3.62	
	1500	3/4" diameter	.067		.11	3.98	
	1600	1" diameter	.073		.19	4.48	
	1700	1-1/4" diameter	.080		.33	5.10	
	1800	1-1/2" diameter	.089		.52	5.95	
	2000	Minimum labor/equipment charge	2	Job		114	
520	0010	**EXPANSION ANCHORS** & shields					520
	0100	Bolt anchors for concrete, brick or stone, no layout and drilling					
	0200	Expansion shields, zinc, 1/4" diameter, 1" long, single	.089	Ea.	.65	6.15	
	0300	1-3/8" long, double	.094		.72	6.60	
	0400	3/8" diameter, 2" long, single	.094		1.08	7.20	
	0500	2" long, double	.100		1.33	7.95	
	0600	1/2" diameter, 2-1/2" long, single	.100		1.78	8.70	
	0700	2-1/2" long, double	.107		1.72	9	
	0800	5/8" diameter, 2-5/8" long, single	.107		2.55	10.45	
	0900	3" long, double	.114		2.55	10.90	
	1000	3/4" diameter, 2-3/4" long, single	.114		3.79	13	
	1100	4" long, double	.123		5.05	15.65	
	1300	1" diameter, 6" long, double	.133		19.45	41	
	1500	Self drilling, steel, 1/4" diameter bolt	.308		.85	19.05	
	1600	3/8" diameter bolt	.348		1.20	22	
	1700	1/2" diameter bolt	.400		1.79	26	
	1800	5/8" diameter bolt	.444		3.23	31	
	1900	3/4" diameter bolt	.500		5.65	38	
	2000	7/8" diameter bolt	.571		8.50	47.50	
	2100	Hollow wall anchors for gypsum board,					
	2200	plaster, tile or wall board					
	2300	1/8" diameter, short		Ea.	.19	.33	
	2400	Long			.20	.34	
	2500	3/16" diameter, short			.43	.73	
	2600	Long			.46	.79	
	2700	1/4" diameter, short			.54	.92	
	2800	Long			.62	1.05	
	3000	Toggle bolts, bright steel, 1/8" diameter, 2" long	.094		.23	5.75	
	3100	4" long	.100		.40	6.40	
	3200	3/16" diameter, 3" long	.100		.29	6.20	
	3300	6" long	.107		.44	6.85	
	3400	1/4" diameter, 3" long	.107		.32	6.65	
	3500	6" long	.114		.48	7.30	
	3600	3/8" diameter, 3" long	.114		.71	7.70	
	3700	6" long	.133		1.04	9.40	
	3800	1/2" diameter, 4" long	.133		2.23	11.40	
	3900	6" long	.160		2.98	14.20	
	4000	Nailing anchors					
	4100	Nylon anchor standard nail, 1/4" diameter, 1" long		C	12.15	20.50	
	4200	1-1/2" long			15.70	27	
	4300	2" long			26	44	
	4400	Zamac anchor stainless nail, 1/4" diameter, 1" long			18.75	32	
	4500	1-1/2" long			22.50	38	
	4600	2" long			31.50	54	
	8000	Wedge anchors, not including layout or drilling					
	8050	Carbon steel, 1/4" diameter, 1-3/4" long	.053	Ea.	.34	3.62	
	8100	3 1/4" long	.055		.52	4.03	
	8150	3/8" diameter, 2-1/4" long	.053		.51	3.91	

050 | Metal Materials, Coatings and Fastenings

050 500 | Metal Fastening

			WORK-HOURS	UNIT	BARE COSTS MAT.	TOTAL INCL. O&P	
520	8200	5" long	.055	Ea.	.89	4.66	520
	8250	1/2" diameter, 2-3/4" long	.057		.91	4.81	
	8300	7" long	.062		1.58	6.20	
	8350	5/8" diameter, 3-1/2" long	.062		1.73	6.45	
	8400	8-1/2" long	.070		2.80	8.70	
	8450	3/4" diameter, 4-1/4" long	.070		2.30	7.85	
	8500	10" long	.080		5.15	13.30	
	8550	1" diameter, 6" long	.080		7.70	17.65	
	8600	12" long	.100		10.75	24	
	8650	1-1/4" diameter, 9" long	.114		14.70	31	
	8700	12" long	.133		16.50	36	
	8750	For type 303 stainless steel, add			350%		
	8800	For type 316 stainless steel, add			450%		
	9000	Minimum labor/equipment charge	2	Job		114	
575	0010	WELDING Field. Cost per welder, no operating engineer	1	Hr.	2.73	97	575
	0200	With 1/2 operating engineer	1.500		2.73	123	
	0300	With 1 operating engineer	2		2.73	150	
	0500	With no operating engineer, minimum	.602	Ton	1.82	58.50	
	0600	Maximum	3.200		7.30	310	
	0800	With one operating engineer per welder, minimum	1.203		1.82	90.50	
	0900	Maximum	6.400		7.30	475	
	1200	Continuous fillet, stick welding, incl. equipment					
	1300	Single pass, 1/8" thick, 0.1#/L.F.	.033	L.F.	.09	3.23	
	1400	3/16" thick, 0.2#/L.F.	.067		.18	6.45	
	1500	1/4" thick, 0.3#/L.F.	.100		.27	9.70	
	1610	5/16" thick, 0.4#/L.F.	.133		.36	12.90	
	1800	3 passes, 3/8" thick, 0.5#/L.F.	.167		.46	16.15	
	2010	4 passes, 1/2" thick, 0.7#/L.F.	.235		.64	23	
	2200	5 to 6 passes, 3/4" thick, 1.3#/L.F.	.421		1.18	40.50	
	2400	8 to 11 passes, 1" thick, 2.4#/L.F.	.800		2.18	77	
	9000	Minimum labor/equipment charge	2	Job		185	

055 | Metal Fabrications

055 100 | Metal Stairs

			WORK-HOURS	UNIT	BARE COSTS MAT.	TOTAL INCL. O&P	
104	0010	STAIR Steel, safety nosing, steel stringers					104
	0020	Grating tread and pipe railing, 3'-6" wide	.914	Riser	70	189	
	0100	4'-0" wide	1.067		76	211	
	0200	Cement fill metal pan and picket rail, 3'-6" wide	.914		60	172	
	0300	4'-0" wide	1.067		65	193	
	0350	Wall rail, both sides, 3'-6" wide	.604		50	132	
	0500	Checkered plate tread, industrial, 3'-6" wide	1.143		84	231	
	0550	Circular for tanks, 3'-0" wide	.970		73	199	
	0800	Custom steel stairs, minimum	.914		85	215	
	0810	Average	1.067		125	296	
	0900	Maximum	1.600		185	440	
	1100	For 4' wide stairs, add			10%		
	1300	For 5' wide stairs, add			20%		
	1500	Landing, steel pan, conventional	.200	S.F.	30	67	

055 | Metal Fabrications

055 100 | Metal Stairs

			WORK-HOURS	UNIT	BARE COSTS MAT.	TOTAL INCL. O&P	
104	1810	Spiral aluminum, 5'-0" diameter, stock units	.711	Riser	210	410	104
	1820	Custom units	.711		320	600	
	1830	Stock units, 4'-0" diameter, safety treads	.640		235	450	
	1840	Oak treads	.640		250	475	
	1850	5'-0" diameter, safety treads	.711		255	490	
	1860	Oak treads	.711		280	535	
	1870	6'-0" diameter, safety treads	.800		295	570	
	1880	Oak treads	.800		330	630	
	1900	Spiral, cast iron, 4'-0" diameter, ornamental, minimum	.711		150	310	
	1920	Maximum	1.280		235	505	
	2000	Spiral, steel, industrial checkered plate, 4' diameter	.711		80	191	
	2200	Stock units, 6'-0" diameter	.800		125	275	
	3900	Industrial ships ladder, 3' W, grating treads, 2 line pipe rail	1.067		55	176	
	4000	Aluminum	1.067	↓	75	210	
	9000	Minimum labor/equipment charge	16	Job		1,225	

055 200 | Handrails & Railings

203	0010	RAILING, PIPE Aluminum, 2 rail, 1-1/4" diam., satin finish	.200	L.F.	10	32.50	203
	0030	Clear anodized	.200		12.50	36.50	
	0040	Dark anodized	.200		14.15	39.50	
	0080	1-1/2" diameter, satin finish	.200		12.10	36	
	0090	Clear anodized	.200		13.50	38	
	0100	Dark anodized	.200		15	41	
	0140	Aluminum, 3 rail, 1-1/4" diam., satin finish	.234		15.50	44	
	0150	Clear anodized	.234		19.30	50.50	
	0160	Dark anodized	.234		21.50	54	
	0200	1-1/2" diameter, satin finish	.234		18.75	49.50	
	0210	Clear anodized	.234		21	53.50	
	0220	Dark anodized	.234		23	57.50	
	0500	Steel, 2 rail, primed, 1-1/4" diameter	.200		7.50	28	
	0520	1-1/2" diameter	.200		8.25	29.50	
	0540	Galvanized, 1-1/4" diameter	.200		10.40	33.50	
	0560	1-1/2" diameter	.200		11.65	35	
	0580	Steel, 3 rail, primed, 1-1/4" diameter	.234		11.10	37	
	0600	1-1/2" diameter	.234		11.85	38	
	0620	Galvanized, 1-1/4" diameter	.234		15.65	45	
	0640	1-1/2" diameter	.234		16.95	47	
	0700	Stainless steel, 2 rail, 1-1/4" diam. #4 finish	.234		24.50	59.50	
	0720	High polish	.234		39	83.50	
	0740	Mirror polish	.234		48	100	
	0760	Stainless steel, 3 rail, 1-1/2" diam., #4 finish	.267		36.50	82.50	
	0770	High polish	.267		60	123	
	0780	Mirror finish	.267		72	144	
	0900	Wall rail, alum. pipe, 1-1/4" diam., satin finish	.150		5.75	21.50	
	0905	Clear anodized	.150		7.05	23.50	
	0910	Dark anodized	.150		8.50	26	
	0915	1-1/2" diameter, satin finish	.150		6.30	22.50	
	0920	Clear anodized	.150		8	25	
	0925	Dark anodized	.150		9.95	28.50	
	0930	Steel pipe, 1-1/4" diameter, primed	.150		4.40	19.10	
	0935	Galvanized	.150		6.35	22.50	
	0940	1-1/2" diameter, primed	.150		4.60	19.40	
	0945	Galvanized	.150		6.50	22.50	
	0955	Stainless steel pipe, 1-1/2" diam., #4 finish	.299		19.65	56	
	0960	High polish	.299	↓	38.50	88.50	

055 | Metal Fabrications

055 200 | Handrails & Railings

			WORK-HOURS	UNIT	BARE COSTS MAT.	TOTAL INCL. O&P	
203	0965	Mirror polish	.299	L.F.	46.50	102	203
	9000	Minimum labor/equipment charge	4	Job		287	
208	0010	RAILINGS, ORNAMENTAL Aluminum, bronze or stainless, minimum	.333	L.F.	74	151	208
	0100	Maximum	.889	"	440	815	
	0200	Aluminum pipe rail, minimum	.533	L.F.	48	121	
	0300	Maximum	1		158	340	
	0400	Hand-forged wrought iron, minimum	.667		78	182	
	0500	Maximum	1		340	655	
	0600	Composite metal and wood or glass, minimum	1.333		150	355	
	0700	Maximum	1.600		395	790	
	9000	Minimum labor/equipment charge	4	Job		287	

061 | Rough Carpentry

061 100 | Wood Framing

			WORK-HOURS	UNIT	BARE COSTS MAT.	TOTAL INCL. O&P	
102	0011	BLOCKING					102
	2600	Miscellaneous, to wood construction					
	2620	2" x 4"	47.059	M.B.F.	545	3,675	
	2660	2" x 8"	29.630	"	575	2,750	
	2720	To steel construction					
	2740	2" x 4"	57.143	M.B.F.	545	4,300	
	2780	2" x 8"	38.095	"	575	3,200	
	9000	Minimum labor/equipment charge	2	Job		117	
114	0010	FRAMING, JOISTS					114
	0020						
	2002	Joists, 2" x 4"	.013	L.F.	.40	1.44	
	2100	2" x 6"	.013		.55	1.70	
	2152	2" x 8"	.015		.85	2.30	
	2202	2" x 10"	.018		1.25	3.18	
	2252	2" x 12"	.018		1.54	3.69	
	2302	2" x 14"	.021		1.76	4.23	
	2352	3" x 6"	.017		1.69	3.89	
	2402	3" x 10"	.021		2.26	5.05	
	2452	3" x 12"	.027		3.39	7.35	
	2502	4" x 6"	.020		2.26	5.05	
	2552	4" x 10"	.027		3.76	8	
	2602	4" x 12"	.036		4.51	9.80	
	2607	Sister joist, 2" x 6"	.020		.61	2.22	
	2612	2" x 8"	.025		.85	2.91	
	2617	2" x 10"	.030		1.25	3.90	
	2622	2" x 12"	.035		1.54	4.68	
	3000	Composite wood joist 9-1/2" deep	17.778	M.L.F.	1,575	3,750	
	3010	11-1/2" deep	18.182		1,625	3,850	
	3020	14" deep	19.512		2,100	4,700	
	3030	16" deep	20.513		2,200	4,975	
	4000	Open web joist 12" deep	18.182		2,125	4,700	
	4010	14" deep	19.512		2,150	4,825	
	4020	16" deep	20.513		2,200	4,975	
	4030	18" deep	21.622		2,350	5,275	

061 | Rough Carpentry

061 100 | Wood Framing

			WORK-HOURS	UNIT	BARE COSTS MAT.	TOTAL INCL. O&P	
114	9000	Minimum labor/equipment charge	2	Job		114	114
116	0010	**FRAMING, MISCELLANEOUS**					116
	0020						
	2002	Firestops, 2" x 4"	.021	L.F.	.40	1.89	
	2102	2" x 6"	.027		.61	2.60	
	5002	Nailers, treated, wood construction, 2" x 4"	.020		.55	2.13	
	5102	2" x 6"	.021		.85	2.71	
	5122	2" x 8"	.023		1.19	3.38	
	5202	Steel construction, 2" x 4"	.021		.55	2.21	
	5222	2" x 6"	.023		.85	2.79	
	5242	2" x 8"	.025		1.19	3.48	
	7002	Rough bucks, treated, for doors or windows, 2" x 6"	.040		.85	3.79	
	7102	2" x 8"	.042		1.19	4.51	
	8000	Stair stringers, 2" x 10"	.123		1.14	9.20	
	8100	2" x 12"	.123		1.40	9.65	
	8150	3" x 10"	.128		2.05	11.05	
	8200	3" x 12"	.128		3.08	12.80	
	9000	Minimum labor/equipment charge	2	Job		114	
118	0010	**FRAMING, COLUMNS**					118
	0020						
	0100	4" x 4"	.041	L.F.	1.37	4.76	
	0150	4" x 6"	.058		2.05	6.90	
	0200	4" x 8"	.073		2.73	8.90	
	0250	6" x 6"	.074		3.90	11	
	0300	6" x 8"	.091		5.20	14.25	
	0350	6" x 10"	.107		6.50	17.35	
	9000	Minimum labor/equipment charge	4	Job		235	
127	0010	**FRAMING, TREATED LUMBER**					127
	0020	water-borne salt, c.c.a., a.c.a., wet, .40 p.c.f. retention					
	0100	2" x 4"		M.B.F.	765	1,300	
	0110	2" x 6"			765	1,300	
	0120	2" x 8"			805	1,375	
	0130	2" x 10"			960	1,625	
	0140	2" x 12"			980	1,675	
	0200	4" x 4"			1,425	2,450	
	0210	4" x 6"			1,425	2,450	
	0220	4" x 8"			1,425	2,450	
	0250	Add for .60 P.C.F. retention			40%		
	0260	Add for 2.5 P.C.F. retention			200%		
	0270	Add for K.D.A.T.			20%		
128	0010	**FRAMING, WALLS**					128
	0020						
	2002	Headers over openings, 2" x 6"	.044	L.F.	.61	3.66	
	2052	2" x 8"	.047		.85	4.22	
	2110	2" x 10"	.050		1.14	4.88	
	2152	2" x 12"	.053		1.54	5.75	
	2202	4" x 12"	.084		4.51	12.65	
	2252	6" x 12"	.114		7.80	20	
	5002	Plates, untreated, 2" x 3"	.019		.30	1.61	
	5022	2" x 4"	.020		.40	1.86	
	5041	2" x 6"	.021		.61	2.30	
	5122	Studs, 8' high wall, 2" x 3"	.013		.30	1.29	

061 | Rough Carpentry

061 100 | Wood Framing

			WORK-HOURS	UNIT	BARE COSTS MAT.	TOTAL INCL. O&P	
128	5142	2" x 4"	.015	L.F.	.40	1.54	128
	5162	2" x 6"	.016		.61	1.98	
	5182	3" x 4"	.020	↓	1.13	3.12	
	9000	Minimum labor/equipment charge	2	Job		114	
138	0010	PARTITIONS Wood stud with single bottom plate and					138
	0020	double top plate, no waste, std. & better lumber					
	0182	2" x 4" studs, 8' high, studs 12" O.C.	.200	L.F.	4.36	19.15	
	0202	16" O.C.	.160		3.56	15.45	
	0302	24" O.C.	.128		2.77	12.25	
	0382	10' high, studs 12" O.C.	.200		5.15	20.50	
	0402	16" O.C.	.160		4.16	16.45	
	0502	24" O.C.	.128		3.17	12.95	
	0582	12' high, studs 12" O.C.	.246		5.95	24.50	
	0602	16" O.C.	.200		4.75	19.85	
	0701	24" O.C.	.160	↓	3.56	15.45	
	0702						
	0782	2" x 6" studs, 8' high, studs 12" O.C.	.229	L.F.	6.65	25	
	0802	16" O.C.	.178		5.45	19.85	
	0902	24" O.C.	.139		4.23	15.45	
	0982	10' high, studs 12" O.C.	.229		7.85	27	
	1002	16" O.C.	.178		6.35	21.50	
	1102	24" O.C.	.139		4.84	16.45	
	1182	12' high, studs 12" O.C.	.291		9.05	32.50	
	1202	16" O.C.	.229		7.25	26	
	1302	24" O.C.	.178		5.45	19.85	
	1402	For horizontal blocking, 2" x 4", add	.027		.40	2.25	
	1502	2" x 6", add	.027		.61	2.60	
	1600	For openings, add	.064	↓		3.77	
	1702	Headers for above openings, material only, add		B.F.	635	1,075	
	9000	Minimum labor/equipment charge	2	Job		114	

061 150 | Sheathing

			WORK-HOURS	UNIT	BARE COSTS MAT.	TOTAL INCL. O&P	
154	0010	SHEATHING Plywood on roof, CDX					154
	0032	5/16" thick	.010	S.F.	.28	1.06	
	0052	3/8" thick	.010		.30	1.12	
	0102	1/2" thick	.011		.37	1.30	
	0202	5/8" thick	.012		.43	1.45	
	0302	3/4" thick	.013		.56	1.74	
	0502	Plywood on walls with exterior CDX, 3/8" thick	.013		.30	1.29	
	0602	1/2" thick	.014		.43	1.57	
	0702	5/8" thick	.015		.43	1.62	
	0802	3/4" thick	.016		.56	1.93	
	1200	For structural 1 exterior plywood, add			10%		
	1402	With boards, on roof 1" x 6" boards, laid horizontal	.022		.70	2.49	
	1502	Laid diagonal	.025		.80	2.81	
	1702	1" x 8" boards, laid horizontal	.018		.80	2.43	
	1802	Laid diagonal	.022		.80	2.66	
	2000	For steep roofs, add					
	2200	For dormers, hips and valleys, add			5%		
	2402	Boards on walls, 1" x 6" boards, laid regular	.025		.80	2.81	
	2502	Laid diagonal	.027		.80	2.97	
	2702	1" x 8" boards, laid regular	.021		.80	2.60	
	2802	Laid diagonal	.025		.80	2.81	
	2852	Gypsum, weatherproof, 1/2" thick	.015		.31	1.42	
	2902	Sealed, 4/10" thick	.015		.29	1.35	
	3000	Wood fiber, regular, no vapor barrier, 1/2" thick	.013	↓	.36	1.40	

061 | Rough Carpentry

061 150 | Sheathing

			WORK-HOURS	UNIT	BARE COSTS MAT.	TOTAL INCL. O&P	
154	3100	5/8" thick	.013	S.F.	.47	1.59	154
	3300	No vapor barrier, in colors, 1/2" thick	.013		.51	1.65	
	3400	5/8" thick	.013		.62	1.84	
	3600	With vapor barrier one side, white, 1/2" thick	.013		.50	1.63	
	3700	Vapor barrier 2 sides	.013		.78	2.12	
	3800	Asphalt impregnated, 25/32" thick	.013		.31	1.31	
	3850	Intermediate, 1/2" thick	.013	↓	.27	1.25	
	9000	Minimum labor/equipment charge	4	Job		227	

061 160 | Subfloor

			WORK-HOURS	UNIT	BARE COSTS MAT.	TOTAL INCL. O&P	
164	0012	SUBFLOOR Plywood, CDX, 1/2" thick	.011	SF Flr.	.36	1.25	164
	0102	5/8" thick	.012		.43	1.44	
	0202	3/4" thick	.013		.56	1.71	
	0302	1-1/8" thick, 2-4-1 including underlayment	.015		1.01	2.62	
	0502	With boards, 1" x 10" S4S, laid regular	.015		.97	2.50	
	0602	Laid diagonal	.018		.97	2.68	
	0802	1" x 8" S4S, laid regular	.016		.97	2.58	
	0902	Laid diagonal	.019		.97	2.74	
	1100	Wood fiber, T&G, 2' x 8' planks, 1" thick	.016		1.05	2.72	
	1200	1-3/8" thick	.018	↓	1.30	3.26	
	9000	Minimum labor/equipment charge	2	Job		117	

062 | Finish Carpentry

062 200 | Millwork Moldings

			WORK-HOURS	UNIT	BARE COSTS MAT.	TOTAL INCL. O&P	
220	0010	MOLDINGS, EXTERIOR					220
	1500	Cornice, boards, pine, 1" x 2"	.024	L.F.	.15	1.64	
	1700	1" x 6"	.040		.50	3.13	
	2000	1" x 12"	.044		1.62	5.30	
	2200	Three piece, built-up, pine, minimum	.100		1.23	7.80	
	2300	Maximum	.123		3.43	12.85	
	3000	Trim, exterior, sterling pine, corner board, 1" x 4"	.040		.31	2.80	
	3100	1" x 6"	.040		.53	3.17	
	3350	Fascia, 1" x 6"	.032		.53	2.72	
	3370	1" x 8"	.036		.59	3.04	
	3400	Moldings, back band	.032		.55	2.77	
	3500	Casing	.032		.30	2.33	
	3600	Crown	.032		1.01	3.54	
	3700	Porch rail with balusters	.364		6	30.50	
	3800	Screen	.020	↓	.22	1.52	
	3850						
	4100	Verge board, sterling pine, 1" x 4"	.040	L.F.	.31	2.80	
	4200	1" x 6"	.040		.53	3.17	
	4300	2" x 6"	.048		.88	4.26	
	4400	2" x 8"	.048		1.17	4.76	
	4700	For redwood trim, add		↓	200%		
	9000	Minimum labor/equipment charge	2	Job		114	
224	0010	MOLDINGS, TRIM					224
	0020						

062 | Finish Carpentry

062 200 | Millwork Moldings

			WORK-HOURS	UNIT	BARE COSTS MAT.	TOTAL INCL. O&P	
224	0200	Astragal, stock pine, 11/16" x 1-3/4"	.031	L.F.	.75	3.08	224
	0250	1-5/16" x 2-3/16"	.033		1.27	4.08	
	0800	Chair rail, stock pine, 5/8" x 2-1/2"	.030		.70	2.89	
	0900	5/8" x 3-1/2"	.033		1.30	4.12	
	1000	Closet pole, stock pine, 1-1/8" diameter	.040		.85	3.73	
	1100	Fir, 1-5/8" diameter	.040		1.10	4.15	
	3300	Half round, stock pine, 1/4" x 1/2"	.030		.40	2.37	
	3350	1/2" x 1"	.031	↓	.50	2.64	
	3400	Handrail, fir, single piece, stock, hardware not included					
	3450	1-1/2" x 1-3/4"	.100	L.F.	.90	7.25	
	3470	Pine, 1-1/2" x 1-3/4"	.100		.90	7.25	
	3500	1-1/2" x 2-1/2"	.105		1.25	8.15	
	3600	Lattice, stock pine, 1/4" x 1-1/8"	.030		.22	2.06	
	3700	1/4" x 1-3/4"	.032		.25	2.25	
	3800	Miscellaneous, custom, pine or cedar, 1" x 1"	.030		.15	1.95	
	3900	Nominal 1" x 3"	.033		.40	2.59	
	4100	Birch or oak, custom, nominal 1" x 1"	.033		.22	2.28	
	4200	Nominal 1" x 3"	.037		.60	3.14	
	4400	Walnut, custom, nominal 1" x 1"	.037		.35	2.72	
	4500	Nominal 1" x 3"	.040		.90	3.81	
	4700	Teak, custom, nominal 1" x 1"	.037		.80	3.48	
	4800	Nominal 1" x 3"	.040		2.10	5.85	
	4900	Quarter round, stock pine, 1/4" x 1/4"	.029		.15	1.93	
	4950	3/4" x 3/4"	.031	↓	.35	2.40	
	5600	Wainscot moldings, 1-1/8" x 9/16", 2' high, minimum	.105	S.F.	5.45	15.30	
	5700	Maximum	.123	"	11	26	
	9000	Minimum labor/equipment charge	2	Job		114	
228	0010	**MOLDINGS, WINDOW AND DOOR**					228
	0020						
	2800	Door moldings, stock, decorative, 1-1/8" wide, plain	.471	Set	25	69.50	
	2900	Detailed	.471	"	55	121	
	3150	Door trim set, 1 head and 2 sides, pine, 2-1/2 wide	1.356	Opng.	11	96	
	3170	4-1/2" wide	1.509	"	20	120	
	3200	Glass beads, stock pine, 1/4" x 11/16"	.028	L.F.	.25	2.04	
	3250	3/8" x 1/2"	.029		.30	2.18	
	3270	3/8" x 7/8"	.030		.35	2.30	
	4850	Parting bead, stock pine, 3/8" x 3/4"	.029		.25	2.10	
	4870	1/2" x 3/4"	.031		.30	2.30	
	5000	Stool caps, stock pine, 11/16" x 3-1/2"	.040		1.25	4.41	
	5100	1-1/16" x 3-1/4"	.053		2.50	7.30	
	5300	Threshold, oak, 3' long, inside, 5/8" x 3-5/8"	.250	Ea.	5.20	23	
	5400	Outside, 1-1/2" x 7-5/8"	.500	"	18	59.50	
	5900	Window trim sets, including casings, header, stops,					
	5910	stool and apron, 2-1/2" wide, minimum	.615	Opng.	13	57.50	
	5950	Average	.800		19	78	
	6000	Maximum	1.333	↓	30	127	
	9000	Minimum labor/equipment charge	2	Job		114	

062 400 | Plastic Laminate

408	0010	**COUNTER TOP** Stock, plastic lam., 24" wide w/backsplash, min.	.267	L.F.	4.75	23.50	408
	0100	Maximum	.320		16	45.50	
	0300	Custom plastic, 7/8" thick, aluminum molding, no splash	.267		15.25	41.50	
	0400	Cove splash	.267		19.90	49	
	0600	1-1/4" thick, no splash	.286		17.95	46.50	
	0700	Square splash	.286	↓	22.50	55	

062 | Finish Carpentry

062 400 | Plastic Laminate

			WORK-HOURS	UNIT	BARE COSTS MAT.	TOTAL INCL. O&P	
408	0900	Square edge, plastic face, 7/8" thick, no splash	.267	L.F.	19.20	47.50	408
	1000	With splash	.267		25	57.50	
	1200	For stainless channel edge, 7/8" thick, add			2.15	3.68	
	1300	1-1/4" thick, add			2.50	4.27	
	1500	For solid color suede finish, add			1.90	3.24	
	1700	For end splash, add		Ea.	12	20.50	
	1900	For cut outs, standard, add, minimum	.250		2.55	18.55	
	2000	Maximum	1		3.05	62.50	
	2100	Postformed, including backsplash and front edge	.267	L.F.	8.30	29.50	
	2110	Mitred, add	.667	Ea.		37.50	
	2200	Built-in place, 25" wide, plastic laminate	.320	L.F.	10.50	36.50	
	2300	Ceramic tile mosaic	.320		24	59.50	
	2500	Marble, stock, with splash, 1/2" thick, minimum	.471		29	76.50	
	2700	3/4" thick, maximum	.615		74	163	
	2900	Maple, solid, laminated, 1-1/2" thick, no splash	.286		30	67.50	
	3000	With square splash	.286		34	74.50	
	3200	Stainless steel	.333	S.F.	71	140	
	3400	Recessed cutting block with trim, 16" x 20" x 1"	1	Ea.	40	125	
	3600	Table tops, plastic laminate, square edge, 7/8" thick	.178	S.F.	6.80	22	
	3700	1-1/8" thick	.200	"	7	23.50	
	9000	Minimum labor/equipment charge	2.133	Job		122	

064 | Architectural Woodwork

064 300 | Stairwork & Handrails

			WORK-HOURS	UNIT	BARE COSTS MAT.	TOTAL INCL. O&P	
306	0011	STAIRS, PREFABRICATED					306
	0100	Box stairs, prefabricated, 3'-0" wide					
	0110	Oak treads, no handrails, 2' high	3.200	Flight	140	420	
	0200	4' high	4		295	730	
	0300	6' high	4.571		480	1,075	
	0400	8' high	5.333		600	1,325	
	0600	With pine treads for carpet, 2' high	3.200		110	370	
	0700	4' high	4		185	545	
	0800	6' high	4.571		280	745	
	0900	8' high	5.333		350	905	
	1100	For 4' wide stairs, add			25%		
	1500	Prefabricated stair rail with balusters, 5 risers	1.067	Ea.	160	335	
	1700	Basement stairs, prefabricated, soft wood,					
	1710	open risers, 3' wide, 8' high	4	Flight	350	825	
	1900	Open stairs, prefabricated prefinished poplar, metal stringers,					
	1910	treads 3'-6" wide, no railings					
	2000	3' high	3.200	Flight	400	865	
	2100	4' high	4		505	1,100	
	2200	6' high	4.571		880	1,750	
	2300	8' high	5.333		1,400	2,725	
	2500	For prefab. 3 piece wood railings & balusters, add for					
	2600	3' high stairs	1.067	Ea.	140	300	
	2700	4' high stairs	1.143		175	365	
	2800	6' high stairs	1.231		275	545	

064 | Architectural Woodwork

064 300 | Stairwork & Handrails

			WORK-HOURS	UNIT	BARE COSTS MAT.	TOTAL INCL. O&P	
306	2900	8' high stairs	1.333	Ea.	340	660	306
	3100	For 3'-6" x 3'-6" platform, add	4	↓	160	500	
	3300	Curved stairways, 3'-3" wide, prefabricated, oak, unfinished,					
	3310	incl. curved balustrade system, open one side					
	3400	9' high	22.857	Flight	6,000	11,600	
	3500	10' high	22.857		6,700	12,800	
	3700	Open two sides, 9' high	32		9,500	18,100	
	3800	10' high	32		10,200	19,200	
	4000	Residential, wood, oak treads, prefabricated	10.667		1,075	2,425	
	4200	Built in place	36.364	↓	1,275	4,225	
	4400	Spiral, oak, 4'-6" diameter, unfinished, prefabricated,					
	4500	incl. railing, 9' high	10.667	Flight	3,700	6,925	
	9000	Minimum labor/equipment charge	5.333	Job		305	
308	0010	**STAIR PARTS** Balusters, turned, 30" high, pine, minimum	.286	Ea.	4	23	308
	0100	Maximum	.308	"	7	29.50	
	0300	30" high birch balusters, minimum	.286	Ea.	6	26.50	
	0400	Maximum	.308		8	31	
	0600	42" high, pine balusters, minimum	.296		6	27	
	0700	Maximum	.320		8	31.50	
	0900	42" high birch balusters, minimum	.296		7	29	
	1000	Maximum	.320	↓	10	35.50	
	1050	Baluster, stock pine, 1-1/16" x 1-1/16"	.033	L.F.	.75	3.19	
	1100	1-5/8" x 1-5/8"	.036	"	1.40	4.46	
	1200	Newels, 3-1/4" wide, starting, minimum	1.143	Ea.	36	127	
	1300	Maximum	1.333		300	590	
	1500	Landing, minimum	1.600		70	210	
	1600	Maximum	2	↓	350	710	
	1800	Railings, oak, built-up, minimum	.133	L.F.	5	16.15	
	1900	Maximum	.145		15	34	
	2100	Add for sub rail	.073		4	10.95	
	2300	Risers, Beech, 3/4" x 7-1/2" high	.125		4.75	15.20	
	2400	Fir, 3/4" x 7-1/2" high	.125		1.35	9.45	
	2600	Oak, 3/4" x 7-1/2" high	.125		4.30	14.45	
	2800	Pine, 3/4" x 7-1/2" high	.121		1.35	9.20	
	2850	Skirt board, pine, 1" x 10"	.145		1.65	11.10	
	2900	1" x 12"	.154	↓	2	12.15	
	3000	Treads, oak, 1-1/16" x 9-1/2" wide, 3' long	.444	Ea.	22	63	
	3100	4' long	.471		29	76.50	
	3300	1-1/16" x 11-1/2" wide, 3' long	.444		28	73.50	
	3400	6' long	.571		56	128	
	3600	Beech treads, add		↓	40%		
	3800	For mitered return nosings, add		L.F.	2.60	4.44	
	9000	Minimum labor/equipment charge	2.667	Job		152	
310	0010	**RAILING** Custom design, architectural grade, hardwood, minimum	.211	L.F.	12	32.50	310
	0100	Maximum	.267	"	40	83	
	0300	Stock interior railing with spindles 6" O.C., 4' long	.200	L.F.	28	59.50	
	0400	8' long	.167	"	26	53.50	
	9000	Minimum labor/equipment charge	2.667	Job		152	

071 | Waterproofing and Dampproofing

071 600 | Bitum. Dampproofing

			WORK-HOURS	UNIT	BARE COSTS MAT.	TOTAL INCL. O&P	
602	0010	**BITUMINOUS ASPHALT COATING** For foundation					602
	0030	Brushed on, below grade, 1 coat	.012	S.F.	.04	.73	
	0100	2 coat	.016		.09	1.05	
	0300	Sprayed on, below grade, 1 coat, 25.6 S.F./gal.	.010		.11	.73	
	0400	2 coat, 20.5 S.F./gal.	.016		.13	1.11	
	0600	Troweled on, asphalt with fibers, 1/16" thick	.016		.12	1.10	
	0700	1/8" thick	.020		.25	1.54	
	1000	1/2" thick	.023		.95	2.91	
	9000	Minimum labor/equipment charge	2.667	Job		149	

073 | Shingles and Roofing Tiles

073 100 | Shingles

			WORK-HOURS	UNIT	BARE COSTS MAT.	TOTAL INCL. O&P	
108	0010	**WOOD** 16" No. 1 red cedar shingles, 5" exposure, on roof	3.200	Sq.	155	445	108
	0200	7-1/2" exposure, on walls	3.902		111	410	
	0300	18" No. 1 red cedar perfections, 5 1/2" exposure, on roof	2.909		163	445	
	0600	Resquared, and rebutted, 5-1/2" exposure, on roof	2.667		168	440	
	0900	7-1/2" exposure, on walls	3.265		133	415	
	1000	Add to above for fire retardant shingles, 16" long			40	68	
	1050	18" long			40	68	
	1060	Preformed ridge shingles	.020	L.F.	1.70	4.04	
	1100	Hand-split red cedar shakes, 1/2" thick x 24" long, 10" exp. on roof	3.200	Sq.	130	405	
	1110	3/4" thick x 24" long	3.556		156	470	
	1200	1/2" thick, 18" long, 8-1/2" exposure	4		102	400	
	1210	3/4" thick x 18" long, 8 1/2" exp. on roof	4.444		131	475	
	1700	Add to above for fire retardant shakes, 24" long			42	71.50	
	1800	18" long			42	71.50	
	1810	Ridge shakes	.023	L.F.	1.65	4.13	
	2000	White cedar shingles, 16" long, extras, 5" exposure, on roof	3.333	Sq.	88	340	
	2100	7-1/2" exposure, on walls	4		63	335	
	2300	For 15# organic felt underlayment on roof, 1 layer, add	.125		2.40	11.20	
	2400	2 layers, add	.250		4.80	22.50	
	2700	Panelized systems, No.1 cedar shingles on 5/16" CDX plywood					
	2800	On walls, 8' strips, 7" or 14" exposure	.023	S.F.	2.12	4.96	
	2900	Matching flush corners	.040	L.F.	2.07	5.90	
	3500	On roofs, 8' strips, 7" or 14" exposure	2.667	Sq.	174	450	
	3600	Matching lap corners	.040	L.F.	1.10	4.15	
	3700	Matching rake corners	.040		1.40	4.66	
	3800	Matching valley sheets	.040		5.70	11.95	
	9000	Minimum labor/equipment charge	2.667	Job		152	

079 | Joint Sealers

079 204 | Sealants & Caulkings

		WORK-HOURS	UNIT	BARE COSTS MAT.	TOTAL INCL. O&P
0010	**CAULKING AND SEALANTS**				
0020	Acoustical sealant, elastomeric		Gal.	25.50	43.50
0032	Backer rod, polyethylene, 1/4" diameter	.017	L.F.	.02	1.05
0052	1/2" diameter	.017		.03	1.06
0072	3/4" diameter	.017		.05	1.10
0092	1" diameter	.017	↓	.07	1.13
0100	Caulking compound, oil base, bulk				
0200	Brilliant white color		Gal.	11	18.75
0300	Aluminum pigment and other colors		"	14	24
0500	Bulk, in place, 1/4" x 1/2", 154 L.F./gal.	.031	L.F.	.07	1.90
0600	1/2" x 1/2", 77 L.F./gal.	.032		.14	2.08
0800	3/4" x 3/4", 34 L.F./gal.	.035		.32	2.56
0900	3/4" x 1", 26 L.F./gal.	.040		.43	3.06
1000	1" x 1", 19 L.F./gal.	.044	↓	.59	3.58
1400	Butyl based, bulk		Gal.	16.20	27.50
1500	Cartridges		"	24.50	42
1700	Bulk, in place 1/4" x 1/2", 154 L.F./gal.	.035	L.F.	.11	2.21
1800	1/2" x 1/2", 77 L.F./gal.	.044	"	.21	2.93
2000	Latex acrylic based, bulk		Gal.	21	35.50
2100	Cartridges		"	24.50	42
2200	Bulk in place, 1/4" x 1/2", 154 L.F./gal.	.035	L.F.	.14	2.25
2250					
2300	Polysulfide compounds, 1 component, bulk		Gal.	44	75
2400	Cartridges		"	48	82
2600	1 or 2 component, in place, 1/4" x 1/4", 308 L.F./gal.	.055	L.F.	.14	3.43
2700	1/2" x 1/4", 154 L.F./gal.	.059		.29	3.92
2900	3/4" x 3/8", 68 L.F./gal.	.062		.65	4.67
3000	1" x 1/2", 38 L.F./gal.	.062	↓	1.16	5.55
3200	Polyurethane, 1 or 2 component, bulk		Gal.	41	70
3300	Cartridges		"	44.50	76
3500	1 or 2 component, in place, 1/4" x 1/4", 308 L.F./gal.	.053	L.F.	.13	3.30
3600	1/2" x 1/4", 154 L.F./gal.	.055		.26	3.65
3800	3/4" x 3/8", 68 L.F./gal.	.062		.59	4.58
3900	1" x 1/2", 38 L.F./gal.	.073	↓	1.05	6
4100	Silicone rubber, bulk		Gal.	33	56.50
4200	Cartridges		"	38.50	66
4300	Bulk in place, 1/4" x 1/2", 154 L.F./gal.	.034	L.F.	.24	2.38
4350					
4400	Neoprene gaskets, closed cell, adhesive, 1/8" x 3/8"	.033	L.F.	.18	2.24
4500	1/4" x 3/4"	.037		.45	2.94
4700	1/2" x 1"	.040		1.30	4.54
4800	3/4" x 1-1/2"	.048	↓	2.70	7.40
5500	Resin epoxy coating, 2 component, heavy duty		Gal.	25	42.50
5802	Tapes, sealant, P.V.C. foam adhesive, 1/16" x 1/4"		L.F.	.05	.09
5902	1/16" x 1/2"			.07	.12
5952	1/16" x 1"			.11	.19
6002	1/8" x 1/2"		↓	.08	.14
6200	Urethane foam, 2 component, handy pack, 0.75 C.F.		Ea.	19.50	33.50
6300	50.0 C.F. pack		C.F.	14	24
9000	Minimum labor/equipment charge	2	Job		116

081 | Metal Doors and Frames

081 100 | Steel Doors & Frames

			WORK-HOURS	UNIT	BARE COSTS MAT.	TOTAL INCL. O&P	
103	0010	**COMMERCIAL STEEL DOORS** Flush, full panel					103
	0020	Hollow metal 1-3/8" thick, 20 ga., 2'-0" x 6'-8"	.800	Ea.	130	267	
	0040	2'-8" x 6'-8"	.889		140	290	
	0060	3'-0" x 6'-8"	.941		147	305	
	0100	3'-0" x 7'-0"	.941		146	305	
	0120	For vision lite, add			44	75	
	0140	For narrow lite, add			49	83.50	
	0160	For bottom louver, add			113	194	
	0230	For baked enamel finish, add			30%		
	0260	For galvanizing, add			15%		
	0320	Half glass, 20 ga., 2'-0" x 6'-8"	.800	Ea.	187	365	
	0340	2'-8" x 6'-8"	.889		195	385	
	0360	3'-0" x 6'-8"	.941		207	410	
	0400	3'-0" x 7'-0"	.941		211	415	
	1020	Hollow core, 1-3/4" thick, full panel, 20 ga., 2'-8" x 6'-8"	.889		166	335	
	1040	3'-0" x 6'-8"	.941		176	355	
	1060	3'-0" x 7'-0"	.941		183	365	
	1080	4'-0" x 7'-0"	1.067		228	450	
	1100	4'-0" x 8'-0"	1.231		267	530	
	1120	18 ga., 2'-8" x 6'-8"	.941		187	375	
	1140	3'-0" x 6'-8"	1		194	390	
	1160	3'-0" x 7'-0"	1		202	405	
	1180	4'-0" x 7'-0"	1.143		252	500	
	1200	4'-0" x 8'-0"	1.143		330	625	
	1220	Half glass, 20 ga., 2'-8" x 6'-8"	.800		215	415	
	1240	3'-0" x 6'-8"	.889		221	430	
	1260	3'-0" x 7'-0"	.889		237	455	
	1280	4'-0" x 7'-0"	1		287	545	
	1300	4'-0" x 8'-0"	1.231		330	630	
	1320	18 ga., 2'-8" x 6'-8"	.889		245	470	
	1340	3'-0" x 6'-8"	.941		253	485	
	1360	3'-0" x 7'-0"	.941		261	500	
	1380	4'-0" x 7'-0"	1.067		315	600	
	1400	4'-0" x 8'-0"	1.143		360	675	
	1720	Insulated, 1-3/4" thick, full panel, 18 ga., 3'-0" x 6'-8"	1.067		210	420	
	1740	2'-8" x 7'-0"	1		214	425	
	1760	3'-0" x 7'-0"	1.067		213	425	
	1800	4'-0" x 8'-0"	1.231		310	600	
	1820	Half glass, 18 ga., 3'-0" x 6'-8"	1		272	520	
	1840	2'-8" x 7'-0"	.941		275	530	
	1860	3'-0" x 7'-0"	1		281	540	
	1900	4'-0" x 8'-0"	1.143		375	700	
	9000	Minimum labor/equipment charge	2	Job		117	
118	0010	**STEEL FRAMES, KNOCK DOWN** 18 ga., up to 5-3/4" deep					118
	0020						
	0025	6'-8" high, 3'-0" wide, single	1	Ea.	65.50	170	
	0040	6'-0" wide, double	1.143		85.50	213	
	0100	7'-0" high, 3'-0" wide, single	1		66.50	172	
	0140	6'-0" wide, double	1.143		87	216	
	2800	18 ga. drywall, up to 4-7/8" deep, 7'-0" high, 3'-0" wide, single	1		68.50	175	
	2840	6'-0" wide, double	1.143		86.50	214	
	3600	16 ga., up to 5-3/4" deep, 7'-0" high, 4'-0" wide, single	1.067		73.50	188	
	3640	8'-0" wide, double	1.333		101	250	
	3700	8'-0" high, 4'-0" wide, single	1.067		85	208	
	3740	8'-0" wide, double	1.333		114	272	

081 | Metal Doors and Frames

081 100 | Steel Doors & Frames

		WORK-HOURS	UNIT	BARE COSTS MAT.	TOTAL INCL. O&P	
118	4000 6-3/4" deep, 7'-0" high, 4'-0" wide, single	1.067	Ea.	80.50	201	118
	4040 8'-0" wide, double	1.333		107	261	
	4100 8'-0" high, 4'-0" wide, single	1.067		94	222	
	4140 8'-0" wide, double	1.333		115	274	
	4400 8-3/4" deep, 7'-0" high, 4'-0" wide, single	1.067		108	247	
	4440 8'-0" wide, double	1.333		139	315	
	4500 8'-0" high, 4'-0" wide, single	1.067		123	272	
	4540 8'-0" wide, double	1.333		153	340	
	4800 16 ga. drywall, up to 3-7/8" deep, 7'-0" high, 3'-0" wide, single	1		72.50	182	
	4840 6'-0" wide, double	1.143		92.50	225	
	4900 For welded frames, add		↓	27	46	
	4902					
	5400 16 ga. "B" label, up to 5-3/4" deep, 7'-0" high, 4'-0" wide, single	1.067	Ea.	93	221	
	5440 8'-0" wide, double	1.333		119	281	
	5800 6-3/4" deep, 7'-0" high, 4'-0" wide, single	1.067		96.50	227	
	5840 8'-0" wide, double	1.333		133	305	
	6200 8-3/4" deep, 7'-0" high, 4'-0" wide, single	1.067		125	275	
	6240 8'-0" wide, double	1.333	↓	158	350	
	6300 For "A" label use same price as "B" label					
	6400 For baked enamel finish, add			30%		
	6500 For galvanizing, add			15%		
	7900 Transom lite frames, fixed, add	.103	S.F.	16.50	34.50	
	8000 Movable, add	.123	"	20	41	
	9000 Minimum labor/equipment charge	2	Job		117	

082 | Wood and Plastic Doors

082 050 | Wood & Plastic Doors

		WORK-HOURS	UNIT	BARE COSTS MAT.	TOTAL INCL. O&P	
054	0010 WOOD FRAMES					054
	0400 Exterior frame, incl. ext. trim, pine, 5/4 x 4-9/16" deep	.043	L.F.	3.48	8.45	
	0420 5-3/16" deep	.043		3.78	8.95	
	0440 6-9/16" deep	.043		4.42	10	
	0600 Oak, 5/4 x 4-9/16" deep	.046		5.95	12.85	
	0620 5-3/16" deep	.046		6.40	13.65	
	0640 6-9/16" deep	.046		7.55	15.55	
	0800 Walnut, 5/4 x 4-9/16" deep	.046		7.75	15.95	
	0820 5-3/16" deep	.046		10.05	19.85	
	0840 6-9/16" deep	.046		11.80	23	
	1000 Sills, 8/4 x 8" deep, oak, no horns	.160		9.75	26	
	1020 2" horns	.160		10.45	27	
	1040 3" horns	.160		11.35	28.50	
	1100 8/4 x 10" deep, oak, no horns	.178		12.65	32.50	
	1120 2" horns	.178		13.85	34	
	1140 3" horns	.178	↓	15.15	36.50	
	1200 For casing, see division 062-212					
	1220					
	2000 Exterior, colonial, frame & trim, 3' opng., in-swing, minimum	.727	Ea.	240	450	
	2020 Maximum	.800		500	905	
	2100 5'-4" opening, in-swing, minimum	.941		415	765	
	2120 Maximum	1.067	↓	775	1,375	

082 | Wood and Plastic Doors

082 050 | Wood & Plastic Doors

			WORK-HOURS	UNIT	BARE COSTS MAT.	TOTAL INCL. O&P	
054	2140	Out-swing, minimum	.941	Ea.	440	810	054
	2160	Maximum	1.067		805	1,425	
	2400	6'-0" opening, in-swing, minimum	1		440	810	
	2420	Maximum	1.600		805	1,475	
	2460	Out-swing, minimum	1		545	985	
	2480	Maximum	1.600		995	1,800	
	2600	For two sidelights, add, minimum	.533	Opng.	155	296	
	2620	Maximum	.800	"	820	1,450	
	2700	Custom birch frame, 3'-0" opening	1	Ea.	148	310	
	2750	6'-0" opening	1	"	210	415	
	3000	Interior frame, pine, 11/16" x 3-5/8" deep	.043	L.F.	3.20	8	
	3020	4-9/16" deep	.043		3.40	8.30	
	3040	5-3/16" deep	.043		3.85	9.10	
	3200	Oak, 11/16" x 3-5/8" deep	.046		4.14	9.75	
	3220	4-9/16" deep	.046		4.14	9.75	
	3240	5-3/16" deep	.046		4.26	9.90	
	3400	Walnut, 11/16" x 3-5/8" deep	.046		6.75	14.25	
	3420	4-9/16" deep	.046		6.75	14.25	
	3440	5-3/16" deep	.046		7	14.65	
	3600	Pocket door frame	1	Ea.	79	193	
	3800	Threshold, oak, 5/8" x 3-5/8" deep	.080	L.F.	2.16	8.40	
	3820	4-5/8" deep	.084		2.80	9.80	
	3840	5-5/8" deep	.089		3.34	10.90	
	4020	For casing, see division 062-212					
	9000	Minimum labor/equipment charge	2	Job		117	
062	0010	**WOOD DOOR, ARCHITECTURAL** Flush, interior, 7 ply, hollow core,					062
	0020	Lauan face, 2'-0" x 6'-8"	.941	Ea.	29	104	
	0040	2'-6" x 6'-8"	.941		31.50	108	
	0080	3'-0" x 6'-8"	.941		35	114	
	0100	4'-0" x 6'-8"	1		57	156	
	0120	Birch face, 2'-0" x 6'-8"	.941		37.50	118	
	0140	2'-6" x 6'-8"	.941		39	121	
	0180	3'-0" x 6'-8"	.941		40.50	124	
	0200	4'-0" x 6'-8"	1		64	167	
	0220	Oak face, 2'-0" x 6'-8"	.941		39	121	
	0240	2'-6" x 6'-8"	.941		43	128	
	0280	3'-0" x 6'-8"	.941		48.50	137	
	0300	4'-0" x 6'-8"	1		74.50	185	
	0320	Walnut face, 2'-0" x 6'-8"	.941		106	235	
	0340	2'-6" x 6'-8"	.941		106	235	
	0380	3'-0" x 6'-8"	.941		123	266	
	0400	4'-0" x 6'-8"	1		166	340	
	0420	For 7'-0" high, add			9.25	15.75	
	0440	For 8'-0" high, add			22.50	38.50	
	0460	For 8'-0" high walnut, add			27.50	47.50	
	0480	For prefinishing, clear, add			26.50	45	
	0500	For prefinishing, stain, add			29.50	50.50	
	0600						
	1320	M.D. overlay on hardboard, 2'-0" x 6'-8"	.941	Ea.	72.50	179	
	1340	2'-6" x 6'-8"	.941		75.50	184	
	1380	3'-0" x 6'-8"	.941		79	190	
	1400	4'-0" x 6'-8"	1		103	235	
	1420	For 7'-0" high, add			5.40	9.25	
	1440	For 8'-0" high, add			13.55	23	
	1720	H.P. plastic laminate, 2'-0" x 6'-8"	1		109	244	

082 | Wood and Plastic Doors

082 050 | Wood & Plastic Doors

		WORK-HOURS	UNIT	BARE COSTS MAT.	TOTAL INCL. O&P	
062	1740	2'-6" x 6'-8"	1	Ea.	109	244	062
	1780	3'-0" x 6'-8"	1.067		123	273	
	1800	4'-0" x 6'-8"	1.143		174	365	
	1820	For 7'-0" high, add			4.40	7.50	
	1840	For 8'-0" high, add			14.30	24.50	
	2020	5 ply particle core, lauan face, 2'-6" x 6'-8"	1.067		53.50	154	
	2040	3'-0" x 6'-8"	1.143		60	169	
	2080	3'-0" x 7'-0"	1.231		58	172	
	2100	4'-0" x 7'-0"	1.333		72.50	202	
	2120	Birch face, 2'-6" x 6'-8"	1.067		58.50	163	
	2140	3'-0" x 6'-8"	1.143		65	178	
	2180	3'-0" x 7'-0"	1.231		63.50	181	
	2200	4'-0" x 7'-0"	1.333		88.50	230	
	2220	Oak face, 2'-6" x 6'-8"	1.067		60.50	166	
	2240	3'-0" x 6'-8"	1.143		70	186	
	2280	3'-0" x 7'-0"	1.231		77	204	
	2300	4'-0" x 7'-0"	1.333		107	261	
	2320	Walnut face, 2'-0" x 6'-8"	1.067		122	270	
	2340	2'-6" x 6'-8"	1.143		122	275	
	2380	3'-0" x 6'-8"	1.231		143	315	
	2400	4'-0" x 6'-8"	1.333		190	400	
	2440	For 8'-0" high, add			19.80	34	
	2460	For 8'-0" high walnut, add			33	56.50	
	2480	For solid wood core, add			27.50	47.50	
	2720	For prefinishing, clear, add			36.50	62	
	2740	For prefinishing, stain, add		↓	53	90	
	2750						
	3320	M.D. overlay on hardboard, 2'-6" x 6'-8"	1.143	Ea.	81.50	206	
	3340	3'-0" x 6'-8"	1.231		90	226	
	3380	3'-0" x 7'-0"	1.333		95.50	241	
	3400	4'-0" x 7'-0"	1.600		118	294	
	3440	For 8'-0" height, add			13.20	22.50	
	3460	For solid wood core, add			26.50	45	
	3720	H.P. plastic laminate, 2'-6" x 6'-8"	1.231		125	286	
	3740	3'-0" x 6'-8"	1.333		143	320	
	3780	3'-0" x 7'-0"	1.455		144	330	
	3800	4'-0" x 7'-0"	2		202	460	
	3840	For 8'-0" height, add			16.50	28	
	3860	For solid wood core, add			24	41	
	4000	Exterior, flush, solid wood core, birch, 1-3/4" x 7'-0" x 2'-6"	1.067		83	204	
	4020	2'-8" wide	1.067		85	208	
	4040	3'-0" wide	1.143		89.50	220	
	4100	Oak faced 1-3/4" x 7'-0" x 2'-6" wide	1.067		103	239	
	4120	2'-8" wide	1.067		109	249	
	4140	3'-0" wide	1.143		113	261	
	4200	Walnut faced 1-3/4" x 7'-0" x 2'-6" wide	1.067		169	350	
	4220	2'-8" wide	1.067		177	365	
	4240	3'-0" wide	1.143		190	390	
	4300	For 6'-8" high door, deduct from 7'-0" door		↓	11	18.75	
	9000	Minimum labor/equipment charge	2	Job		117	

285

084 | Entrances and Storefronts

084 100 | Aluminum

		Description	WORK-HOURS	UNIT	BARE COSTS MAT.	TOTAL INCL. O&P	
105	0010	STOREFRONT SYSTEMS Aluminum frame, clear 3/8" plate glass,					105
	0020	incl. 3' x 7' door with hardware (400 sq. ft. max. wall)					
	0500	Wall height to 12' high, commercial grade	.107	S.F.	11	24.50	
	0600	Institutional grade	.123		14.60	32	
	0700	Monumental grade	.139		21	43.50	
	1000	6' x 7' door with hardware, commercial grade	.119		11.25	25.50	
	1100	Institutional grade	.139		15.40	34	
	1200	Monumental grade	.160		21.50	45	
	1500	For bronze anodized finish, add			18%		
	1600	For black anodized finish, add			36%		
	1700	For stainless steel framing, add to monumental		↓	75%		
	2000	For individual doors see division 081					
	9000	Minimum labor/equipment charge	16	Job		880	
107	0010	SWING DOORS Alum. entrance, 6' x 7', incl. hdwre & oper.	22.857	Opng.	4,000	8,450	107
	0020	For anodized finish, add		"	200	340	
	9000	Minimum labor/equipment charge	16	Job		1,150	

084 300 | Stainless Steel

		Description	WORK-HOURS	UNIT	BARE COSTS MAT.	TOTAL INCL. O&P	
301	0010	STAINLESS STEEL AND GLASS Entrance unit, narrow stiles					301
	0020	3' x 7' opening, including hardware, minimum	10	Opng.	3,500	6,675	
	0050	Average	11.429		3,600	6,950	
	0100	Maximum	13.333		4,100	7,950	
	1000	For solid bronze entrance units, statuary finish, add			60%		
	1100	Without statuary finish, add		↓	45%		
	2000	Balanced doors, 3' x 7', economy	17.778	Ea.	5,025	9,850	
	2100	Premium	22.857	"	8,650	16,400	
	9000	Minimum labor/equipment charge	8	Job		575	

084 600 | Automatic Doors

		Description	WORK-HOURS	UNIT	BARE COSTS MAT.	TOTAL INCL. O&P	
602	0010	SLIDING ENTRANCE 12' x 7'-6" opng., 5' x 7' door, 2 way traf.,					602
	0020	mat activated, panic pushout, incl. operator & hardware,					
	0030	not including glass or glazing	22.857	Opng.	5,100	9,975	
	9000	Minimum labor/equipment charge	22.857	Job		1,275	

087 | Hardware

087 100 | Finish Hardware

		Description	WORK-HOURS	UNIT	BARE COSTS MAT.	TOTAL INCL. O&P	
108	0010	DEADLOCKS Mortise, heavy duty, outside key	.889	Ea.	94	211	108
	0020	Double cylinder	.889		105	231	
	0100	Medium duty, outside key	.800		73.50	171	
	0110	Double cylinder	.800		93	204	
	1000	Tubular, standard duty, outside key	.800		40.50	115	
	1010	Double cylinder	.800		51.50	133	
	1200	Night latch, outside key	.800	↓	50.50	132	
110	0010	DOORSTOPS Holder and bumper, floor or wall	.333	Ea.	12.10	39.50	110
	1300	Wall bumper	.333	↓	4.28	26.50	

087 | Hardware

087 100 | Finish Hardware

			WORK-HOURS	UNIT	BARE COSTS MAT.	TOTAL INCL. O&P	
110	1600	Floor bumper, 1" high	.333	Ea.	3.54	25	110
	1900	Plunger type, door mounted	.333	↓	22	57	
	9000	Minimum labor/equipment charge	1.333	Job		76	
112	0010	**ENTRANCE LOCKS** Cylinder, grip handle, deadlocking latch	.889	Ea.	93	209	112
	0020	Deadbolt	1		111	246	
	0100	Push and pull plate, dead bolt	1		103	232	
	0900	For handicapped lever, add		↓	117	200	
116	0011	**HINGES, MAT'L. ONLY**, full mort., avg.freq., stl.base, 4-1/2"x 4-1/2",USP					116
	0100	5" x 5", USP		Pr.	23.50	40.50	
	0200	6" x 6", USP			48.50	83	
	0400	Brass base, 4-1/2" x 4-1/2", US10			30	51	
	0500	5" x 5", US10			42	71.50	
	0600	6" x 6", US10			70	119	
	0800	Stainless steel base, 4-1/2" x 4-1/2", US32		↓	50	85.50	
	0900	For non removable pin, add		Ea.	2	3.41	
	0910	For floating pin, driven tips, add			3.75	6.40	
	0930	For hospital type tip on pin, add			11	18.75	
	0940	For steeple type tip on pin, add		↓	7.35	12.55	
	1000	Full mortise, high frequency, steel base, 4-1/2" x 4-1/2", USP		Pr.	36.50	62	
	1100	5" x 5", USP			42	71.50	
	1200	6" x 6", USP			78	133	
	1400	Brass base, 4-1/2" x 4-1/2", US10			54	92.50	
	1500	5" x 5", US10			63	108	
	1600	6" x 6", US10			100	171	
	1800	Stainless steel base, 4-1/2" x 4-1/2", US32		↓	86	147	
	1930	For hospital type tip on pin, add		Ea.	14.45	24.50	
	2000	Full mortise low frequency, steel base, 4-1/2" x 4-1/2", USP		Pr.	6.60	11.25	
	2100	5" x 5", USP			18.60	32	
	2200	6" x 6", USP			34	58	
	2400	Brass base, 4-1/2" x 4-1/2", US10			25	42.50	
	2500	5" x 5", US10			37	63	
	2800	Stainless steel base, 4-1/2" x 4-1/2", US32		↓	41	70	
120	0010	**LOCKSET** Standard duty, cylindrical, with sectional trim					120
	0020	Non-keyed, passage	.667	Ea.	32.50	93	
	0100	Privacy	.667		39.50	105	
	0400	Keyed, single cylinder function	.800		63	153	
	0420	Hotel	1		78.50	190	
	1000	Heavy duty with sectional trim, non-keyed, passages	.667		94.50	200	
	1100	Privacy	.667		119	242	
	1400	Keyed, single cylinder function	.800		141	286	
	1420	Hotel	1		210	415	
	1600	Communicating	.800		163	325	
	1690	For re-core cylinder, add			23.50	40.50	
	1700	Residential, interior door, minimum	.500		11.35	48	
	1720	Maximum	1		30.50	109	
	1800	Exterior, minimum	.571		26	77	
	1820	Maximum	1	↓	106	238	
	9000	Minimum labor/equipment charge	1.333	Job		76	
125	0010	**MORTISE LOCKSET** Comm., wrought knobs & full escutcheon trim					125
	0020	Non-keyed, passage, minimum	.889	Ea.	120	256	
	0030	Maximum	1		194	390	
	0040	Privacy, minimum	.889		128	270	
	0050	Maximum	1		210	415	
	0100	Keyed, office/entrance/apartment, minimum	1	↓	147	310	

087 | Hardware

087 100 | Finish Hardware

			WORK-HOURS	UNIT	BARE COSTS MAT.	TOTAL INCL. O&P	
125	0110	Maximum	1.143	Ea.	252	500	125
	0120	Single cylinder, typical, minimum	1		126	272	
	0130	Maximum	1.143		233	460	
	0200	Hotel, minimum	1.143		156	330	
	0210	Maximum	1.333		254	510	
	0300	Communication, double cylinder, minimum	1		156	325	
	0310	Maximum	1.143		202	410	
	1000	Wrought knobs and sectional trim, non-keyed, passage, minimum	.800		82	185	
	1010	Maximum	.889		162	325	
	1040	Privacy, minimum	.800		96	210	
	1050	Maximum	.889		173	345	
	1100	Keyed, entrance,office/apartment, minimum	.889		142	293	
	1110	Maximum	1		206	410	
	1120	Single cylinder, typical, minimum	.889		137	285	
	1130	Maximum	1		200	395	
	2000	Cast knobs and full escutcheon trim					
	2010	Non-keyed, passage, minimum	.889	Ea.	174	345	
	2020	Maximum	1		284	540	
	2040	Privacy, minimum	.889		210	410	
	2050	Maximum	1		305	575	
	2120	Keyed, single cylinder, typical, minimum	1		210	415	
	2130	Maximum	1.143		330	630	
	2200	Hotel, minimum	1.143		252	500	
	2210	Maximum	1.333		420	790	
	2220						
	3000	Cast knob and sectional trim, non-keyed, passage, minimum	.800	Ea.	132	270	
	3010	Maximum	.800		263	490	
	3040	Privacy, minimum	.800		150	300	
	3050	Maximum	.800		263	490	
	3100	Keyed, office/entrance/apartment, minimum	.889		175	350	
	3110	Maximum	.889		276	525	
	3120	Single cylinder, typical, minimum	.889		175	350	
	3130	Maximum	.889		330	615	
	3190	For re-core cylinder, add			23	39.50	
	3200						
	4000	Keyless, pushbutton type, with deadbolt, standard	.889	Ea.	81	189	
	4100	Heavy duty	.889		121	257	
	4150	Card type, 1 time zone, minimum			265	455	
	4200	Maximum			900	1,525	
	4250	3 time zones, minimum			685	1,175	
	4300	Maximum			1,575	2,675	
	4350	System with printer, and control console, 3 zones		Total	7,675	13,100	
	4400	6 zones		"	10,100	17,200	
	4450	For each door, minimum, add		Ea.	1,125	1,925	
	4500	Maximum, add		"	1,675	2,875	
	4510						
127	0010	**PANIC DEVICE** For rim locks, single door, exit only	1.333	Ea.	269	535	127
	0020	Outside key and pull	1.600		320	635	
	0200	Bar and vertical rod, exit only	1.600		410	790	
	0210	Outside key and pull	2		490	945	
	0400	Bar and concealed rod	2		445	875	
	0600	Touch bar, exit only	1.333		310	605	
	0610	Outside key and pull	1.600		395	765	
	0700	Touch bar and vertical rod, exit only	1.600		445	850	
	0710	Outside key and pull	2		520	1,000	
	1000	Mortise, bar, exit only	2		335	685	

087 | Hardware

087 100 | Finish Hardware

			WORK-HOURS	UNIT	BARE COSTS MAT.	TOTAL INCL. O&P	
127	1600	Touch bar, exit only	2	Ea.	420	825	127
	2000	Narrow stile, rim mounted, bar, exit only	1.333		440	830	
	2010	Outside key and pull	1.600		480	905	
	2200	Bar and vertical rod, exit only	1.600		440	845	
	2210	Outside key and pull	2		520	1,000	
	2400	Bar and concealed rod, exit only	2.667		495	995	
	3000	Mortise, bar, exit only	2		345	705	
	3600	Touch bar, exit only	2	↓	505	975	
	4000	Double doors, exit only	4	Pr.	565	1,200	
	4500	Exit & entrance	4	"	595	1,250	
	9000	Minimum labor/equipment charge	3.200	Job		182	
129	0010	PUSH-PULL Push plate, pull plate, aluminum	.667	Ea.	52	126	129
	0500	Bronze	.667		65	149	
	1500	Pull handle and push bar, aluminum	.727		96.50	206	
	2000	Bronze	.800		126	261	
	3000	Push plate both sides, aluminum	.571		11.90	53	
	3500	Bronze	.615		30	86	
	4000	Door pull, designer style, cast aluminum, minimum	.667		55.50	132	
	5000	Maximum	1		259	500	
	6000	Cast bronze, minimum	.667		60	140	
	7000	Maximum	1		280	535	
	8000	Walnut, minimum	.667		46	116	
	9000	Maximum	1	↓	257	500	
	9800	Minimum labor/equipment charge	1.600	Job		91	
132	0011	SPECIAL HINGES, MATERIAL ONLY, Paumelle, high frequency					132
	0020	Steel base, 6" x 4-1/2", US10		Pr.	100	171	
	0100	Bronze base, 5" x 4-1/2", US10			124	212	
	0200	Paumelle, average frequency, steel base, 4-1/2" x 3-1/2", US10			104	177	
	0400	Olive knuckle, low frequency, bronze base, 6" x 4-1/2", US10		↓	108	185	
	1000	Electric hinge with concealed conductor, average frequency					
	1010	Steel base, 4-1/2" x 4-1/2", US26D		Pr.	198	340	
	1100	Bronze base, 4-1/2" x 4-1/2", US26D		"	208	355	
	1200	Electric hinge with concealed conductor, high frequency					
	1210	Steel base, 4-1/2" x 4-1/2", US26D		Pr.	148	253	
	1600	Double weight, 800 lb., steel base, removable pin, 5" x 6", USP			100	171	
	1700	Steel base-welded pin, 5" x 6", USP			111	189	
	1800	Triple weight, 2000 lb., steel base, welded pin, 5" x 6", USP			115	197	
	2000	Pivot reinf., high frequency, steel base, 7-3/4" door plate, USP			108	185	
	2200	Bronze base, 7-3/4" door plate, US10		↓	155	265	
	3000	Swing clear, full mortise, full or half surface, high frequency,					
	3010	Steel base, 5" high, USP		Pr.	91	155	
	3200	Swing clear, full mortise, average frequency					
	3210	Steel base, 4-1/2" high, USP		Pr.	72	123	
	4000	Wide throw, average frequency, steel base, 4-1/2" x 6", USP			84	143	
	4200	High frequency, steel base, 4-1/2" x 6", USP		↓	147	251	
	4600	Spring hinge, single acting, 6" flange, steel		Ea.	37	63	
	4700	Brass			64	109	
	4900	Double acting, 6" flange, steel			66	112	
	4950	Brass		↓	108	185	
	8000	Continuous hinge, steel	.250	L.F.	4	21	

087 200 | Operators

202	0010	AUTOMATIC OPENERS Swing doors, single	20	Ea.	2,000	4,575	202
	0100	Single operating pair	32	Pr.	3,700	8,200	

087 | Hardware

087 200 | Operators

			WORK-HOURS	UNIT	BARE COSTS MAT.	TOTAL INCL. O&P	
202	0400	For double simultaneous doors, one way, add	13.333	Pr.	315	1,325	202
	0500	Two way, add	17.778	↓	390	1,700	
	1000	Sliding doors, 3' wide, including track & hanger, single	26.667	Opng.	3,400	7,350	
	1300	Bi-parting	32		4,025	8,750	
	1450	Activating carpet, single door, one way, add	7.273		635	1,500	
	1550	Two way, add	12.308	↓	910	2,250	
	1750	Handicap opener, button operating	2	Ea.	1,025	1,850	
206	0010	**DOOR CLOSER** Rack and pinion	1.231	Ea.	90.50	225	206
	0020	Adjustable backcheck, 3 way mount, all sizes, regular arm	1.333		95.50	240	
	0040	Hold open arm	1.333		104	254	
	0100	Fusible link	1.231		81	209	
	0200	Non sized, regular arm	1.333		90.50	230	
	0240	Hold open arm	1.333		112	269	
	0400	4 way mount, non sized, regular arm	1.333		124	288	
	0440	Hold open arm	1.333	↓	133	305	
	1950						
	2000	Backcheck and adjustable power, hinge face mount					
	2010	All sizes, regular arm	1.231	Ea.	114	265	
	2040	Hold open arm	1.231		122	279	
	2400	Top jamb mount, all sizes, regular arm	1.333		114	271	
	2440	Hold open arm	1.333		122	285	
	2800	Top face mount, all sizes, regular arm	1.231		113	265	
	2840	Hold open arm	1.231		122	279	
	4000	Backcheck, overhead concealed, all sizes, regular arm	1.455		120	289	
	4040	Concealed arm	1.600		128	310	
	4400	Compact overhead, concealed, all sizes, regular arm	1.455		219	455	
	4440	Concealed arm	1.600		228	480	
	4800	Concealed in door, all sizes, regular arm	1.455		81	222	
	4840	Concealed arm	1.600		87	240	
	4900	Floor concealed, all sizes, single acting	3.636		103	385	
	4940	Double acting	3.636		132	435	
	5000	For cast aluminum cylinder, deduct			12	20.50	
	5040	For delayed action, add			20	34	
	5080	For fusible link arm, add			9	15.35	
	5120	For shock absorbing arm, add			25	42.50	
	5160	For spring power adjustment, add			19	32.50	
	6000	Closer-holder, hinge face mount, all sizes, exposed arm	1.231		90	225	
	7000	Electronic closer-holder, hinge facemount, concealed arm	1.600		138	325	
	7400	With built-in detector	1.600	↓	430	830	
	9000	Minimum labor/equipment charge	2	Job		114	

087 300 | Weatherstripping/Seals

304	0010	**THRESHOLD** 3' long door saddles, aluminum, minimum	.400	Ea.	6	33	304
	0100	Maximum	.667		44	113	
	0500	Bronze, minimum	.400		24.50	64.50	
	0600	Maximum	.667		99	207	
	0700	Rubber, 1/2" thick, 5-1/2" wide	.400		27.50	70	
	0800	2-3/4" wide	.400	↓	12.75	44.50	
	9000	Minimum labor/equipment charge	2	Job		114	

092 | Lath, Plaster and Gypsum Board

092 600 | Gypsum Board Systems

			WORK-HOURS	UNIT	BARE COSTS MAT.	TOTAL INCL. O&P
602	0010	**BLUEBOARD** For use with thin coat				
	0100	plaster application (see division 092-154)				
	1000	3/8" thick, on walls or ceilings, standard, no finish included	.008	S.F.	.17	.78
	1100	With thin coat plaster finish	.018		.24	1.44
	1400	On beams, columns, or soffits, standard, no finish included	.024		.25	1.78
	1450	With thin coat plaster finish	.034		.31	2.45
	3000	1/2" thick, on walls or ceilings, standard, no finish included	.008		.18	.79
	3100	With thin coat plaster finish	.018		.25	1.47
	3300	Fire resistant, no finish included	.008		.19	.81
	3400	With thin coat plaster finish	.018		.27	1.50
	3450	On beams, columns, or soffits, standard, no finish included	.024		.28	1.83
	3500	With thin coat plaster finish	.034		.35	2.52
	3700	Fire resistant, no finish included	.024		.31	1.88
	3800	With thin coat plaster finish	.034		.40	2.60
	5000	5/8" thick, on walls or ceilings, fire resistant, no finish included	.008		.18	.79
	5100	With thin coat plaster finish	.018		.28	1.52
	5500	On beams, columns, or soffits, no finish included	.024		.28	1.83
	5600	With thin coat plaster finish	.034		.35	2.52
	6000	For high ceilings, over 8' high, add	.005		.09	.45
	6500	For over 3 stories high, add per story	.003		.04	.22
	9000	Minimum labor/equipment charge	4	Job		227
608	0010	**DRYWALL** Gypsum plasterboard, nailed or screwed to studs,				
	0100	unless otherwise noted				
	0150	3/8" thick, on walls, standard, no finish included	.008	S.F.	.15	.72
	0200	On ceilings, standard, no finish included	.009		.15	.77
	0250	On beams, columns, or soffits, no finish included	.024		.24	1.75
	0270					
	0300	1/2" thick, on walls, standard, no finish included	.008	S.F.	.15	.72
	0350	Taped and finished	.017		.20	1.28
	0400	Fire resistant, no finish included	.008		.19	.78
	0450	Taped and finished	.017		.23	1.33
	0500	Water resistant, no finish included	.008		.23	.84
	0550	Taped and finished	.017		.28	1.42
	0600	Prefinished, vinyl, clipped to studs	.018		.50	1.86
	0650					
	1000	On ceilings, standard, no finish included	.009	S.F.	.16	.79
	1050	Taped and finished	.021		.22	1.57
	1100	Fire resistant, no finish included	.009		.19	.84
	1150	Taped and finished	.021		.24	1.60
	1200	Water resistant, no finish included	.009		.23	.90
	1250	Taped and finished	.021		.27	1.66
	1500	On beams, columns, or soffits, standard, no finish included	.024		.32	1.89
	1550	Taped and finished	.034		.37	2.55
	1600	Fire resistant, no finish included	.024		.29	1.85
	1650	Taped and finished	.034		.33	2.48
	1700	Water resistant, no finish included	.024		.34	1.92
	1750	Taped and finished	.034		.38	2.57
	2000	5/8" thick, on walls, standard, no finish included	.008		.19	.78
	2050	Taped and finished	.017		.24	1.34
	2100	Fire resistant, no finish included	.008		.21	.81
	2150	Taped and finished	.017		.26	1.39
	2200	Water resistant, no finish included	.008		.28	.94
	2250	Taped and finished	.017		.33	1.50
	2300	Prefinished, vinyl, clipped to studs	.018		.58	2
	3000	On ceilings, standard, no finish included	.009		.19	.84

092 | Lath, Plaster and Gypsum Board

092 600 | Gypsum Board Systems

			WORK-HOURS	UNIT	BARE COSTS MAT.	TOTAL INCL. O&P	
608	3050	Taped and finished	.021	S.F.	.24	1.60	608
	3100	Fire resistant, no finish included	.009		.21	.87	
	3150	Taped and finished	.021		.26	1.64	
	3200	Water resistant, no finish included	.009		.28	.99	
	3250	Taped and finished	.021		.33	1.75	
	3500	On beams, columns, or soffits, standard, no finish included	.024		.28	1.83	
	3550	Taped and finished	.034		.34	2.49	
	3600	Fire resistant, no finish included	.024		.30	1.86	
	3650	Taped and finished	.034		.36	2.54	
	3700	Water resistant, no finish included	.024		.37	1.99	
	3750	Taped and finished	.034		.42	2.63	
	4000	Fireproofing, beams or columns, 2 layers, 1/2" thick, incl finish	.048		.48	3.58	
	4050	5/8" thick	.053		.54	3.96	
	4100	3 layers, 1/2" thick	.071		.72	5.30	
	4150	5/8" thick	.076		.78	5.70	
	5050	For 1" thick coreboard on columns	.033		.50	2.76	
	5100	For foil-backed board, add			.07	.12	
	5200	For high ceilings, over 8' high, add	.005		.08	.44	
	5270	For textured spray, add	.010		.10	.71	
	5300	For over 3 stories high, add per story	.003		.04	.22	
	5350	For finishing corners, inside or outside, add	.015	L.F.	.05	.92	
	5500	For acoustical sealant, add per bead	.016	"	.02	.94	
	5550	Sealant, 1 quart tube		Ea.	4	6.80	
	5600	Sound deadening board, 1/4" gypsum	.009	S.F.	.15	.77	
	5650	1/2" wood fiber	.009	"	.20	.85	
	9000	Minimum labor/equipment charge	4	Job		227	
612	0010	METAL STUDS, DRYWALL Partitions, 10' high, with runners					612
	1800						
	2000	Non-load bearing, galvanized, 25 ga. 1-5/8", 16" O.C.	.018	S.F.	.23	1.40	
	2100	24" O.C.	.015		.18	1.19	
	2200	2-1/2" wide, 16" O.C.	.018		.25	1.47	
	2250	24" O.C.	.016		.21	1.25	
	2300	3-5/8" wide, 16" O.C.	.019		.29	1.56	
	2350	24" O.C.	.016		.25	1.33	
	2400	4" wide, 16" O.C.	.019		.34	1.65	
	2450	24" O.C.	.016		.27	1.39	
	2500	6" wide, 16" O.C.	.020		.44	1.85	
	2550	24" O.C.	.017		.33	1.51	
	2600	20 ga. studs, 1-5/8" wide, 16" O.C.	.018		.41	1.71	
	2650	24" O.C.	.015		.33	1.44	
	2700	2-1/2" wide, 16" O.C.	.018		.46	1.83	
	2750	24" O.C.	.016		.37	1.53	
	2800	3-5/8" wide, 16" O.C.	.019		.55	2.01	
	2850	24" O.C.	.016		.44	1.65	
	2900	4" wide, 16" O.C.	.019		.58	2.07	
	2950	24" O.C.	.016		.46	1.71	
	3000	6" wide, 16" O.C.	.020		.73	2.35	
	3050	24" O.C.	.017		.58	1.94	
	4000	LB studs, light ga. structural, galv., 18 ga., 2-1/2", 16" O.C.	.040		.85	3.73	
	4100	24" O.C.	.033		.69	3.08	
	4200	3-5/8" wide, 16" O.C.	.042		.99	4.09	
	4250	24" O.C.	.035		.79	3.33	
	4300	4" wide, 16" O.C.	.044		1.03	4.28	
	4350	24" O.C.	.036		.83	3.49	
	4400	6" wide, 16" O.C.	.047		1.26	4.84	
	4450	24" O.C.	.038		1.01	3.90	

092 | Lath, Plaster and Gypsum Board

092 600 | Gypsum Board Systems

			WORK-HOURS	UNIT	BARE COSTS MAT.	TOTAL INCL. O&P	
612	4600	16 ga. studs, 2-1/2", 16" O.C.	.044	S.F.	1.07	4.36	612
	4650	24" O.C.	.036		.86	3.55	
	4700	3-5/8" wide, 16" O.C.	.047		1.13	4.61	
	4750	24" O.C.	.038		.98	3.85	
	4800	4" wide, 16" O.C.	.050		1.30	5.05	
	4850	24" O.C.	.040		1.05	4.06	
	4900	6" wide, 16" O.C.	.053		1.60	5.75	
	4950	24" O.C.	.042		1.28	4.59	
	9000	Minimum labor/equipment charge	2	Job		114	

092 800 | Drywall Accessories

			WORK-HOURS	UNIT	BARE COSTS MAT.	TOTAL INCL. O&P	
804	0010	ACCESSORIES, DRYWALL Casing bead, galvanized steel	2.759	C.L.F.	13.85	181	804
	0100	Vinyl	2.667		18.80	185	
	0300	Corner bead, galvanized steel, 1" x 1"	2		8.15	128	
	0400	1-1/4" x 1-1/4"	2.286		10.40	148	
	0600	Vinyl corner bead	2		13.10	136	
	0700	Door casing, vinyl, for 2" wall systems	3.200		22.50	221	
	0900	Furring channel, galv. steel, 7/8" deep, standard	3.077		15.35	201	
	1000	Resilient	3.137		15.95	207	
	1100	J trim, galvanized steel, 1/2" wide	2.667		14.10	177	
	1120	5/8" wide	2.712		14.35	180	
	1500	Z stud, galvanized steel, 1-1/2" wide	3.077		23.50	215	
	9000	Minimum labor/equipment charge	2.667	Job		152	

093 | Tile

093 100 | Ceramic Tile

			WORK-HOURS	UNIT	BARE COSTS MAT.	TOTAL INCL. O&P	
102	0011	CERAMIC TILE					102
	0020						
	0600	Cove base, 4-1/4" x 4-1/4" high, mud set	.176	L.F.	2.61	12.95	
	0700	Thin set	.125		2.40	10.15	
	0900	6" x 4-1/4" high, mud set	.160		2.55	12.10	
	1000	Thin set	.117		2.38	9.70	
	1200	Sanitary cove base, 6" x 4-1/4" high, mud set	.172		2.60	12.80	
	1300	Thin set	.129		2.42	10.35	
	1500	6" x 6" high, mud set	.190		3.15	14.60	
	1600	Thin set	.137		2.91	11.65	
	2400	Bullnose trim, 4-1/4" x 4-1/4", mud set	.195		2.70	14.05	
	2500	Thin set	.125		2.42	10.20	
	2700	6" x 4-1/4" bullnose trim, mud set	.190		3.45	15.15	
	2800	Thin set	.129		3.25	11.85	
	3000	Floors, natural clay, random or uniform, thin set, color group 1	.087	S.F.	2.81	9.05	
	3100	Color group 2	.087		3	9.40	
	3300	Porcelain type, 1 color, color group 2, 1" x 1"	.087		3.26	9.85	
	3400	2" x 2" or 2" x 1", thin set	.084		3.69	10.40	
	3600	For random blend, 2 colors, add			.60	1.02	
	3700	4 colors, add			.85	1.46	
	4300	Specialty tile, 4-1/4" x 4-1/4" x 1/2", decorator finish	.087		5.50	13.65	
	4310						

093 | Tile

093 100 | Ceramic Tile

			WORK-HOURS	UNIT	BARE COSTS MAT.	TOTAL INCL. O&P
102	4500	Add for epoxy grout, 1/16" joint, 1" x 1" tile	.020	S.F.	.43	1.70
	4600	2" x 2" tile	.020	"	.38	1.60
	4800	Pregrouted sheets, walls, 4-1/4" x 4-1/4", 6" x 4-1/4"				
	4810	and 8-1/2" x 4-1/4", 4 S.F. sheets, silicone grout	.067	S.F.	3.23	8.80
	5100	Floors, unglazed, 2 S.F. sheets,				
	5110	urethane adhesive	.089	S.F.	3.28	9.90
	5400	Walls, interior, thin set, 4-1/4" x 4-1/4" tile	.084		1.90	7.35
	5500	6" x 4-1/4" tile	.084		1.95	7.40
	5700	8-1/2" x 4-1/4" tile	.084		2.75	8.80
	5800	6" x 6" tile	.080		2.25	7.70
	6000	Decorated wall tile, 4-1/4" x 4-1/4", minimum	.059		2.40	7
	6100	Maximum	.089		13.30	27
	6600	Crystalline glazed, 4-1/4" x 4-1/4", mud set, plain	.160		2.95	12.80
	6700	4-1/4" x 4-1/4", scored tile	.160		3.20	13.20
	6900	1-3/8" squares	.172		3.20	13.80
	7000	For epoxy grout, 1/16" joints, 4-1/4" tile, add	.020		.45	1.74
	7200	For tile set in dry mortar, add	.009			.45
	7300	For tile set in portland cement mortar, add	.055			2.69
	9500	Minimum labor/equipment charge	4.923	Job		239

093 300 | Quarry Tile

304	0010	QUARRY TILE Base, cove or sanitary, 2" or 5" high, mud set				
	0100	1/2" thick	.145	L.F.	2.65	11.60
	0300	Bullnose trim, red, mud set, 6" x 6" x 1/2" thick	.133		2.68	11.05
	0400	4" x 4" x 1/2" thick	.145		2.68	11.60
	0600	4" x 8" x 1/2" thick, using 8" as edge	.123		2.64	10.45
	0610					
	0700	Floors, mud set, 1,000 S.F. lots, red, 4" x 4" x 1/2" thick	.133	S.F.	2.89	11.45
	0900	6" x 6" x 1/2" thick	.114		2.74	10.20
	1000	4" x 8" x 1/2" thick	.123		2.87	10.85
	1300	For waxed coating, add			.31	.53
	1500	For colors other than green, add			.49	.84
	1600	For abrasive surface, add			.41	.70
	1800	Brown tile, imported, 6" x 6" x 3/4"	.133		3.60	12.60
	1900	8" x 8" x 1"	.145		4.10	14.10
	2100	For thin set mortar application, deduct	.023			1.11
	2500					
	2700	Stair tread & riser, 6" x 6" x 3/4", plain	.320	S.F.	3.15	21
	2800	Abrasive	.340		3.55	22.50
	3000	Wainscot, 6" x 6" x 1/2", thin set, red	.152		2.70	11.95
	3100	Colors other than green	.152		3.20	12.80
	3300	Window sill, 6" wide, 3/4" thick	.178	L.F.	3.10	13.95
	3400	Corners	.200	Ea.	2.75	14.45
	9000	Minimum labor/equipment charge	4.923	Job		239

095 | Acoustical Treatment and Wood Flooring

095 100 | Acoustical Ceilings

		WORK-HOURS	UNIT	BARE COSTS MAT.	TOTAL INCL. O&P	
104	0010 **SUSPENDED ACOUSTIC CEILING BOARDS** Not including					104
	0100 suspension system					
	0300 Fiberglass boards, film faced, 2' x 2' or 2' x 4', 5/8" thick	.013	S.F.	.39	1.39	
	0400 3/4" thick	.013		.46	1.54	
	0500 3" thick, thermal, R11	.018		.97	2.67	
	0600 Glass cloth faced fiberglass, 3/4" thick	.016		1.26	3.07	
	0700 1" thick	.016		1.40	3.33	
	0820 1-1/2" thick, nubby face	.017		1.70	3.87	
	0900 Mineral fiber boards, 5/8" thick, aluminum faced, 24" x 24"	.013		1.06	2.57	
	0930 24" x 48"	.012		1.04	2.47	
	0960 Standard face	.012		.45	1.46	
	1000 Plastic coated face	.020		.70	2.33	
	1200 Mineral fiber, 2 hour rating, 5/8" thick	.012		.62	1.74	
	1300 Mirror faced panels, 15/16" thick, 2' x 2'	.016		4.55	8.70	
	1900 Eggcrate, acrylic, 1/2" x 1/2" x 1/2" cubes	.016		1.07	2.74	
	2100 Polystyrene eggcrate, 3/8" x 3/8" x 1/2" cubes	.016		.88	2.40	
	2200 1/2" x 1/2" x 1/2" cubes	.016		1.20	2.96	
	2210					
	2400 Luminous panels, prismatic, acrylic	.020	S.F.	1.27	3.31	
	2500 Polystyrene	.020		.66	2.27	
	2700 Flat white acrylic	.020		2.23	4.94	
	2800 Polystyrene	.020		1.52	3.73	
	3000 Drop pan, white, acrylic	.020		3.28	6.75	
	3100 Polystyrene	.020		2.72	5.75	
	3600 Perforated aluminum sheets, .024" thick, corrugated, painted	.016		1.32	3.17	
	3700 Plain	.016		1.18	2.93	
	3720 Mineral fiber, 24" x 24" or 48", reveal edge, painted, 5/8" thick	.013		.72	1.98	
	3740 3/4" thick	.014		1.20	2.84	
	3750 Wood fiber in cementitious binder, 2' x 2' or 4', painted, 1" thick	.013		.90	2.29	
	3760 2" thick	.015		1.47	3.34	
	3770 2-1/2" thick	.016		2	4.32	
	3780 3" thick	.018		2.30	4.93	
	9000 Minimum labor/equipment charge	2	Job		114	
106	0010 **SUSPENDED CEILINGS, COMPLETE** Including standard					106
	0100 suspension system but not incl. 1-1/2" carrier channels					
	0600 Fiberglass ceiling board, 2' x 4' x 5/8", plain faced,	.016	S.F.	.79	2.26	
	0700 Offices, 2' x 4' x 3/4"	.021		.88	2.70	
	1800 Tile, Z bar suspension, 5/8" mineral fiber tile	.053		1.12	4.95	
	1900 3/4" mineral fiber tile	.053		1.20	5.10	
	9000 Minimum labor/equipment charge	4	Job		227	

096 | Flooring and Carpet

096 600 | Resilient Tile Flooring

		WORK-HOURS	UNIT	BARE COSTS MAT.	TOTAL INCL. O&P	
601	0010 **RESILIENT** Asphalt tile, on concrete, 1/8" thick					601
	0050 Color group B	.020	S.F.	.82	2.48	
	0100 Color group C & D	.020		.88	2.58	
	0300 For wood subfloor, add to above for felt underlayment			.15	.26	

096 | Flooring and Carpet

096 600 | Resilient Tile Flooring

			WORK-HOURS	UNIT	BARE COSTS MAT.	TOTAL INCL. O&P	
601	0500	For less than 500 S.F., add	.016	S.F.		.87	601
	0600	For over 5000 S.F., deduct	.005	↓		.27	
	0800	Base, cove, rubber or vinyl, .080" thick					
	1100	Standard colors, 2-1/2" high	.025	L.F.	.42	2.08	
	1150	4" high	.025		.54	2.28	
	1200	6" high	.025		.79	2.71	
	1450	1/8" thick, standard colors, 2-1/2" high	.025		.48	2.19	
	1500	4" high	.025		.70	2.56	
	1550	6" high	.025	↓	.72	2.59	
	1600	Corners, 2-1/2" high	.025	Ea.	.93	2.95	
	1630	4" high	.025		.97	3.02	
	1660	6" high	.025	↓	1.24	3.47	
	1700	Conductive flooring, rubber tile, 1/8" thick	.025	S.F.	2.20	5.10	
	1800	Homogeneous vinyl tile, 1/8" thick	.025		3	6.50	
	2200	Cork tile, standard finish, 1/8" thick	.025		2.02	4.81	
	2250	3/16" thick	.025		2.52	5.65	
	2300	5/16" thick	.025		3.42	7.20	
	2350	1/2" thick	.025		4.03	8.25	
	2500	Urethane finish, 1/8" thick	.025		3.08	6.60	
	2550	3/16" thick	.025		4.37	8.80	
	2600	5/16" thick	.025		5.60	10.90	
	2650	1/2" thick	.025		7.65	14.40	
	3700	Polyethylene, in rolls, no base incl., landscape surfaces	.029		1.80	4.63	
	3800	Nylon action surface, 1/8" thick	.029		2	4.98	
	3900	1/4" thick	.029		2.85	6.45	
	4000	3/8" thick	.029		3.60	7.70	
	5500	Polyvinyl chloride, sheet goods for gyms, 1/4" thick	.100		2.90	10.35	
	5600	3/8" thick	.133		3.30	12.85	
	5900	Rubber, sheet goods, 36" wide, 1/8" thick	.067		2.60	8.05	
	5950	3/16" thick	.080		3.65	10.55	
	6000	1/4" thick	.089		4.25	12.05	
	6050	Tile, marbleized colors, 12" x 12", 1/8" thick	.020		2.60	5.50	
	6100	3/16" thick	.020		3.70	7.40	
	6300	Special tile, plain colors, 1/8" thick	.020		3.28	6.70	
	6350	3/16" thick	.020		4.40	8.60	
	7000	Vinyl composition tile, 12" x 12", 1/16" thick	.016		.62	1.92	
	7050	Embossed	.016		.77	2.19	
	7100	Marbleized	.016		.77	2.19	
	7150	Solid	.016		.88	2.37	
	7200	3/32" thick, embossed	.016		.82	2.26	
	7250	Marbleized	.016		.88	2.37	
	7300	Solid	.016		1.30	3.09	
	7350	1/8" thick, marbleized	.016		.98	2.54	
	7400	Solid	.016		1.53	3.47	
	7500	Vinyl tile, 12" x 12", .050" thick, minimum	.016		1.40	3.26	
	7550	Maximum	.016		2.75	5.55	
	7600	1/8" thick, minimum	.016		1.75	3.86	
	7650	Solid colors	.016		2.20	4.62	
	7700	Marbleized or Travertine pattern	.016		2.80	5.65	
	7750	Florentine pattern	.016		3.30	6.50	
	7800	Maximum	.016		6.65	12.20	
	8000	Vinyl sheet goods, backed, .065" thick, minimum	.032		1.15	3.69	
	8050	Maximum	.040		2.10	5.75	
	8100	.080" thick, minimum	.035		1.35	4.19	
	8150	Maximum	.040		2.40	6.25	
	8200	.125" thick, minimum	.035	↓	1.50	4.44	

096 | Flooring and Carpet

096 600 | Resilient Tile Flooring

			WORK-HOURS	UNIT	BARE COSTS MAT.	TOTAL INCL. O&P	
601	8250	Maximum	.040	S.F.	3.30	7.80	601
	8300	.250" thick, minimum	.035		2.35	5.90	
	8350	Maximum	.040		4.30	9.50	
	8700	Adhesive cement, 1 gallon does 200 to 300 S.F.		Gal.	12.20	21	
	8800	Asphalt primer, 1 gallon per 300 S.F.			7.65	13.05	
	8900	Emulsion, 1 gallon per 140 S.F.			7.65	13.05	
	8950	Latex underlayment			23.50	40.50	
	9500	Minimum labor/equipment charge	2	Job		108	

096 780 | Resilient Accessories

			WORK-HOURS	UNIT	BARE COSTS MAT.	TOTAL INCL. O&P	
781	0010	STAIR TREADS AND RISERS See index for materials other					781
	0100	than rubber and vinyl					
	0300	Rubber, molded tread, 12" wide, 5/16" thick, black	.070	L.F.	5.55	13.25	
	0400	Colors	.070		5.70	13.45	
	0600	1/4" thick, black	.070		5.20	12.60	
	0700	Colors	.070		5.65	13.40	
	0900	Grip strip safety tread, colors, 5/16" thick	.070		8	17.40	
	1000	3/16" thick	.067		6.10	14	
	1200	Landings, smooth sheet rubber, 1/8" thick	.067	S.F.	2.75	8.30	
	1300	3/16" thick	.067	"	3.55	9.65	
	1500	Nosings, 1-1/2" deep, 3" wide, black	.057	L.F.	1.50	5.65	
	1600	Colors	.057		2.05	6.60	
	1800	Risers, 7" high, 1/8" thick, flat	.032		1.50	4.28	
	1900	Coved	.032		1.75	4.71	
	2100	Vinyl, molded tread, 12" wide, colors, 1/8" thick	.070		2.65	8.25	
	2200	1/4" thick	.070		3.80	10.20	
	2300	Landing material, 1/8" thick	.040	S.F.	2.35	6.15	
	2400	Riser, 7" high, 1/8" thick, coved	.046	L.F.	1.38	4.82	
	2500	Tread and riser combined, 1/8" thick	.100	"	3.50	11.35	
	9000	Minimum labor/equipment charge	2.667	Job		144	

096 850 | Sheet Carpet

			WORK-HOURS	UNIT	BARE COSTS MAT.	TOTAL INCL. O&P	
852	0010	CARPET Commercial grades, direct cement					852
	0700	Nylon, level loop, 26 oz., light to medium traffic	.140	S.Y.	12.25	28.50	
	0900	32 oz., medium traffic	.140		14.60	32.50	
	1100	40 oz., medium to heavy traffic	.140		18.50	39.50	
	2100	Nylon, plush, 20 oz., light traffic	.140		7.50	20.50	
	2800	24 oz., light to medium traffic	.140		8.75	22.50	
	2900	30 oz., medium traffic	.140		10.90	26	
	3000	36 oz., medium traffic	.140		13.25	30.50	
	3100	42 oz., medium to heavy traffic	.160		16.10	36.50	
	3200	46 oz., medium to heavy traffic	.160		19.10	41	
	3300	54 oz., heavy traffic	.160		21.50	45.50	
	3500	Olefin, 15 oz., light traffic	.140		3.94	14.35	
	3650	22 oz., light traffic	.140		4.94	16.05	
	4500	50 oz., medium to heavy traffic, level loop	.140		32.50	63	
	4700	32 oz., medium to heavy traffic, patterned	.163		32.50	63.50	
	4900	48 oz., heavy traffic, patterned	.163		41	78.50	
	5600	For bound carpet baseboard, add	.027	L.F.	.85	2.89	
	5610	For stairs, not incl. price of carpet, add	.267	Riser		14.35	
	8950	For tackless, stretched installation, add padding to above					
	9000	Sponge rubber pad, minimum	.053	S.Y.	2.15	6.55	
	9100	Maximum	.053		5.90	12.85	
	9200	Felt pad, minimum	.053		2.50	7.15	
	9300	Maximum	.053		4.20	10.05	
	9400	Bonded urethane pad, minimum	.053		2.65	7.40	

096 | Flooring and Carpet

096 850 | Sheet Carpet

		WORK-HOURS	UNIT	BARE COSTS MAT.	TOTAL INCL. O&P	
9500	Maximum	.053	S.Y.	4.60	10.70	
9600	Prime urethane pad, minimum	.053		1.70	5.75	
9700	Maximum	.053	↓	3	7.95	
9900	Carpet cleaning machine, rent		Day		49.70	
9910	Minimum labor/equipment charge	2.667	Job		144	

099 | Painting and Wall Coverings

099 100 | Exterior Painting

		WORK-HOURS	UNIT	BARE COSTS MAT.	TOTAL INCL. O&P
0010	**DOORS AND WINDOWS**				
0100	Door frames & trim, only				
0110	Brushwork, primer	.011	L.F.	.01	.57
0120	Finish coat, exterior latex	.011		.01	.57
0130	Primer & 1 coat, exterior latex	.019		.03	1.01
0140	Primer & 2 coats, exterior latex	.025	↓	.04	1.36
0150	Doors, flush, both sides, incl. frame & trim				
0160	Roll & brush, primer	.400	Ea.	1.85	24
0170	Finish coat, exterior latex	.333		1.61	19.90
0180	Primer & 1 coat, exterior latex	.727		3.46	44
0190	Primer & 2 coats, exterior latex	.941		4.93	57
0200	Brushwork, stain, sealer & 2 coats polyurethane	.800	↓	8.50	56
0210	Doors, French, both sides, 10-15 lite, incl. frame & trim				
0220	Brushwork, primer	.842	Ea.	.37	44
0230	Finish coat, exterior latex	.696		.32	37
0240	Primer & 1 coat, exterior latex	1.524		.69	80.50
0250	Primer & 2 coats, exterior latex	2		.99	105
0260	Brushwork, stain, sealer & 2 coats polyurethane	1.600	↓	1.70	85.50
0270	Doors, louvered, both sides, incl. frame & trim				
0280	Brushwork, primer	.889	Ea.	2.03	49.50
0290	Finish coat, exterior latex	.744		1.77	41.50
0300	Primer & 1 coat, exterior latex	1.600		3.80	88.50
0310	Primer & 2 coats, exterior latex	2.133		5.40	119
0320	Brushwork, stain, sealer & 2 coats polyurethane	1.778	↓	9.35	108
0330	Doors, panel, both sides, incl. frame & trim				
0340	Roll & brush, primer	.571	Ea.	1.94	32.50
0350	Finish coat, exterior latex	.500		1.69	28.50
0360	Primer & 1 coat, exterior latex	1.067		3.63	61
0370	Primer & 2 coats, exterior latex	1.455		5.20	84
0380	Brushwork, stain, sealer & 2 coats polyurethane	1.143	↓	8.95	74.50
0410	1 to 6 lite				
0420	Brushwork, primer	.242	Ea.	1.49	15.05
0430	Finish coat, exterior latex	.364		1.30	21
0440	Primer & 1 coat, exterior latex	.571		2.80	34
0450	Primer & 2 coats, exterior latex	.889		3.99	53
0460	Stain, sealer & 1 coat varnish	.485	↓	2.88	30.50
0470	7 to 10 lite				
0480	Brushwork, primer	.444	Ea.	1.85	26
0490	Finish coat, exterior latex	.667		1.61	37
0500	Primer & 1 coat, exterior latex	1.067	↓	3.46	61

099 | Painting and Wall Coverings

099 100 | Exterior Painting

			WORK-HOURS	UNIT	BARE COSTS MAT.	TOTAL INCL. O&P	
104	0510	Primer & 2 coats, exterior latex	1.600	Ea.	4.94	91	104
	0520	Stain, sealer & 1 coat varnish	.727	↓	3.56	44	
	0530	12 lite					
	0540	Brushwork, primer	.615	Ea.	2.10	36	
	0550	Finish coat, exterior latex	.800		1.83	44.50	
	0560	Primer & 1 coat, exterior latex	1.391		3.93	79	
	0570	Primer & 2 coats, exterior latex	2		5.60	112	
	0580	Stain, sealer & 1 coat varnish	.889	↓	4.04	53	
	0590	For oil base paint, add		Ea.	10%		
106	0010	**SIDING** Exterior, Alkyd (oil base)					106
	0020	Labor cost includes protection of adjacent items not painted					
	0500	Spray	.003	S.F.	.05	.23	
	0800	Paint 2 coats, brushwork	.009		.13	.66	
	1000	Spray	.004		.20	.57	
	1200	Stucco, rough, oil base, paint 2 coats, brushwork	.015		.11	.94	
	1400	Roller	.009		.13	.67	
	1600	Spray	.004		.11	.41	
	1800	Texture 1-11 or clapboard, oil base, primer coat, brushwork	.005		.06	.38	
	2000	Spray	.003		.05	.26	
	2400	Paint 2 coats, brushwork	.009		.13	.67	
	2600	Spray	.006		.12	.51	
	3400	Stain 2 coats, brushwork	.013		.11	.85	
	4000	Spray	.006		.10	.48	
	4200	Wood shingles, oil base primer coat, brushwork	.011		.06	.66	
	4400	Spray	.007		.05	.43	
	5000	Paint 2 coats, brushwork	.020		.09	1.19	
	5200	Spray	.011		.09	.74	
	6500	Stain 2 coats, brushwork	.021		.11	1.25	
	7000	Spray	.006		.10	.47	
	8000	For latex paint, deduct		↓	10%		
120	0010	**TRIM**					120
	0100	Door frames & trim (see Doors, interior or exterior)					
	0110	Fascia, latex paint, one coat coverage					
	0120	1" x 4", brushwork	.010	L.F.	.01	.53	
	0130	Roll	.005		.01	.27	
	0140	Spray	.003		.01	.17	
	0150	1" x 6" to 1" x 10", brushwork	.011		.02	.60	
	0160	Roll	.006		.02	.33	
	0170	Spray	.003		.02	.20	
	0180	1" x 12", brushwork	.012		.03	.69	
	0190	Roll	.008		.03	.44	
	0200	Spray	.004	↓	.03	.23	
	0210	Gutters & downspouts, metal, zinc chromate paint					
	0220	Brushwork, gutters, 5", first coat	.011	L.F.	.03	.62	
	0230	Second coat	.009		.02	.50	
	0240	Third coat	.008		.02	.43	
	0250	Downspouts, 4", first coat	.029		.02	1.51	
	0260	Second coat	.017		.02	.88	
	0270	Third coat	.013	↓	.02	.67	
	0280	Gutters & downspouts, wood					
	0290	Brushwork, gutters, 5", primer	.013	L.F.	.03	.69	
	0300	Finish coat, exterior latex	.010		.02	.54	
	0310	Primer & 1 coat exterior latex	.020		.05	1.13	
	0320	Primer & 2 coats exterior latex	.028	↓	.07	1.57	

099 | Painting and Wall Coverings

099 100 | Exterior Painting

		WORK-HOURS	UNIT	BARE COSTS MAT.	TOTAL INCL. O&P
0330	Downspouts, 4", primer	.012	L.F.	.02	.67
0340	Finish coat, exterior latex	.010		.02	.53
0350	Primer & 1 coat exterior latex	.019		.04	1.04
0360	Primer & 2 coats exterior latex	.027		.06	1.49
0370	Molding, exterior, up to 14" wide				
0380	Brushwork, primer	.005	L.F.	.08	.40
0390	Finish coat, exterior latex	.005		.07	.38
0400	Primer & 1 coat exterior latex	.011		.16	.83
0410	Primer & 2 coats exterior latex	.017		.22	1.25
0420	Stain & fill	.008		.04	.46
0430	Shellac	.004		.05	.32
0440	Varnish	.006		.07	.45

099 200 | Interior Painting

		WORK-HOURS	UNIT	BARE COSTS MAT.	TOTAL INCL. O&P
0010	**DOORS & WINDOWS, LATEX**				
0100	Doors flush, both sides, incl. frame & trim				
0110	Roll & brush, primer	.400	Ea.	1.50	23
0120	Finish coat, latex	.333		1.78	20
0130	Primer & 1 coat latex	.727		3.28	43
0140	Primer & 2 coats latex	.941		4.96	57
0160	Spray, both sides, primer	.083		1.35	6.60
0170	Finish coat, latex	.067		1.61	6.20
0180	Primer & 1 coat latex	.148		2.96	12.70
0190	Primer & 2 coats latex	.200		4.79	18.45
0200	Doors, French, both sides, 10-15 lite, incl. frame & trim				
0210	Roll & brush, primer	.842	Ea.	.30	44
0220	Finish coat, latex	.696		.36	37
0230	Primer & 1 coat latex	1.524		.66	79.50
0240	Primer & 2 coats latex	2		.99	105
0260	Doors, louvered, both sides, incl. frame & trim				
0270	Roll & brush, primer	.889	Ea.	1.65	49.50
0280	Finish coat, latex	.744		1.95	41.50
0290	Primer & 1 coat, latex	1.600		3.60	88.50
0300	Primer & 2 coats, latex	2.133		5.45	119
0320	Spray, both sides, primer	.167		1.48	11.15
0330	Finish coat, latex	.125		1.77	9.50
0340	Primer & 1 coat, latex	.286		3.25	20.50
0350	Primer & 2 coats, latex	.381		5.25	28.50
0360	Doors, panel, both sides, incl. frame & trim				
0370	Roll & brush, primer	.571	Ea.	1.57	31.50
0380	Finish coat, latex	.500		1.86	28.50
0390	Primer & 1 coat, latex	1.067		3.44	61
0400	Primer & 2 coats, latex	1.455		5.20	84.50
0420	Spray, both sides, primer	.100		1.42	7.60
0430	Finish coat, latex	.083		1.69	7.20
0440	Primer & 1 coat, latex	.182		3.10	14.65
0450	Primer & 2 coats, latex	.250		5.05	21.50
0460	Windows, per interior side, base on 15 SF				
0470	1 to 6 lite				
0480	Brushwork, primer	.242	Ea.	1.21	14.60
0490	Finish coat, enamel	.364		1.63	21.50
0500	Primer & 1 coat enamel	.571		2.84	34
0510	Primer & 2 coats enamel	.889		4.39	54
0530	7 to 10 lite				
0540	Brushwork, primer	.444	Ea.	1.50	25.50
0550	Finish coat, enamel	.667		1.83	37

099 | Painting and Wall Coverings

099 200 | Interior Painting

			WORK-HOURS	UNIT	BARE COSTS MAT.	TOTAL INCL. O&P	
214	0560	Primer & 1 coat enamel	1.067	Ea.	3.33	60	214
	0570	Primer & 2 coats enamel	1.600	↓	5	91	
	0590	12 lite					
	0600	Brushwork, primer	.615	Ea.	1.70	35	
	0610	Finish coat, enamel	.800		2.07	45	
	0620	Primer & 1 coat enamel	1.391		3.78	78	
	0630	Primer & 2 coats enamel	2	↓	5.65	113	
	0650	for oil base paint, add		S.F.	10%		
224	0010	**WALLS AND CEILINGS**					224
	0020	Labor cost includes protection of adjacent items not painted					
	0100	Concrete, dry wall or plaster, oil base, primer or sealer coat					
	0200	Smooth finish, brushwork	.006	S.F.	.03	.36	
	0240	Roller	.004		.03	.25	
	0300	Sand finish, brushwork	.007		.06	.45	
	0340	Roller	.005		.06	.34	
	0380	Spray	.003		.04	.22	
	0800	Paint 2 coats, smooth finish, brushwork	.012		.07	.74	
	0840	Roller	.007		.06	.45	
	0880	Spray	.005		.07	.37	
	0900	Sand finish, brushwork	.013		.10	.85	
	0940	Roller	.008		.09	.55	
	0980	Spray	.005		.08	.38	
	1600	Glaze coating, 5 coats, spray, clear	.009		.51	1.32	
	1640	Multicolor	.009	↓	.68	1.62	
	2000	Masonry or concrete block, oil base, primer or sealer coat					
	2100	Smooth finish, brushwork	.007	S.F.	.04	.40	
	2180	Spray	.003		.05	.26	
	2200	Sand finish, brushwork	.007		.08	.52	
	2280	Spray	.003		.10	.34	
	2800	Paint 2 coats, smooth finish, brushwork	.011		.06	.65	
	2880	Spray	.006		.05	.39	
	2900	Sand finish, brushwork	.012		.11	.80	
	2980	Spray	.006		.11	.48	
	3600	Glaze coating, 5 coats, spray, clear	.009		.51	1.32	
	3620	Multicolor	.009		.68	1.62	
	4000	Block filler, 1 coat, brushwork	.015		.10	.97	
	4100	Silicone, water repellent, 2 coats, spray	.004		.21	.57	
	4120	For latex paint, deduct		↓	10%		

101 | Visual Display Boards, Compartments and Cubicles

101 600 | Toilet Compartments

			WORK-HOURS	UNIT	BARE COSTS MAT.	TOTAL INCL. O&P	
602	0010	**PARTITIONS, TOILET**					602
	0100	Cubicles, ceiling hung, marble	8	Ea.	1,100	2,325	
	0200	Painted metal	4		350	825	
	0300	Plastic laminate on particle board	4		480	1,050	
	0400	Porcelain enamel	4		715	1,450	
	0500	Stainless steel	4		860	1,675	
	0600	For handicap units, incl. 52" grab bars, add		↓	172	293	
	0700						

101 | Visual Display Boards, Compartments and Cubicles

101 600 | Toilet Compartments

		Description	WORK-HOURS	UNIT	BARE COSTS MAT.	TOTAL INCL. O&P
602	0800	Floor & ceiling anchored, marble	6.400	Ea.	1,200	2,400
	1000	Painted metal	3.200		355	795
	1100	Plastic laminate on particle board	3.200		485	1,025
	1200	Porcelain enamel	3.200		735	1,450
	1300	Stainless steel	3.200		1,000	1,875
	1400	For handicap units, incl. 52" grab bars, add			172	293
	1600	Floor mounted, marble	5.333		710	1,525
	1700	Painted metal	2.286		320	670
	1800	Plastic laminate on particle board	2.286		450	895
	1900	Porcelain enamel	2.286		715	1,350
	2000	Stainless steel	2.286		970	1,800
	2100	For handicap units, incl. 52" grab bars, add			172	293
	2200	For juvenile units, deduct		↓	33	56.50
	2300					
	2400	Floor mounted, headrail braced, marble	5.333	Ea.	825	1,725
	2500	Painted metal	2.667		330	715
	2600	Plastic laminate on particle board	2.667		480	970
	2700	Porcelain enamel	2.667		725	1,400
	2800	Stainless steel	2.667		980	1,800
	2900	For handicap units, incl. 52" grab bars, add			170	290
	3000	Wall hung partitions, painted metal	2.286		405	820
	3200	Porcelain enamel	2.286		730	1,375
	3300	Stainless steel	2.286		965	1,775
	3400	For handicap units, incl. 52" grab bars, add		↓	172	293
	4000	Screens, entrance, floor mounted, 58" high, 48" wide				
	4100	Marble	1.778	Ea.	450	875
	4200	Painted metal	1.067		148	315
	4300	Plastic laminate on particle board	1.067		305	585
	4400	Porcelain enamel	1.067		290	560
	4500	Stainless steel	1.067		535	980
	4600	Urinal screen, 18" wide, ceiling braced, marble	2.667		450	910
	4700	Painted metal	2		158	385
	4800	Plastic laminate on particle board	2		244	530
	4900	Porcelain enamel	2		277	585
	5000	Stainless steel	2	↓	375	755
	5050					
	5100	Floor mounted, head rail braced				
	5200	Marble	2.667	Ea.	420	855
	5300	Painted metal	2		153	375
	5400	Plastic laminate on particle board	2		205	465
	5500	Porcelain enamel	2		291	610
	5600	Stainless steel	2		445	875
	5700	Pilaster, flush, marble	1.778		520	975
	5800	Painted metal	1.600		199	430
	5900	Plastic laminate on particle board	1.600		281	575
	6000	Porcelain enamel	1.600		240	505
	6100	Stainless steel	1.600	↓	330	660
	6150					
	6200	Post braced, marble	1.778	Ea.	505	955
	6300	Painted metal	1.600		199	430
	6400	Plastic laminate on particle board	1.600		286	580
	6500	Porcelain enamel	1.600	↓	240	505
	6600	Stainless steel	1.600		330	650
	6700	Wall hung, bracket supported				
	6800	Painted metal	1.600	Ea.	204	440
	6900	Plastic laminate on particle board	1.600	↓	281	575

101 | Visual Display Boards, Compartments and Cubicles

101 600 | Toilet Compartments

			WORK-HOURS	UNIT	BARE COSTS MAT.	TOTAL INCL. O&P	
602	7000	Porcelain enamel	1.600	Ea.	245	510	602
	7100	Stainless steel	1.600	↓	299	605	
	7200						
	7400	Flange supported, painted metal	1.600	Ea.	143	335	
	7500	Plastic laminate on particle board	1.600		167	375	
	7600	Porcelain enamel	1.600		296	595	
	7700	Stainless steel	1.600		350	690	
	7800	Wedge type, painted metal	1.600		173	385	
	8000	Porcelain enamel	1.600		250	520	
	8100	Stainless steel	1.600	↓	355	705	
	9000	Minimum labor/equipment charge	3.200	Job		182	

104 | Identifying and Pedestrian Control Devices

104 100 | Directories

			WORK-HOURS	UNIT	BARE COSTS MAT.	TOTAL INCL. O&P	
104	0010	DIRECTORY BOARDS Plastic, glass covered, 30" x 20"	5.333	Ea.	197	640	104
	0100	36" x 48"	8	↓	570	1,425	
	0900	Outdoor, weatherproof, black plastic, 36" x 24"	8		600	1,475	
	1000	36" x 36"	10.667	↓	700	1,800	
	9000	Minimum labor/equipment charge	8	Job		455	

104 300 | Signs

304	0011	SIGNS, Plaques, 20" x 30", up to 450 letters, cast alum.	4	Ea.	620	1,275	304
	4000	Cast bronze	4		960	1,850	
	4200	30" x 36", up to 900 letters cast aluminum	5.333		1,300	2,525	
	4300	Cast bronze	5.333		1,725	3,275	
	5100	Exit signs, 24 ga. alum., 14" x 12" surface mounted	.267		8.50	29.50	
	5200	10" x 7"	.400		6.50	33.50	
	6400	Replacement sign faces, 6" or 8"	.160	↓	19.85	41	
	9000	Minimum labor/equipment charge	2	Job		114	

105 | Lockers, Protective Covers and Postal Specialties

105 380 | Canopies

			WORK-HOURS	UNIT	BARE COSTS MAT.	TOTAL INCL. O&P	
384	0010	CANOPIES Wall hung, aluminum, prefinished, 8' x 10'	18.462	Ea.	980	3,050	384
	0300	8' x 20'	21.818	↓	1,900	4,850	
	1000	12' x 20'	24	↓	2,800	6,575	
	1050						
	2300	Aluminum entrance canopies, flat soffit					
	2500	3'-6" x 4'-0", clear anodized	4	Ea.	425	955	
	4700	Canvas awnings, including canvas, frame & lettering					
	5000	Minimum	.160	S.F.	11	28	

105 | Lockers, Protective Covers and Postal Specialties

105 380	Canopies	WORK-HOURS	UNIT	BARE COSTS MAT.	TOTAL INCL. O&P	
384	5300 Average	.178	S.F.	17	39.50	384
	5500 Maximum	.200	↓	33	68	
	9000 Minimum labor/equipment charge	8	Job		455	

107 | Telephone Specialties

107 550	Telephone Enclosures	WORK-HOURS	UNIT	BARE COSTS MAT.	TOTAL INCL. O&P	
551	0010 TELEPHONE ENCLOSURE					551
	0020					
	0300 Shelf type, wall hung, minimum	3.200	Ea.	660	1,300	
	0400 Maximum	3.200		2,150	3,850	
	0600 Booth type, painted steel, indoor or outdoor, minimum	10.667		2,600	5,025	
	0700 Maximum (stainless steel)	10.667		8,650	15,400	
	1900 Outdoor, drive-up type, wall mounted	4		680	1,400	
	2000 Post mounted, stainless steel posts	5.333	↓	1,050	2,150	
	9000 Minimum labor/equipment charge	4	Job		227	

108 | Toilet and Bath Accessories and Scales

108 200	Bath Accessories	WORK-HOURS	UNIT	BARE COSTS MAT.	TOTAL INCL. O&P	
204	0010 BATHROOM ACCESSORIES					204
	0020					
	0200 Curtain rod, stainless steel, 5' long, 1" diameter	.615	Ea.	23	74.50	
	0300 1-1/4" diameter	.615	"	24	76	
	0500 Dispenser units, combined soap & towel dispensers,					
	0510 mirror and shelf, flush mounted	.800	Ea.	255	475	
	0600 Towel dispenser and waste receptacle,					
	0610 18 gallon capacity	.800	Ea.	305	570	
	0800 Grab bar, straight, 1-1/4" diameter, stainless steel, 18" long	.333		28	67.50	
	1100 36" long	.400		42	94	
	3000 Mirror with stainless steel, 3/4" square frame, 18" x 24"	.400		45	99.50	
	3300 72" x 24"	1.333		159	345	
	3500 Mirror with 5" stainless steel shelf, 3/4" sq. frame, 18" x 24"	.400		67.50	138	
	3800 72" x 24"	1.333		360	690	
	4200 Napkin/tampon dispenser, recessed	.533		360	640	
	4300 Robe hook, single, regular	.222		9.10	28	
	4400 Heavy duty, concealed mounting	.222		12.70	34.50	
	4600 Soap dispenser, chrome, surface mounted, liquid	.400		38.50	88.50	
	5600 Shelf, stainless steel, 5" wide, 18 ga., 24" long	.333		35	79	
	6100 Toilet tissue dispenser, surface mounted, S.S., single roll	.267		9.60	32	
	6200 Double roll	.333		21	55	
	6290 Toilet seat	.200	↓	14.60	37.50	

108 | Toilet and Bath Accessories and Scales

108 200 | Bath Accessories

			WORK-HOURS	UNIT	BARE COSTS MAT.	TOTAL INCL. O&P	
204	6400	Towel bar, stainless steel, 18" long	.348	Ea.	24	61	204
	6500	30" long	.381		23.50	62.50	
	6700	Towel dispenser, stainless steel, surface mounted	.500		28.50	77.50	
	6800	Flush mounted, recessed	.800		120	250	
	7400	Tumbler holder, tumbler only	.267		12.40	36.50	
	7500	Soap, tumbler & toothbrush	.267		16	43	
	7700	Wall urn ash receiver, surface mount, 11" long	.667		68	155	
	8000	Waste receptacles, stainless steel, with top, 13 gallon	.800	↓	144	291	
	9000	Minimum labor/equipment charge	1.600	Job		91	

142 | Elevators

142 010 | Elevators

			WORK-HOURS	UNIT	BARE COSTS MAT.	TOTAL INCL. O&P	
011	0012	**ELEVATORS**					011
	5000	Passenger, pre-engineered, 5 story, hydraulic, 2,500 lb. cap.	800	Ea.	20,800	86,000	
	5100	For less than 5 stops, deduct	110	Stop	7,100	19,200	
	5200	For 4,000 lb. capacity, general purpose, add		Ea.	5,100	8,675	
	5800						
	7000	Residential, cab type, 1 floor, 2 stop, minimum	80	Ea.	6,400	16,000	
	7100	Maximum	160		10,800	28,500	
	7200	2 floor, 3 stop, minimum	133		9,500	24,700	
	7300	Maximum	266		15,500	43,300	
	7700	Stair climber (chair lift), single seat, minimum	16		3,000	6,125	
	7800	Maximum	80		4,100	12,000	
	8000	Wheelchair, porch lift, minimum	16		4,250	8,250	
	8500	Maximum	32		10,100	19,200	
	8700	Stair lift, minimum	16		8,400	15,300	
	8900	Maximum	80	↓	13,300	27,600	

151 | Pipe and Fittings

151 100 | Miscellaneous Fittings

			WORK-HOURS	UNIT	BARE COSTS MAT.	TOTAL INCL. O&P	
125	0010	**DRAINS**					125
	0020						
	0140	Cornice, C.I., 45° or 90° outlet					
	0200	3" and 4" pipe size	1.333	Ea.	74	202	
	0260	For galvanized body, add		↓	10.65	18.15	
	0280	For polished bronze dome, add			9.25	15.80	
	0400	Deck, auto park, C.I., 13" top					
	0440	3", 4", 5", and 6" pipe size	2	Ea.	235	515	
	0480	For galvanized body, add		"	110	188	
	0500						

151 | Pipe and Fittings

151 100 | Miscellaneous Fittings

			WORK-HOURS	UNIT	BARE COSTS MAT.	TOTAL INCL. O&P	
125	2000	Floor, medium duty, C.I., deep flange, 7" top					125
	2040	2" and 3" pipe size	1.333	Ea.	34	134	
	2080	For galvanized body, add			13.60	23	
	2120	For polished bronze top, add		↓	22	37.50	
	2500	Heavy duty, cleanout & trap w/bucket, C.I., 15" top					
	2540	2", 3", and 4" pipe size	2.667	Ea.	805	1,525	
	2560	For galvanized body, add			240	410	
	2580	For polished bronze top, add		↓	280	480	
	3860	Roof, flat metal deck, C.I. body, 10" aluminum dome					
	3890	3" pipe size	1.143	Ea.	120	270	
	3900	4" pipe size	1.231	"	135	300	
	3980	Precast plank deck, C.I. body, aluminum dome					
	4100	10" top, 3" pipe size	1.231	Ea.	66	183	
	4120	13" top, 4" pipe size	1.333		85	221	
	4160	16" top, 6" pipe size	2		120	320	
	4220	For galvanized body, add		↓	46	78.50	
	4620	Main, all aluminum, 12" low profile dome					
	4640	2", 3" and 4" pipe size	1.143	Ea.	175	365	
	9200	Minimum labor/equipment charge	4.267	Job		243	
141	0010	**FAUCETS/FITTINGS**					141
	0020						
	0150	Bath, faucets, diverter spout combination, sweat	1	Ea.	58	163	
	0200	For integral stops, IPS unions, add			30	51	
	0500	Drain, central lift, 1-1/2" IPS male	.400		33	82	
	0600	Trip lever, 1-1/2" IPS male	.400	↓	34	83.50	
	0810	Bidet					
	0812	Fitting, over the rim, swivel spray/pop-up drain	1	Ea.	112	255	
	0840	Flush valves, with vacuum breaker					
	0850	Water closet					
	0860	Exposed, rear spud	1	Ea.	86	210	
	0870	Top spud	1		79	199	
	0880	Concealed, rear spud	1		114	258	
	0890	Top spud	1		114	258	
	0900	Wall hung	1	↓	100	235	
	0910						
	0920	Urinal					
	0930	Exposed, stall	1	Ea.	83	205	
	0940	Wall, (washout)	1		83	205	
	0950	Pedestal, top spud	1		80	200	
	0960	Concealed, stall	1		91	219	
	0970	Wall (washout)	1		85	208	
	1000	Kitchen sink faucets, top mount, cast spout	.800		47	130	
	1100	For spray, add	.333		10	38.50	
	2000	Laundry faucets, shelf type, IPS or copper unions	.667		47	122	
	2100	Lavatory faucet, centerset, without drain	.800		30	102	
	2200	For pop-up drain, add	.500		12	52	
	2800	Self-closing, center set	.800		85	195	
	3000	Service sink faucet, cast spout, pail hook, hose end	.571		59	137	
	4000	Shower by-pass valve with union	.444		42	100	
	4200	Shower thermostatic mixing valve, concealed	1		186	380	
	4300	For inlet strainer, check, and stops, add			56	95.50	
	5000	Sillcock, compact, brass, IPS or copper to hose	.333	↓	4.50	29	
	9000	Minimum labor/equipment charge	2	Job		127	

151 | Pipe and Fittings

151 100 | Miscellaneous Fittings

		WORK-HOURS	UNIT	BARE COSTS MAT.	TOTAL INCL. O&P	
401	0010 **PIPE, COPPER** Solder joints					401
	0020 Type K tubing, couplings & clevis hangers 10' O.C.					
	1100 1/4" diameter	.095	L.F.	1.15	8	
	1200 1" diameter	.121		2.95	12.70	
	1260 2" diameter	.200	▼	6.90	24.50	
	2000 Type L tubing, couplings & hangers 10' O.C.					
	2100 1/4" diameter	.091	L.F.	.90	7.30	
	2120 3/8" diameter	.095		1.13	8	
	2140 1/2" diameter	.099		1.31	8.50	
	2160 5/8" diameter	.101		1.35	8.70	
	2180 3/4" diameter	.105		1.90	9.95	
	2200 1" diameter	.118		2.55	11.80	
	2220 1-1/4" diameter	.138		3.36	14.50	
	2240 1-1/2" diameter	.154		4.16	16.85	
	2260 2" diameter	.190		6.15	22.50	
	2280 2-1/2" diameter	.258		8.60	29.50	
	2300 3" diameter	.286		11.55	36	
	2320 3-1/2" diameter	.372		15.35	47.50	
	2340 4" diameter	.410		19.25	56	
	2360 5" diameter	.471		44	102	
	2380 6" diameter	.600	▼	55	129	
	3000 Type M tubing, couplings & hangers 10' O.C.					
	3140 1/2" diameter	.095	L.F.	1.06	7.85	
	3180 3/4" diameter	.103		1.45	8.95	
	3200 1" diameter	.114	▼	1.94	10.60	
	4000 Type DWV tubing, couplings & hangers 10' O.C.					
	4100 1-1/4" diameter	.133	L.F.	2.62	12.95	
	4120 1-1/2" diameter	.148		3.18	14.80	
	4140 2" diameter	.182		4.16	18.65	
	4160 3" diameter	.276		7	27.50	
	4180 4" diameter	.400		12.45	43.50	
	4200 5" diameter	.444		30	77	
	4220 6" diameter	.571	▼	42.50	106	
	9000 Minimum labor/equipment charge	2	Job		127	
430	0010 **PIPE, COPPER, FITTINGS,** Wrought unless otherwise noted					430
	0040 Solder joints, copper x copper					
	0070 90° Elbow, 1/4"	.364	Ea.	.83	24.50	
	0100 1/2"	.400		.29	26	
	0120 3/4"	.421		.63	27.50	
	0130 1"	.500		1.52	34.50	
	0250 45° Elbow, 1/4"	.364		1.64	26	
	0270 3/8"	.364		1.33	25.50	
	0280 1/2"	.400		.51	26	
	0290 5/8"	.421		2.52	31	
	0300 3/4"	.421		.86	28	
	0310 1"	.500		2.21	36	
	0320 1-1/4"	.533		3.03	39.50	
	0330 1-1/2"	.615		3.63	45.50	
	0340 2"	.727		6.05	56.50	
	0350 2-1/2"	1.231		12.90	92	
	0360 3"	1.231		19.20	103	
	0370 3-1/2"	1.600		33	148	
	0380 4"	1.778		40	171	
	0390 5"	2.667		145	400	
	0400 6"	2.667		225	545	
	0450 Tee, 1/4"	.571	▼	1.83	39.50	

151 | Pipe and Fittings

151 100 | Miscellaneous Fittings

			WORK-HOURS	UNIT	BARE COSTS MAT.	TOTAL INCL. O&P	
430	0470	3/8"	.571	Ea.	1.39	38.50	430
	0480	1/2"	.615		.48	40	
	0490	5/8"	.667		3.05	47.50	
	0500	3/4"	.667		1.19	44	
	0510	1"	.800		3.37	56	
	0520	1-1/4"	.889		5.35	65.50	
	0530	1-1/2"	1		7.40	75.50	
	0540	2"	1.143		11.55	92	
	0550	2-1/2"	2		23	154	
	0560	3"	2.286		36	191	
	0570	3-1/2"	2.667		88	305	
	0580	4"	3.200		77	315	
	0590	5"	4		230	625	
	0600	6"	4		299	745	
	0650	Coupling, 1/4"	.333		.21	21.50	
	0670	3/8"	.333		.27	21.50	
	0680	1/2"	.364		.21	23.50	
	0690	5/8"	.381		.62	25	
	0700	3/4"	.381		.42	25	
	0710	1"	.444		.86	30	
	0720	1-1/4"	.471		1.50	32	
	0730	1-1/2"	.533		1.97	37	
	0740	2"	.615		3.27	44.50	
	0750	2-1/2"	1.067		7	73	
	0760	3"	1.231		10.70	89	
	0770	3-1/2"	2		18.30	145	
	0780	4"	2.286		21	166	
	0790	5"	2.667		46	231	
	0800	6"	3		72	300	
	2000	DWV, solder joints, copper x copper					
	2030	90° Elbow, 1-1/4"	.615	Ea.	2.68	43.50	
	2050	1-1/2"	.667		3.60	48	
	2070	2"	.800		5.20	59	
	2090	3"	1.600		12.60	113	
	2100	4"	1.778		55	195	
	2250	Tee, Sanitary, 1-1/4"	.889		4.99	64.50	
	2270	1-1/2"	1		6.60	75	
	2290	2"	1.143		7.70	85.50	
	2310	3"	2.286		25	173	
	2330	4"	2.667		60	254	
	2400	Coupling, 1-1/4"	.571		1.17	38.50	
	2420	1-1/2"	.615		1.55	41.50	
	2440	2"	.727		2.14	50	
	2460	3"	1.455		3.60	89.50	
	2480	4"	1.600		9.10	107	
	9000	Minimum labor/equipment charge	2	Job		127	

151 550 | Plastic Pipe

551	0010	**PIPE, PLASTIC**					551
	0020	Fiberglass reinforced, couplings 10' O.C., hangers 3 per 10'					
	0240	2" diameter	.276	L.F.	11.75	36	
	0280	4" diameter	.340		18.30	50	
	0300	6" diameter	.421		27	70	
	1800	PVC, couplings 10' O.C., hangers 3 per 10'					
	1820	Schedule 40					
	1860	1/2" diameter	.148	L.F.	1.37	11.70	

151 | Pipe and Fittings

151 550 | Plastic Pipe

			WORK-HOURS	UNIT	BARE COSTS MAT.	TOTAL INCL. O&P	
551	1870	3/4" diameter	.157	L.F.	1.41	12.35	551
	1880	1" diameter	.174		1.62	13.80	
	1890	1-1/4" diameter	.190		1.73	15.05	
	1900	1-1/2" diameter	.222		1.83	17.20	
	1910	2" diameter	.271		2.24	19.25	
	1920	2-1/2" diameter	.286		2.81	21	
	1930	3" diameter	.302		3.29	23	
	1940	4" diameter	.333		4.29	26.50	
	1950	5" diameter	.372		6.15	31.50	
	1960	6" diameter	.410	↓	7.25	36	
	4100	DWV type, schedule 40, couplings 10' O.C., hangers 3 per 10'					
	4120	ABS					
	4140	1-1/4" diameter	.190	L.F.	1.93	15.35	
	4150	1-1/2" diameter	.222		2.03	17.55	
	4160	2" diameter	.271	↓	2.43	19.55	
	4400	PVC					
	4410	1-1/4" diameter	.190	L.F.	1.76	15.10	
	4460	2" diameter	.271		2.13	19.05	
	4470	3" diameter	.302		3.07	22.50	
	4480	4" diameter	.333		3.93	25.50	
	4490	6" diameter	.410	↓	7.70	36.50	
	5360	CPVC, couplings 10' O.C., hangers 3 per 10'					
	5380	Schedule 40					
	5460	1/2" diameter	.148	L.F.	2.35	13.40	
	5470	3/4" diameter	.157		2.88	14.85	
	5480	1" diameter	.174		3.47	16.90	
	5490	1-1/4" diameter	.190		3.94	18.85	
	5500	1-1/2" diameter	.222		4.25	21.50	
	5510	2" diameter	.271		5.20	24.50	
	5520	2-1/2" diameter	.286		8.05	30.50	
	5530	3" diameter	.302	↓	9.70	34	
	5560						
	9900	Minimum labor/equipment charge	2	Job		127	
558	0010	**PIPE, PLASTIC, FITTINGS**					558
	0500	PVC, high impact/pressure, Schedule 40					
	0530	90° Elbow, 1/2"	.400	Ea.	.28	26	
	0550	3/4"	.421		.30	27	
	0560	1"	.500		.55	33	
	0570	1-1/4"	.533		.94	35.50	
	0580	1-1/2"	.571		1.04	38	
	0590	2"	.696		1.63	43	
	0600	3"	1		4.92	65.50	
	0610	4"	1.231		5.90	80.50	
	0620	6"	2		33.50	171	
	0800	Tee, 1/2"	.615		.40	40	
	0820	3/4"	.667		.45	42.50	
	0830	1"	.727		.75	47.50	
	0840	1-1/4"	.800		1.15	53	
	0850	1-1/2"	.800		1.50	53	
	0860	2"	.941		2.10	57	
	0870	3"	1.600		9	106	
	0880	4"	2		16	141	
	0890	6"	3.200	↓	55	277	
	2700	PVC (white), Schedule 40, socket joints					
	2760	90° Elbow, 1/2"	.364	Ea.	.18	23.50	

151 | Pipe and Fittings

151 550 | Plastic Pipe

		WORK-HOURS	UNIT	BARE COSTS MAT.	TOTAL INCL. O&P	
2810	2"	.571	Ea.	1.10	34.50	
4500	DWV, ABS, non pressure, socket joints					
4540	1/4 Bend, 1-1/4"	.471	Ea.	2.02	33	
4560	1-1/2"	.500	↓	1.23	34.50	
4570	2"	.571	↓	1.76	36	
4800	Tee, sanitary					
4820	1-1/4"	.727	Ea.	2.10	50	
4830	1-1/2"	.800		2.10	54.50	
4840	2"	.941	↓	3.16	58.50	
5000	PVC, Schedule 40, socket joints					
5040	1/4 Bend, 1-1/4" diameter	.471	Ea.	2.50	34.50	
5060	1-1/2"	.500		1.05	33.50	
5070	2"	.571		1.35	35	
5080	3"	.941		4.10	60.50	
5090	4"	1.143		8.65	80	
5100	6"	2		55	208	
5250	Tee, sanitary 1-1/4"	.727		3.15	51.50	
5270	1-1/2"	.800		1.85	53.50	
5280	2"	.941		2.65	58	
5290	3"	1.455		7.05	95.50	
5300	4"	1.778	↓	13.50	125	
5500	CPVC, Schedule 80, threaded joints					
5540	90° Elbow, 1/4"	.400	Ea.	4.43	32.50	
5560	1/2"	.444		3.06	33	
5570	3/4"	.471		4.58	37.50	
5580	1"	.533		6.45	45	
5590	1-1/4"	.571		12.35	57	
5600	1-1/2"	.615		13.30	61.50	
5610	2"	.727		17.85	71.50	
5620	2-1/2"	.889		55.50	145	
5630	3"	1.143		59.50	168	
6000	Coupling, 1/4"	.400		6.70	36.50	
6020	1/2"	.444		5.55	38	
6030	3/4"	.471		8.95	45.50	
6040	1"	.533		10.15	51	
6050	1-1/4"	.571		10.75	55	
6060	1-1/2"	.615		10.65	57	
6070	2"	.727		13.60	64.50	
6080	2-1/2"	.800		24.50	87.50	
6090	3"	.842	↓	28.50	97	
9900	Minimum labor/equipment charge	2	Job		127	

152 | Plumbing Fixtures

152 100 | Fixtures

		WORK-HOURS	UNIT	BARE COSTS MAT.	TOTAL INCL. O&P	
0010	**BATHS**					
0100	Tubs, recessed porcelain enamel on cast iron, with trim					
0180	48" x 42"	4	Ea.	1,000	1,925	
0220	72" x 36"	5.333	↓	1,075	2,150	

152 | Plumbing Fixtures

152 100 | Fixtures

			Work-Hours	Unit	Bare Costs Mat.	Total Incl. O&P	
104	0300	Mat bottom, 4' long	2.909	Ea.	855	1,625	104
	0380	5' long	3.636		325	765	
	0560	Corner 48" x 44"	3.636		1,150	2,200	
	2000	Enameled formed steel, 4'-6" long	2.759		236	560	
	2200	5' long	2.909		219	545	
	4600	Module tub & showerwall surround, molded fiberglass					
	4610	5' long x 34" wide x 76" high	4	Ea.	575	1,200	
	4750	Handicap with 1-1/2" OD grab bar, antiskid bottom					
	4760	60" x 32-3/4" x 72" high	4	Ea.	610	1,275	
	4770	75" x 40" x 76" high with molded seat	4.571	"	690	1,450	
	6000	Whirlpool, bath with vented overflow, molded fiberglass					
	6100	66" x 48" x 24"	16	Ea.	1,950	4,250	
	6400	72" x 36" x 24"	16		1,850	4,100	
	6500	60" x 30" x 21"	16		1,450	3,400	
	6600	72" x 42" x 22"	16		2,325	4,925	
	6700	83" x 65"	53.333		3,550	9,100	
	6710	For color add			10%		
	6711	For designer colors and trim add			25%		
	7000	Redwood hot tub system					
	7050	4' diameter x 4' deep	16	Ea.	955	2,550	
	7150	6' diameter x 4' deep	20		1,525	3,725	
	7200	8' diameter x 4' deep	20		2,100	4,700	
	9000	Minimum labor/equipment charge	5.333	Job		305	
	9600	Rough-in, supply, waste and vent, for all above tubs, add	7.729	Ea.	122	650	
116	0010	**DRINKING FOUNTAIN** For connection to cold water supply					116
	1000	Wall mounted, non-recessed					
	2700	Stainless steel, single bubbler, no back	2	Ea.	645	1,225	
	2740	With back	2		400	810	
	2780	Dual handle & wheelchair projection type	2		360	740	
	2820	Dual level for handicapped type	2.500		720	1,375	
	3980	For rough-in, supply and waste, add	3.620		49.50	315	
	4000	Wall mounted, semi-recessed					
	4200	Poly-marble, single bubbler	2	Ea.	465	920	
	4600	Stainless steel, satin finish, single bubbler	2	"	490	965	
	6000	Wall mounted, fully recessed					
	6400	Poly-marble, single bubbler	2	Ea.	530	1,025	
	6800	Stainless steel, single bubbler	2		415	840	
	7580	For rough-in, supply and waste, add	4.372		49.50	360	
	7590						
	7600	Floor mounted, pedestal type					
	8600	Enameled iron, heavy duty service, 2 bubblers	4	Ea.	785	1,600	
	8880	For freeze-proof valve system, add	4		231	645	
	8900	For rough-in, supply and waste, add	4.372		49.50	360	
	9000	Minimum labor/equipment charge	4	Job		253	
136	0010	**LAVATORIES** With trim, white unless noted otherwise					136
	0020						
	0500	Vanity top, porcelain enamel on cast iron					
	0600	20 x 18"	2.500	Ea.	153	405	
	0640	26" x 18" oval	2.500	"	255	575	
	0860	For color, add			25%		
	1000	Cultured marble, 19" x 17", single bowl	2.500	Ea.	82.50	283	
	1120	25" x 22", single bowl	2.500		104	320	
	1900	Stainless steel, self-rimming, 25" x 22", single bowl, ledge	2.500		139	380	
	1960	17" x 22", single bowl	2.500		134	370	
	2600	Steel, enameled, 20" x 17", single bowl	2.759		81	296	
	2900	Vitreous china, 20" x 16", single bowl	2.963		152	430	

152 | Plumbing Fixtures

152 100 | Fixtures

			WORK-HOURS	UNIT	BARE COSTS MAT.	TOTAL INCL. O&P	
136	2960	20" x 17", single bowl	2.963	Ea.	146	420	136
	3580	Rough-in, supply, waste and vent for all above lavatories	6.957	↓	78	530	
	4000	Wall hung					
	4040	Porcelain enamel on cast iron, 16" x 14", single bowl	2	Ea.	310	640	
	4180	20" x 18", single bowl	2		172	405	
	4580	For color, add			30%		
	6000	Vitreous china, 18" x 15", single bowl with backsplash	2.286		165	415	
	6960	Rough-in, supply, waste and vent for above lavatories	9.639	↓	128	765	
	9000	Minimum labor/equipment charge	2.667	Job		169	
148	0011	**SHOWERS**, Stall, with door and trim					148
	0020						
	1520	32" square	8	Ea.	325	1,025	
	1540	Terrazzo receptor, 32" square	8		690	1,625	
	1580	36" corner angle	8.889		755	1,775	
	3000	Fiberglass, one piece, with 3 walls, 32" x 32" square	6.667		330	945	
	3100	36" x 36" square	6.667		380	1,025	
	4960	Rough-in, supply, waste and vent for above showers	7.805		76.50	580	
	5500	Head, water economizer, 3.0 GPM	.333	↓	48	104	
	9000	Minimum labor/equipment charge	4	Job		227	
168	0010	**URINALS**					168
	0020						
	3000	Wall hung, vitreous china, with hanger & self-closing valve	5.333	Ea.	380	950	
	3300	Rough-in, supply, waste & vent	5.654		84.50	465	
	5000	Stall type, vitreous china, includes valve	6.400		465	1,150	
	6980	Rough-in, supply, waste and vent	8.040	↓	119	655	
	9000	Minimum labor/equipment charge	4	Job		227	
176	0010	**WASH FOUNTAINS** Rigging not included					176
	1900	Group, foot control					
	2000	Precast terrazzo, circular, 36" diam., 5 or 6 persons	8	Ea.	1,450	2,950	
	2100	54" diameter for 8 or 10 persons	9.600		1,825	3,675	
	2400	Semi-circular, 36" diam. for 3 persons	8		1,350	2,750	
	2500	54" diam. for 4 or 5 persons	9.600	↓	1,625	3,350	
	5610	Group, infrared control, barrier free					
	5614	Precast terrazzo					
	5620	Semi-circular 36" diam. for 3 persons	8	Ea.	2,275	4,350	
	5630	46" diam. for 4 persons	8.571		2,475	4,725	
	5640	Circular, 54" diam. for 8 persons, button control	9.600		3,750	6,975	
	5700	Rough-in, supply, waste and vent for above wash fountains	8.791	↓	112	700	
	9000	Minimum labor/equipment charge	8	Job		475	
180	0010	**WATER CLOSETS**					180
	0100						
	0150	Tank type, vitreous china, incl. seat, supply pipe w/stop					
	0200	Wall hung, one piece	3.019	Ea.	700	1,375	
	0400	Two piece, close coupled	3.019		390	835	
	0960	For rough-in, supply, waste, vent and carrier	5.861		161	605	
	1000	Floor mounted, one piece	3.019		505	1,025	
	1100	Two piece, close coupled, water saver	3.019		145	420	
	1960	For color, add			30%		
	1980	For rough-in, supply, waste and vent	5.246	↓	103	475	
	3000	Bowl only, with flush valve, seat					
	3100	Wall hung	2.759	Ea.	315	690	
	3200	For rough-in, supply, waste and vent, single WC	6.250	"	171	650	
	9000	Minimum labor/equipment charge	4	Job		227	

153 | Plumbing Appliances

153 100 | Water Appliances

			WORK-HOURS	UNIT	BARE COSTS MAT.	TOTAL INCL. O&P	
105	0010	**WATER COOLER**					105
	0100	Wall mounted, non-recessed					
	0140	4 GPH	4	Ea.	335	800	
	1000	Dual height, 8.2 GPH	4.211		545	1,175	
	1040	14.3 GPH	4.211		560	1,200	
	3300	Semi-recessed, 8.1 GPH	4	↓	545	1,150	
	4600	Floor mounted, flush-to-wall					
	4640	4 GPH	2.667	Ea.	355	785	
	4980	For stainless steel cabinet, add			75	128	
	5000	Dual height, 8.2 GPH	4	↓	590	1,275	
	9000	Minimum labor/equipment charge	4	Job		253	

160 | Raceways

160 200 | Conduits

			WORK-HOURS	UNIT	BARE COSTS MAT.	TOTAL INCL. O&P	
205	0010	**CONDUIT** To 15' high, includes 2 terminations, 2 elbows and					205
	0020	11 beam clamps per 100 L.F.					
	2500	Steel, intermediate conduit (IMC), 1/2" diameter	.080	L.F.	1.01	6.60	
	2530	3/4" diameter	.089		1.22	7.50	
	2550	1" diameter	.114		1.65	9.75	
	2570	1-1/4" diameter	.123		2.10	11.05	
	2600	1-1/2" diameter	.133		2.65	12.60	
	2630	2" diameter	.160		3.25	15.30	
	2650	2-1/2" diameter	.200		5.60	21.50	
	2670	3" diameter	.267		7.45	29	
	2700	3-1/2" diameter	.296		10.10	35.50	
	2730	4" diameter	.320		12	40	
	5000	Electric metallic tubing (EMT), 1/2" diameter	.047		.38	3.51	
	5020	3/4" diameter	.062		.54	4.65	
	5040	1" diameter	.070		.79	5.55	
	5060	1-1/4" diameter	.080		1.16	6.85	
	5080	1-1/2" diameter	.089		1.39	7.80	
	5100	2" diameter	.100		1.85	9.20	
	5120	2-1/2" diameter	.133		4.10	15.10	
	5140	3" diameter	.160		5.15	18.50	
	5160	3-1/2" diameter	.178		6.95	22.50	
	5180	4" diameter	.200	↓	8.15	26	
	5200	Field bends, 45° to 90°, 1/2" diameter	.090	Ea.		5.45	
	5220	3/4" diameter	.100			6.10	
	5240	1" diameter	.110			6.65	
	5260	1-1/4" diameter	.211			12.80	
	5280	1-1/2" diameter	.222			13.50	
	5300	2" diameter	.308			18.70	
	5320	Offsets, 1/2" diameter	.123			7.45	
	5340	3/4" diameter	.129			7.80	
	5360	1" diameter	.151			9.15	
	5380	1-1/4" diameter	.267			16.20	
	5400	1-1/2" diameter	.286			17.35	
	7600	EMT, "T" fittings with covers, 1/2" diameter, set screw	.500	↓	6.95	42	

160 | Raceways

160 200 | Conduits

			WORK-HOURS	UNIT	BARE COSTS MAT.	TOTAL INCL. O&P	
205	9000	Minimum labor/equipment charge	2	Job		122	205
260	0010	**CUTTING AND DRILLING**					260
	0100	Hole drilling to 10' high, concrete wall					
	0110	8" thick, 1/2" pipe size	.667	Ea.		40.50	
	0120	3/4" pipe size	.667			40.50	
	0130	1" pipe size	.842			51	
	0140	1-1/4" pipe size	.842			51	
	0150	1-1/2" pipe size	.842			51	
	0160	2" pipe size	1.818			110	
	0170	2-1/2" pipe size	1.818			110	
	0180	3" pipe size	1.818			110	
	0190	3-1/2" pipe size	2.424			148	
	0200	4" pipe size	2.424			148	
	0500	12" thick, 1/2" pipe size	.851			52	
	0520	3/4" pipe size	.851			52	
	0540	1" pipe size	1.096			67	
	0560	1-1/4" pipe size	1.096			67	
	0570	1-1/2" pipe size	1.096			67	
	0580	2" pipe size	2.222			135	
	0590	2-1/2" pipe size	2.222			135	
	0600	3" pipe size	2.222			135	
	0610	3-1/2" pipe size	2.857			173	
	0630	4" pipe size	3.200			195	
	0650	16" thick, 1/2" pipe size	1.053			64	
	0670	3/4" pipe size	1.143			69.50	
	0690	1" pipe size	1.333			81	
	0710	1-1/4" pipe size	1.455			88	
	0730	1-1/2" pipe size	1.455			88	
	0750	2" pipe size	2.667			162	
	0770	2-1/2" pipe size	2.963			181	
	0790	3" pipe size	3.200			195	
	0810	3-1/2" pipe size	3.478			212	
	0830	4" pipe size	4			243	
	0850	20" thick, 1/2" pipe size	1.250			76	
	0870	3/4" pipe size	1.333			81	
	0890	1" pipe size	1.600			97.50	
	0910	1-1/4" pipe size	1.667			102	
	0930	1-1/2" pipe size	1.739			106	
	0950	2" pipe size	2.963			181	
	0970	2-1/2" pipe size	3.333			203	
	0990	3" pipe size	3.636			220	
	1010	3-1/2" pipe size	4			243	
	1030	4" pipe size	4.706			286	
	1050	24" thick, 1/2" pipe size	1.455			88	
	1070	3/4" pipe size	1.569			95.50	
	1090	1" pipe size	1.860			113	
	1110	1-1/4" pipe size	2			122	
	1130	1-1/2" pipe size	2			122	
	1150	2" pipe size	3.333			203	
	1170	2-1/2" pipe size	3.636			220	
	1190	3" pipe size	4			243	
	1210	3-1/2" pipe size	4.444			270	
	1230	4" pipe size	5.333			325	
	1500	Brick wall, 8" thick, 1/2" pipe size	.444			27	
	1520	3/4" pipe size	.444	▼		27	

160 | Raceways

160 200 | Conduits

			WORK-HOURS	UNIT	BARE COSTS MAT.	TOTAL INCL. O&P	
260	1540	1" pipe size	.602	Ea.		36.50	260
	1560	1-1/4" pipe size	.602			36.50	
	1580	1-1/2" pipe size	.602			36.50	
	1600	2" pipe size	1.404			85.50	
	1620	2-1/2" pipe size	1.404			85.50	
	1640	3" pipe size	1.404			85.50	
	1660	3-1/2" pipe size	1.818			110	
	1680	4" pipe size	2			122	
	1700	12" thick, 1/2" pipe size	.552			33.50	
	1720	3/4" pipe size	.552			33.50	
	1740	1" pipe size	.727			44	
	1760	1-1/4" pipe size	.727			44	
	1780	1-1/2" pipe size	.727			44	
	1800	2" pipe size	1.600			97.50	
	1820	2-1/2" pipe size	1.600			97.50	
	1840	3" pipe size	1.600			97.50	
	1860	3-1/2" pipe size	2.105			128	
	1880	4" pipe size	2.424			148	
	1900	16" thick, 1/2" pipe size	.650			40	
	1920	3/4" pipe size	.650			40	
	1940	1" pipe size	.860			52.50	
	1960	1-1/4" pipe size	.860			52.50	
	1980	1-1/2" pipe size	.860			52.50	
	2000	2" pipe size	1.818			110	
	2010	2-1/2" pipe size	1.818			110	
	2030	3" pipe size	1.818			110	
	2050	3-1/2" pipe size	2.424			148	
	2070	4" pipe size	2.667			162	
	2090	20" thick, 1/2" pipe size	.748			45.50	
	2110	3/4" pipe size	.748			45.50	
	2130	1" pipe size	1			61	
	2150	1-1/4" pipe size	1			61	
	2170	1-1/2" pipe size	1			61	
	2190	2" pipe size	2			122	
	2210	2-1/2" pipe size	2			122	
	2230	3" pipe size	2			122	
	2250	3-1/2" pipe size	2.667			162	
	2270	4" pipe size	2.963			181	
	2290	24" thick, 1/2" pipe size	.851			52	
	2310	3/4" pipe size	.851			52	
	2330	1" pipe size	1.127			68	
	2350	1-1/4" pipe size	1.127			68	
	2370	1-1/2" pipe size	1.127			68	
	2390	2" pipe size	2.222			135	
	2410	2-1/2" pipe size	2.222			135	
	2430	3" pipe size	2.222			135	
	2450	3-1/2" pipe size	2.857			173	
	2470	4" pipe size	3.200			195	
	2480						
	3000	Knockouts to 8' high, metal boxes & enclosures					
	3020	With hole saw, 1/2" pipe size	.151	Ea.		9.15	
	3040	3/4" pipe size	.170			10.40	
	3050	1" pipe size	.200			12.15	
	3060	1-1/4" pipe size	.222			13.50	
	3070	1-1/2" pipe size	.250			15.20	
	3080	2" pipe size	.296			18	

160 | Raceways

160 200 | Conduits

			WORK-HOURS	UNIT	BARE COSTS MAT.	TOTAL INCL. O&P	
260	3090	2-1/2" pipe size	.400	Ea.		24.50	260
	4010	3" pipe size	.500			30.50	
	4030	3-1/2" pipe size	.615			37.50	
	4050	4" pipe size	.727			44	
	4070	With hand punch set, 1/2" pipe size	.200			12.15	
	4090	3/4" pipe size	.250			15.20	
	4110	1" pipe size	.267			16.20	
	4130	1-1/4" pipe size	.286			17.35	
	4150	1-1/2" pipe size	.308			18.70	
	4170	2" pipe size	.400			24.50	
	4190	2-1/2" pipe size	.471			28.50	
	4200	3" pipe size	.533			32.50	
	4220	3-1/2" pipe size	.667			40.50	
	4240	4" pipe size	.800			48.50	
	4260	With hydraulic punch, 1/2" pipe size	.182			11	
	4280	3/4" pipe size	.211			12.80	
	4300	1" pipe size	.211			12.80	
	4320	1-1/4" pipe size	.211			12.80	
	4340	1-1/2" pipe size	.211			12.80	
	4360	2" pipe size	.250			15.20	
	4380	2-1/2" pipe size	.296			18	
	4400	3" pipe size	.348			21	
	4420	3-1/2" pipe size	.400			24.50	
290	0010	**WIREMOLD RACEWAY**					290
	0100	No. 500	.080	L.F.	.55	5.80	
	0110	No. 700	.080		.63	5.95	
	0400	No. 1500, small pancake	.089		1.03	7.15	
	0600	No. 2000, base & cover	.089		1.05	7.20	
	0800	No. 3000, base & cover	.107		2.10	10.05	
	1000	No. 4000, base & cover	.123		3.65	13.65	
	1200	No. 6000, base & cover	.160		5.65	19.35	
	2400	Fittings, elbows, No. 500	.200	Ea.	.95	13.80	
	2800	Elbow cover, No. 2000	.200		1.90	15.40	
	3000	Switch box, No. 500	.500		8.35	44.50	
	3400	Telephone outlet, No. 1500	.500		6.95	42	
	3600	Junction box, No. 1500	.500		4.85	39	
	3800	Plugmold wired sections, No. 2000					
	4000	1 circuit, 6 outlets, 3 ft. long	1	Ea.	19.10	93.50	
	4100	2 circuits, 8 outlets, 6 ft. long	1.509		26	136	
	4200	Tele-power poles, aluminum, 4 outlets	2.963		125	395	
	9990	Minimum labor/equipment charge	1.600	Job		97.50	

161 | Conductors and Grounding

161 100 | Conductors

			WORK-HOURS	UNIT	BARE COSTS MAT.	TOTAL INCL. O&P	
145	0010	**NON-METALLIC SHEATHED CABLE** 600 volt					145
	0100	Copper with ground wire, (Romex)					
	0152	#14, 2 wire	3.200	C.L.F.	15.50	221	
	0202	3 wire	3.478		29	261	

161 | Conductors and Grounding

161 100	Conductors	WORK-HOURS	UNIT	BARE COSTS MAT.	TOTAL INCL. O&P	
145	0252	#12, 2 wire	3.636	C.L.F.	24	262	145
	0302	3 wire	4		40	310	
	0352	#10, 2 wire	4		41	315	
	0402	3 wire	5.714		61	450	
	0452	#8, 3 wire	6.154		130	595	
	0502	#6, 3 wire	6.667	↓	181	715	
	0550	SE type SER aluminum cable, 3 RHW and					
	0602	1 bare neutral, 3 #8 & 1 #8	5.333	C.L.F.	68	445	
	0652	3 #6 & 1 #6	6.154		75	500	
	0702	3 #4 & 1 #6	7.273		95	605	
	0752	3 #2 & 1 #4	8		130	710	
	0802	3 #1/0 & 1 #2	8.889		190	865	
	0852	3 #2/0 & 1 #1	10		225	1,000	
	0902	3 #4/0 & 1 #2/0	11.429	↓	295	1,200	
	6500	Service entrance cap for copper SEU					
	6700	150 amp	.800	Ea.	11.20	67.50	
	6800	200 amp	1	"	16	88.50	
	9000	Minimum labor/equipment charge	2	Job		122	
165	0010	**WIRE**					165
	0020	600 volt type THW, copper solid, #14	.615	C.L.F.	4.70	45.50	
	0030	#12	.727		6.50	55	
	0040	#10	.800	↓	9.90	65.50	
	0051	Wire, 600 volt, stranded					
	0160	#6	1.231	C.L.F.	25.50	118	
	0180	#4	1.509		39.50	159	
	0200	#3	1.600		48	180	
	0220	#2	1.778		61.50	213	
	0240	#1	2		83	263	
	0260	1/0	2.424		95	310	
	0280	2/0	2.759		115	365	
	0300	3/0	3.200		145	445	
	0350	4/0	3.636		185	540	
	0400	250 MCM(kcmil)	4		220	620	
	0420	300 MCM	4.211		270	715	
	0450	350 MCM	4.444		300	780	
	0480	400 MCM	4.706		360	895	
	0490	500 MCM	5		420	1,025	
	0510	750 MCM	7.273		705	1,625	
	0530	Aluminum, stranded, #8	.889		10.80	72.50	
	0540	#6	1		13.50	84.50	
	0560	#4	1.231		17	103	
	0580	#2	1.509		23	131	
	0600	#1	1.778		34	167	
	0620	1/0	2		39	189	
	0640	2/0	2.222		46	214	
	0680	3/0	2.424		57.50	247	
	0700	4/0	2.581		64	266	
	0720	250 MCM	2.759		77	300	
	0740	300 MCM	2.963		105	360	
	0760	350 MCM	3.200		109	380	
	0780	400 MCM	3.478		125	425	
	0800	500 MCM	4		140	480	
	0850	600 MCM	4.211		178	560	
	0880	700 MCM	4.706	↓	200	625	
	9000	Minimum labor/equipment charge	2	Job		122	

162 | Boxes and Wiring Devices

162 300 | Wiring Devices

			WORK-HOURS	UNIT	BARE COSTS MAT.	TOTAL INCL. O&P	
320	0010	**WIRING DEVICES**					320
	0200	Toggle switch, quiet type, single pole, 15 amp	.200	Ea.	3.65	18.35	
	0500	20 amp	.296		5.45	27.50	
	0550	Rocker, 15 amp	.200		4.10	19.15	
	0560	20 amp	.296		13.25	41	
	0600	3 way, 15 amp	.348		5.65	30.50	
	0850	Rocker, 15 amp	.348		5.75	30.50	
	0860	20 amp	.444		19.20	59.50	
	0900	4 way, 15 amp	.533		17	61.50	
	1030	Rocker, 15 amp	.533		15.80	60	
	1040	20 amp	.727		39.50	112	
	1650	Dimmer switch, 120 volt, incandescent, 600 watt, 1 pole	.500		9.50	47	
	2200	Receptacle, duplex, 120 V grounded, 15 amp	.200		2	15.55	
	2300	20 amp	.296		8.10	31.50	
	2400	Dryer, 30 amp	.533		10.35	50.50	
	2500	Range, 50 amp	.727		12.50	66	
	2600	Wall plates, stainless steel, 1 gang	.100		1.85	9.20	
	2800	2 gang	.151		4.60	17	
	3200	Lampholder, keyless	.308		2.65	23	
	3400	Pullchain with receptacle	.364		7.20	34.50	
	9000	Minimum labor/equipment charge	2	Job		122	

166 | Lighting

166 100 | Lighting

			WORK-HOURS	UNIT	BARE COSTS MAT.	TOTAL INCL. O&P	
110	0010	**EXIT AND EMERGENCY LIGHTING**					110
	0080	Exit light ceiling or wall mount, incandescent, single face	1	Ea.	50	146	
	0100	Double face	1.194	"	56	169	
	0300	Emergency light units, battery operated					
	0350	Twin sealed beam light, 25 watt, 6 volt each					
	0500	Lead battery operated	2	Ea.	250	545	
	0700	Nickel cadmium battery operated	2	"	460	905	
	9000	Minimum labor/equipment charge	2	Job		122	
115	0010	**EXTERIOR FIXTURES** With lamps					115
	0400	Quartz, 500 watt	1.509	Ea.	87	240	
	0800	Wall pack, mercury vapor, 175 watt	2		250	545	
	1000	250 watt	2		270	585	
	1100	Low pressure sodium, 35 watt	2		200	460	
	1150	55 watt	2		270	585	
	1160	High pressure sodium, 70 watt	2		290	615	
	1170	150 watt	2		315	655	
	1180	Metal Halide, 175 watt	2		245	540	
	1190	250 watt	2		330	685	
	1200	Floodlights with ballast and lamp,					
	1400	pole mounted pole not included					
	2250	Low pressure sodium, 55 watt	2.963	Ea.	460	960	
	2270	90 watt	4	"	510	1,100	
	9000	Minimum labor/equipment charge	2.133	Job		129	

168 | Special Systems

168 100 | Special Systems

			WORK-HOURS	UNIT	BARE COSTS MAT.	TOTAL INCL. O&P	
120	0010	**DETECTION SYSTEMS**, not including wires & conduits					120
	0020						
	0100	Burglar alarm, battery operated, mechanical trigger	2	Ea.	227	510	
	0200	Electrical trigger	2		271	585	
	0400	For outside key control, add	1		64	171	
	0600	For remote signaling circuitry, add	1		102	235	
	0800	Card reader, flush type, standard	2.963		765	1,475	
	1000	Multi-code	2.963		985	1,850	
	1200	Door switches, hinge switch	1.509		48	175	
	1400	Magnetic switch	1.509		57	189	
	1600	Exit control locks, horn alarm	2		282	600	
	1800	Flashing light alarm	2		320	665	
	2000	Indicating panels, 1 channel	2.963		299	690	
	2200	10 channel	5		1,025	2,075	
	2400	20 channel	8		2,000	3,900	
	2600	40 channel	14.035		3,650	7,100	
	2800	Ultrasonic motion detector, 12 volt	3.478		188	530	
	3000	Infrared photoelectric detector	3.478		155	475	
	3200	Passive infrared detector	3.478		232	610	
	3400	Glass break alarm switch	1		38.50	127	
	3420	Switchmats, 30" x 5'	1.509		69.50	210	
	3440	30" x 25'	2		166	405	
	3460	Police connect panel	2		200	460	
	3480	Telephone dialer	1.509		315	630	
	3500	Alarm bell	2		63	229	
	3520	Siren	2		120	325	
	3540	Microwave detector, 10' to 200'	4		550	1,175	
	3560	10' to 350'	4		1,575	2,925	
	3600	Fire, sprinkler & standpipe alarm control panel, 4 zone	4		835	1,675	
	3800	8 zone	8		1,150	2,475	
	4000	12 zone	12.121	▼	1,650	3,575	
	4150						
	4200	Battery and rack	2	Ea.	630	1,200	
	4400	Automatic charger	1		400	745	
	4600	Signal bell	1		45	138	
	4800	Trouble buzzer or manual station	1		33	118	
	5000	Detector, rate of rise	1		31	114	
	5100	Fixed temperature	1		25.50	105	
	5200	Smoke detector, ceiling type	1.290		58.50	179	
	5400	Duct type	2.500		232	545	
	5600	Light and horn	1.509		96	257	
	5800	Fire alarm horn	1.194		33	129	
	6000	Door holder, electro-magnetic	2		71	243	
	6200	Combination holder and closer	2.500		395	825	
	6400	Code transmitter	2		630	1,200	
	6600	Drill switch	1		79	196	
	6800	Master box	2.963		1,900	3,425	
	7000	Break glass station	1		45.50	139	
	7800	Remote annunciator, 8 zone lamp	4.444		221	650	
	8000	12 zone lamp	6.154		282	860	
	8200	16 zone lamp	7.273		345	1,025	
	8400	Standpipe or sprinkler alarm, alarm device	1	▼	114	255	
	8600	Actuating device	1	Ea.	266	515	
	9410	Minimum labor/equipment charge	2	Job		122	

Part Five
Location Adjustment Factors

Costs shown in *Means ADA Compliance Pricing Guide* are based on national averages for materials and installation. To adjust these costs to a specific location, simply multiply the total project cost by the factor for that city.

The data is arranged alphabetically by state and postal zip code numbers. For a city not listed, use the factor for a nearby city with similar economic characteristics.

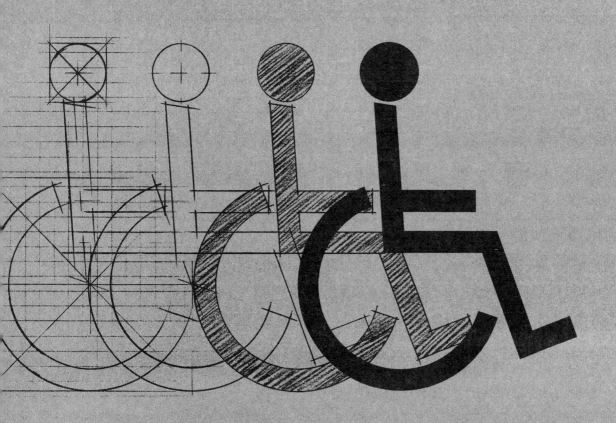

Location Adjustment Factors

STATE/ZIP	CITY	Residential	Commercial
ALABAMA			
350-352	Birmingham	.81	.82
354	Tuscaloosa	.80	.78
355	Jasper	.74	.75
356	Decatur	.82	.83
357-358	Huntsville	.81	.82
359	Gadsden	.79	.80
360-361	Montgomery	.82	.80
362	Anniston	.72	.73
363	Dothan	.81	.79
364	Evergreen	.82	.80
365-366	Mobile	.83	.84
367	Selma	.81	.79
368	Phenix City	.81	.79
369	Butler	.81	.79
ALASKA			
995-996	Anchorage	1.33	1.32
997	Fairbanks	1.30	1.29
998	Juneau	1.32	1.31
999	Ketchikan	1.37	1.36
ARIZONA			
850,853	Phoenix	.93	.90
852	Mesa/Tempe	.89	.86
855	Globe	.94	.91
856-857	Tucson	.93	.90
859	Show Low	.95	.91
860	Flagstaff	.97	.92
863	Prescott	.94	.90
864	Kingman	.94	.90
865	Chambers	.94	.90
ARKANSAS			
716	Pine Bluff	.81	.81
717	Camden	.74	.74
718	Texarkana	.78	.77
719	Hot Springs	.73	.73
720-722	Little Rock	.82	.82
723	West Memphis	.84	.84
724	Jonesboro	.83	.83
725	Batesville	.80	.80
726	Harrison	.81	.81
727	Fayetteville	.72	.69
728	Russellville	.82	.79
729	Fort Smith	.83	.80
749	Poteau	.83	.79
CALIFORNIA			
900-902	Los Angeles	1.15	1.15
903-905	Inglewood	1.12	1.12
906-908	Long Beach	1.13	1.13
910-912	Pasadena	1.11	1.11
913-916	Van Nuys	1.14	1.14
917-918	Alhambra	1.13	1.13
919-921	San Diego	1.15	1.10
922	Palm Springs	1.16	1.12
923-924	San Bernardino	1.14	1.10
925	Riverside	1.18	1.13
926-927	Santa Ana	1.15	1.12
928	Anaheim	1.17	1.15
930	Oxnard	1.19	1.14
931	Santa Barbara	1.16	1.13
932-933	Bakersfield	1.16	1.10
934	San Luis Obispo	1.26	1.13
935	Mojave	1.13	1.09
936-938	Fresno	1.16	1.12
939	Salinas	1.14	1.14
940-941	San Francisco	1.26	1.29
942,956-958	Sacramento	1.13	1.12
943	Palo Alto	1.17	1.20
944	San Mateo	1.18	1.21
945	Vallejo	1.14	1.17
946	Oakland	1.19	1.22
947	Berkeley	1.17	1.20
948	Richmond	1.17	1.20
949	San Rafael	1.28	1.21
950	Santa Cruz	1.18	1.16
951	San Jose	1.25	1.23
952	Stockton	1.17	1.13
953	Modesto	1.16	1.11
954	Santa Rosa	1.19	1.22
955	Eureka	1.14	1.13
959	Marysville	1.14	1.13
960	Redding	1.14	1.13
961	Susanville	1.14	1.13

STATE/ZIP	CITY	Residential	Commercial
COLORADO			
800-802	Denver	.97	.93
803	Boulder	.89	.86
804	Golden	.91	.87
805	Fort Collins	.95	.89
806	Greeley	.91	.85
807	Fort Morgan	.94	.88
808-809	Colorado Springs	.93	.91
810	Pueblo	.94	.92
811	Alamosa	.90	.88
812	Salida	.90	.88
813	Durango	.90	.88
814	Montrose	.88	.86
815	Grand Junction	.92	.87
816	Glenwood Springs	.92	.87
CONNECTICUT			
060	New Britain	1.08	1.09
061	Hartford	1.07	1.08
062	Willimantic	1.09	1.10
063	New London	1.10	1.09
064	Meriden	1.09	1.10
065	New Haven	1.09	1.10
066	Bridgeport	1.06	1.09
067	Waterbury	1.11	1.11
068	Norwalk	1.05	1.10
069	Stamford	1.08	1.12
DISTRICT OF COLUMBIA			
200-205	Washington	.93	.95
DELAWARE			
197	Newark	1.00	1.01
198	Wilmington	1.00	1.01
199	Dover	1.00	1.01
FLORIDA			
320,322	Jacksonville	.86	.85
321	Daytona Beach	.89	.88
323	Tallahassee	.79	.81
324	Panama City	.72	.74
325	Pensacola	.88	.86
326	Gainesville	.86	.83
327-328,347	Orlando	.89	.87
329	Melbourne	.89	.88
330-332,340	Miami	.84	.86
333	Fort Lauderdale	.85	.87
334,349	West Palm Beach	.88	.85
335-336,346	Tampa	.85	.87
337	St. Petersburg	.72	.73
338	Lakeland	.84	.86
339	Fort Myers	.84	.84
342	Sarasota	.83	.85
GEORGIA			
300-303,399	Atlanta	.81	.86
304	Statesboro	.66	.67
305	Gainesville	.61	.65
306	Athens	.71	.76
307	Dalton	.67	.66
308-309	Augusta	.77	.79
310-312	Macon	.80	.80
313-314	Savannah	.80	.81
315	Waycross	.75	.75
316	Valdosta	.76	.76
317	Albany	.77	.79
318-319	Columbus	.77	.77
HAWAII			
967	Hilo	1.25	1.21
968	Honolulu	1.25	1.21
STATES & POSS.			
969	Guam	.88	.85
IDAHO			
832	Pocatello	.92	.91
833	Twin Falls	.83	.82
834	Idaho Falls	.85	.84
835	Lewiston	1.10	1.01
836-837	Boise	.92	.91
838	Coeur d'Alene	.99	.91

STATE/ZIP	CITY	Residential	Commercial
ILLINOIS			
600-603	North Suburban	1.07	1.06
604	Joliet	1.07	1.06
605	South Suburban	1.04	1.03
606	Chicago	1.09	1.08
609	Kankakee	.96	.96
610-611	Rockford	1.00	.99
612	Rock Island	.99	.91
613	La Salle	1.03	.96
614	Galesburg	1.02	.94
615-616	Peoria	1.04	.98
617	Bloomington	.98	.94
618-619	Champaign	1.02	.98
620-622	East St. Louis	.97	.97
623	Quincy	.93	.91
624	Effingham	.95	.92
625	Decatur	.98	.94
626-627	Springfield	.97	.94
628	Centralia	.95	.95
629	Carbondale	.91	.91
INDIANA			
424	Henderson	.91	.89
460	Anderson	.92	.90
461-462	Indianapolis	.95	.93
463-464	Gary	.98	.96
465-466	South Bend	.92	.90
467-468	Fort Wayne	.90	.91
469	Kokomo	.89	.88
470	Lawrenceburg	.89	.86
471	New Albany	.90	.86
472	Columbus	.90	.88
473	Muncie	.91	.90
474	Bloomington	.92	.89
475	Washington	.90	.90
476-477	Evansville	.93	.93
478	Terre Haute	.94	.93
479	Lafayette	.89	.89
IOWA			
500-503,509	Des Moines	.94	.90
504	Mason City	.89	.84
505	Fort Dodge	.84	.79
506-507	Waterloo	.90	.84
508	Creston	.93	.88
510-511	Sioux City	.89	.83
512	Sibley	.83	.81
513	Spencer	.85	.83
514	Carroll	.90	.85
515	Council Bluffs	.94	.88
516	Shenandoah	.78	.73
520	Dubuque	.95	.85
521	Decorah	.94	.84
522-524	Cedar Rapids	.99	.90
525	Ottumwa	.96	.87
526	Burlington	.85	.79
527-528	Davenport	.92	.90
KANSAS			
660-662	Kansas City	.95	.93
664-666	Topeka	.87	.86
667	Fort Scott	.88	.86
668	Emporia	.79	.78
669	Belleville	.92	.86
670-672	Wichita	.88	.85
673	Independence	.83	.80
674	Salina	.87	.83
675	Hutchinson	.81	.77
676	Hays	.90	.86
677	Colby	.90	.86
678	Dodge City	.89	.86
679	Liberal	.82	.79
KENTUCKY			
400-402	Louisville	.92	.89
403-405	Lexington	.89	.86
406	Frankfort	.93	.87
407-409	Corbin	.82	.76
410	Covington	.94	.92
411-412	Ashland	.90	.91
413-414	Campton	.80	.77
415-416	Pikeville	.85	.86
417-418	Hazard	.80	.76
420	Paducah	.93	.88
421-422	Bowling Green	.91	.86
423	Owensboro	.90	.88
425-426	Somerset	.79	.75
427	Elizabethtown	.90	.86

STATE/ZIP	CITY	Residential	Commercial
LOUISIANA			
700-701	New Orleans	.88	.87
703	Thibodaux	.87	.87
704	Hammond	.84	.83
705	Lafayette	.86	.84
706	Lake Charles	.88	.88
707-708	Baton Rouge	.87	.86
710-711	Shreveport	.83	.83
712	Monroe	.81	.81
713-714	Alexandria	.80	.80
MAINE			
039	Kittery	.82	.84
040-041	Portland	.90	.92
042	Lewiston	.91	.92
043	Augusta	.83	.83
044	Bangor	.94	.94
045	Bath	.82	.82
046	Machias	.73	.73
047	Houlton	.84	.84
048	Rockland	.88	.88
049	Waterville	.84	.83
MARYLAND			
206	Waldorf	.88	.88
207-208	College Park	.91	.91
209	Silver Spring	.90	.90
210-212	Baltimore	.92	.92
214	Annapolis	.91	.92
215	Cumberland	.87	.88
216	Easton	.67	.68
217	Hagerstown	.88	.87
218	Salisbury	.79	.80
219	Elkton	.80	.81
MASSACHUSETTS			
010-011	Springfield	1.10	1.08
012	Pittsfield	1.05	1.05
013	Greenfield	1.07	1.04
014	Fitchburg	1.12	1.08
015-016	Worcester	1.13	1.09
017	Framingham	1.10	1.11
018	Lowell	1.13	1.13
019	Lawrence	1.13	1.13
020-022	Boston	1.19	1.20
023-024	Brockton	1.08	1.11
025	Buzzards Bay	1.06	1.08
026	Hyannis	1.08	1.09
027	New Bedford	1.07	1.08
MICHIGAN			
480,483	Royal Oak	1.03	1.02
481	Ann Arbor	1.03	1.02
482	Detroit	1.06	1.05
484-485	Flint	.98	1.00
486	Saginaw	.97	.98
487	Bay City	.95	.96
488-489	Lansing	.97	.94
490	Battle Creek	.98	.92
491	Kalamazoo	.98	.92
492	Jackson	.93	.90
493,495	Grand Rapids	.90	.87
494	Muskegan	.96	.93
496	Traverse City	.93	.90
497	Gaylord	.93	.94
498-499	Iron Mountain	.97	.94
MINNESOTA			
540	New Richmond	.94	.87
550-551	Saint Paul	1.06	1.04
553-554	Minneapolis	1.12	1.09
556-558	Duluth	.97	.99
559	Rochester	1.04	1.01
560	Mankato	.96	.95
561	Windom	.81	.80
562	Willmar	.84	.83
563	St. Cloud	1.05	.97
564	Brainerd	1.02	.95
565	Detroit Lakes	.87	.93
566	Bemidji	.87	.93
567	Thief River Falls	.83	.90
MISSISSIPPI			
386	Clarksdale	.70	.67
387	Greenville	.82	.78
388	Tupelo	.72	.72
389	Greenwood	.72	.69
390-392	Jackson	.83	.79
393	Meridian	.77	.76
394	Laurel	.73	.69
395	Biloxi	.83	.79
396	Mccomb	.69	.67
397	Columbus	.71	.72

STATE/ZIP	CITY	Residential	Commercial
MISSOURI			
630-631	St. Louis	.96	.99
633	Bowling Green	.91	.94
634	Hannibal	1.00	.93
635	Kirksville	.85	.89
636	Flat River	.92	.95
637	Cape Girardeau	.91	.94
638	Sikeston	1.14	.87
639	Poplar Bluff	1.14	.87
640-641	Kansas City	.96	.93
644-645	St. Joseph	.86	.90
646	Chillicothe	.80	.84
647	Harrisonville	.93	.91
648	Joplin	.85	.87
650-651	Jefferson City	.95	.89
652	Columbia	.94	.88
653	Sedalia	.92	.86
654-655	Rolla	.89	.83
656-658	Springfield	.83	.85
MONTANA			
590-591	Billings	.99	.97
592	Wolf Point	.94	.92
593	Miles City	.95	.93
594	Great Falls	.98	.97
595	Havre	.92	.91
596	Helena	.95	.94
597	Butte	.93	.92
598	Missoula	.96	.94
599	Kalispell	.96	.94
NEBRASKA			
680-681	Omaha	.89	.88
683-685	Lincoln	.87	.81
686	Columbus	.76	.75
687	Norfolk	.86	.85
688	Grand Island	.85	.81
689	Hastings	.85	.81
690	Mccook	.76	.72
691	North Platte	.85	.81
692	Valentine	.80	.76
693	Alliance	.77	.73
NEVADA			
889-891	Las Vegas	1.03	1.02
893	Ely	.96	.97
894-895	Reno	.94	.99
897	Carson City	.96	.99
898	Elko	.93	.96
NEW HAMPSHIRE			
030	Nashua	.94	.95
031	Manchester	.94	.95
032-033	Concord	.94	.95
034	Keene	.84	.85
035	Littleton	.86	.87
036	Charleston	.82	.83
037	Claremont	.81	.82
038	Portsmouth	.92	.91
NEW JERSEY			
070-071	Newark	1.09	1.07
072	Elizabeth	1.08	1.06
073	Jersey City	1.09	1.08
074-075	Paterson	1.08	1.08
076	Hackensack	1.07	1.07
077	Long Branch	1.07	1.05
078	Dover	1.08	1.06
079	Summit	1.07	1.05
080,083	Vineland	1.04	1.01
081	Camden	1.05	1.02
082,084	Atlantic city	1.07	1.04
085-086	Trenton	1.07	1.05
087	Point Pleasant	1.07	1.06
088-089	New Brunswick	1.07	1.05
NEW MEXICO			
870-872	Albuquerque	.87	.89
873	Gallup	.89	.91
874	Farmington	.88	.90
875	Santa Fe	.87	.89
877	Las Vegas	.88	.90
878	Socorro	.89	.91
879	Truth/Consequences	.88	.88
880	Las Cruces	.83	.83
881	Clovis	.89	.89
882	Roswell	.90	.90
883	Carrizozo	.91	.91
884	Tucumcari	.90	.90

STATE/ZIP	CITY	Residential	Commercial
NEW YORK			
100-102	New York	1.32	1.32
103	Staten Island	1.27	1.27
104	Bronx	1.26	1.26
105	Mount Vernon	1.25	1.25
106	White Plains	1.21	1.21
107	Yonkers	1.24	1.24
108	New Rochelle	1.25	1.25
109	Suffern	1.11	1.11
110	Queens	1.25	1.25
111	Long Island City	1.26	1.26
112	Brooklyn	1.26	1.26
113	Flushing	1.27	1.27
114	Jamaica	1.25	1.25
115,117,118	Hicksville	1.16	1.16
116	Far Rockaway	1.27	1.27
119	Riverhead	1.16	1.16
120-122	Albany	.99	.99
123	Schenectady	1.00	1.00
124	Kingston	1.12	1.10
125-126	Poughkeepsie	1.13	1.11
127	Monticello	1.11	1.09
128	Glens Falls	.98	.96
130-132	Syracuse	1.01	.99
133-135	Utica	.90	.93
136	Watertown	.93	.96
137-139	Binghamton	.95	.95
140-142	Buffalo	1.05	1.02
143	Niagara Falls	1.04	1.00
144-146	Rochester	1.00	1.01
147	Jamestown	.95	.91
148-149	Elmira	.95	.93
NO. CAROLINA			
270,272-274	Greensboro	.78	.79
271	Winston-Salem	.78	.79
275-276	Raleigh	.78	.78
277	Durham	.78	.79
278	Rocky Mount	.64	.64
279	Elizabeth City	.67	.67
280	Gastonia	.77	.78
281-282	Charlotte	.77	.78
283	Fayetteville	.77	.77
284	Wilmington	.75	.77
285	Kinston	.66	.66
286	Hickory	.64	.65
287-288	Asheville	.75	.77
289	Murphy	.64	.65
NO. DAKOTA			
580-581	Fargo	.80	.85
582	Grand Forks	.80	.85
583	Devils Lake	.81	.86
584	Jamestown	.80	.86
585	Bismarck	.81	.85
586	Dickinson	.82	.86
587	Minot	.81	.85
588	Williston	.81	.85
OHIO			
430-432	Columbus	.95	.93
433	Marion	.90	.91
434-436	Toledo	.98	.97
437-438	Zanesville	.90	.89
439	Steubenville	.94	.94
440	Lorain	1.00	.93
441	Cleveland	1.09	1.02
442-443	Akron	1.01	1.00
444-445	Youngstown	.99	.96
446-447	Canton	.96	.95
448-449	Mansfield	.93	.91
450	Hamilton	.96	.90
451-452	Cincinnati	.98	.92
453-454	Dayton	.91	.90
455	Springfield	.89	.87
456	Chillicothe	.97	.91
457	Athens	.91	.90
458	Lima	.89	.88
OKLAHOMA			
730-731	Oklahoma City	.81	.83
734	Ardmore	.81	.80
735	Lawton	.82	.81
736	Clinton	.79	.81
737	Enid	.81	.80
738	Woodward	.81	.80
739	Guymon	.71	.70
740-741	Tulsa	.87	.84
743	Miami	.83	.81
744	Muskogee	.78	.75
745	Mcalester	.77	.78
746	Ponca City	.80	.79
747	Durant	.78	.80
748	Shawnee	.79	.81

STATE/ZIP	CITY	Residential	Commercial
OREGON			
970-972	Portland	1.10	1.08
973	Salem	1.07	1.06
974	Eugene	1.07	1.06
975	Medford	1.08	1.07
976	Klamath Falls	1.08	1.07
977	Bend	1.07	1.06
978	Pendleton	1.04	1.02
979	Vale	1.00	.98
PENNSYLVANIA			
150-152	Pittsburgh	1.01	.99
153	Washington	.96	.94
154	Uniontown	.95	.93
155	Bedford	.99	.92
156	Greensburg	.96	.94
157	Indiana	.99	.93
158	Dubois	.99	.93
159	Johnstown	1.02	.96
160	Butler	.96	.93
161	New Castle	.99	.96
162	Kittanning	.97	.94
163	Oil City	.88	.93
164-165	Erie	.95	.94
166	Altoona	1.04	.95
167	Bradford	.96	.94
168	State College	.96	.96
169	Wellsboro	.91	.92
170-171	Harrisburg	.97	.96
172	Chambersburg	.94	.93
173-174	York	.95	.93
175-176	Lancaster	.95	.93
177	Williamsport	.92	.91
178	Sunbury	.93	.92
179	Pottsville	.93	.92
180	Lehigh Valley	.99	.98
181	Allentown	1.03	1.02
182	Hazleton	.95	.94
183	Stroudsburg	.99	.98
184-185	Scranton	.96	.99
186-187	Wilkes-Barre	.94	.97
188	Montrose	.92	.95
189	Doylestown	.91	1.03
190-191	Philadelphia	1.11	1.09
193	Westchester	1.02	1.00
194	Norristown	1.05	1.03
195-196	Reading	.97	.98
RHODE ISLAND			
028	Newport	1.05	1.07
029	Providence	1.05	1.07
SOUTH CAROLINA			
290-292	Columbia	.74	.77
293	Spartanburg	.73	.76
294	Charleston	.76	.78
295	Florence	.72	.74
296	Greenville	.73	.76
297	Rock Hill	.65	.68
298	Aiken	.65	.68
299	Beaufort	.69	.71
SOUTH DAKOTA			
570-571	Sioux Falls	.87	.81
572	Watertown	.86	.80
573	Mitchell	.86	.80
574	Aberdeen	.87	.81
575	Pierre	.87	.81
576	Mobridge	.87	.80
577	Rapid City	.87	.81
TENNESSEE			
370-372	Nashville	.82	.82
373-374	Chattanooga	.86	.85
375,380-381	Memphis	.86	.86
376	Johnson City	.81	.80
377-379	Knoxville	.82	.82
382	Mckenzie	.70	.70
383	Jackson	.68	.75
384	Columbia	.76	.76
385	Cookeville	.69	.69

STATE/ZIP	CITY	Residential	Commercial
TEXAS			
750	Mckinney	.93	.86
751	Waxahackie	.83	.83
752-753	Dallas	.91	.87
754	Greenville	.82	.77
755	Texarkana	.91	.80
756	Longview	.87	.76
757	Tyler	.92	.81
758	Palestine	.76	.76
759	Lufkin	.81	.81
760-761	Fort Worth	.86	.85
762	Denton	.90	.81
763	Wichita Falls	.83	.83
764	Eastland	.79	.78
765	Temple	.82	.81
766-767	Waco	.84	.83
768	Brownwood	.77	.76
769	San Angelo	.82	.78
770-772	Houston	.89	.90
773	Huntsville	.77	.77
774	Wharton	.79	.80
775	Galveston	.88	.89
776-777	Beaumont	.87	.89
778	Bryan	.84	.85
779	Victoria	.85	.85
780	Laredo	.81	.82
781-782	San Antonio	.84	.85
783-784	Corpus Christi	.83	.82
785	Mc Allen	.83	.81
786-787	Austin	.81	.84
788	Del Rio	.72	.72
789	Giddings	.78	.77
790-791	Amarillo	.83	.83
792	Childress	.79	.82
793-794	Lubbock	.82	.84
795-796	Abilene	.81	.81
797	Midland	.83	.84
798-799,885	El Paso	.81	.80
UTAH			
840-841	Salt Lake City	.88	.87
842,844	Ogden	.89	.87
843	Logan	.90	.88
845	Price	.83	.82
846-847	Provo	.89	.88
VERMONT			
050	White River Jct.	.77	.76
051	Bellows Falls	.77	.76
052	Bennington	.73	.72
053	Brattleboro	.78	.77
054	Burlington	.86	.87
056	Montpelier	.84	.85
057	Rutland	.88	.87
058	St. Johnsbury	.78	.79
059	Guildhall	.77	.78
129	Plattsburgh	.94	.92
VIRGINIA			
220-221	Fairfax	.88	.89
222	Arlington	.89	.90
223	Alexandria	.90	.91
224-225	Fredericksburg	.83	.83
226	Winchester	.81	.82
227	Culpeper	.79	.80
228	Harrisonburg	.77	.78
229	Charlottesville	.84	.82
230-232	Richmond	.85	.83
233-235	Norfolk	.82	.82
236	Newport News	.83	.82
237	Portsmouth	.80	.80
238	Petersburg	.85	.83
239	Farmville	.76	.74
240-241	Roanoke	.80	.79
242	Bristol	.81	.76
243	Pulaski	.73	.72
244	Staunton	.75	.73
245	Lynchburg	.82	.78
246	Grundy	.72	.72
WASHINGTON			
980-981,987	Seattle	1.01	1.07
982	Everett	.97	1.03
983-984	Tacoma	1.07	1.05
985	Olympia	1.06	1.04
986	Vancouver	1.09	1.03
988	Wenatchee	.99	1.03
989	Yakima	1.04	1.02
990-992	Spokane	1.03	1.02
993	Richland	1.03	1.02
994	Clarkston	1.03	1.02

STATE/ZIP	CITY	Residential	Commercial
WEST VIRGINIA			
247-248	Bluefield	.83	.83
249	Lewisburg	.84	.84
250-253	Charleston	.91	.91
254	Martinsburg	.77	.78
255-257	Huntington	.90	.92
258-259	Beckley	.83	.83
260	Wheeling	.90	.92
261	Parkersburg	.89	.90
262	Buckhannon	.94	.91
263-264	Clarksburg	.94	.91
265	Morgantown	.95	.92
266	Gassaway	.92	.92
267	Romney	.87	.87
268	Petersburg	.93	.90
WISCONSIN			
530,532	Milwaukee	.97	.96
531	Kenosha	.94	.93
534	Racine	.98	.93
535	Beloit	.91	.89
537	Madison	.93	.91
538	Lancaster	.91	.89
539	Portage	.88	.86
541-543	Green Bay	.96	.93
544	Wausau	.90	.87
545	Rhinelander	.95	.91
546	La Crosse	.91	.88
547	Eau Claire	.96	.89
548	Superior	.99	.93
549	Oshkosh	.93	.90
WYOMING			
820	Cheyenne	.87	.82
821	Yellow. Nat'l Park	.81	.78
822	Wheatland	.84	.79
823	Rawlins	.85	.80
824	Worland	.82	.79
825	Riverton	.83	.79
826	Casper	.87	.83
827	Newcastle	.82	.79
828	Sheridan	.86	.83
829-831	Rock Springs	.85	.80
Canadian Factors (reflect Canadian currency)			
ALBERTA	Calgary	1.07	1.04
	Edmonton	1.07	1.03
BRITISH COLUMBIA	Vancouver	1.07	1.08
	Victoria	1.06	1.07
MANITOBA	Winnipeg	1.03	1.02
NEW BRUNSWICK	Moncton	.99	.97
	Saint John	1.02	.99
NEWFOUNDLAND	St. John's	1.00	.99
NOVA SCOTIA	Halifax	.99	.98
ONTARIO	Hamilton	1.18	1.14
	Kitchener	1.08	1.06
	London	1.13	1.11
	Oshawa	1.13	1.11
	Ottawa	1.14	1.12
	St. Catherines	1.09	1.08
	Sudbury	1.09	1.07
	Thunder Bay	1.10	1.08
	Toronto	1.16	1.15
	Windsor	1.10	1.08
PRINCE EDWARD ISLAND	Charlottetown	.97	.95
QUEBEC	Chicoutimi	1.06	1.05
	Montreal	1.13	1.06
	Quebec	1.14	1.06
SASKATCHEWAN	Regina	.95	.95
	Saskatoon	.95	.95

Part Six
Appendix, Glossary and Index

Table of Contents

Resources	331
Cross References: Pricing Guide/ADAAG	337
Cross References: ADAAG/Pricing Guide	339
Abbreviations	341
Glossary	345
Index	347

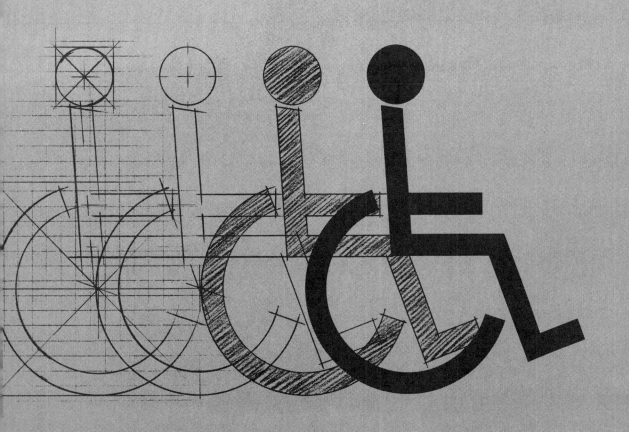

Resources

The following is a brief annotated list of useful sources of information and assistance available to help facility owners, managers, designers, and users understand and respond to the requirements of ADA. Many of the listed organizations and agencies produce or distribute publications relating to the ADA; most of these materials are available in accessible formats upon request.

Note: In addition to reviewing the Americans with Disabilities Act Guidelines, it is also important to contact your state or local access board or building inspector to know what other accessibility legislation might apply to your project.

Accessibility Standards

Accessible and Usable Buildings and Facilities (CABO/ANSI 117.1–1992)

National nonbinding guidelines for accessible design. Incorporated into national model building codes, and adopted by some states as their accessibility standards.
   Council of American Building
      Officials
   5203 Leesburg Pike, #708
   Falls Church, VA 22041
   703-931-4533

Americans with Disabilities Act Accessibility Guidelines (ADAAG)

Minimum technical requirements for the ADA Standards for Accessible Design, included as Appendix A in the ADA Title III Final Rule.
   Department of Justice (DOJ)
   Civil Rights Division, Public Access
      Section
   P.O. Box 66738
   Washington, DC 20035-9998
   202-514-0301 (voice)
   202-514-0383 (TDD)

Uniform Federal Accessibility Standards (UFAS)

Standards for accessible design that currently apply to all federal facilities, those built with federal funds and those that receive federal funds.
   U.S. Architectural and
      Transportation Barriers
      Compliance Board (ATBCB)
   1331 F Street, N.W., Suite 1000
   Washington, DC 20004-1111
   1-800-USA-ABLE (voice)

UFAS Retrofit Guide: Accessibility Modifications for Existing Buildings, 1993

An illustrated design guide for removing architectural barriers in retrofit situations. Designed to be used in conjunction with Uniform Federal Accessibility Standards (UFAS).
   Van Nostrand Reinhold
   7625 Empire Drive
   Florence, KY 41042
   1-800-842-3636

Accessibility Guidelines

Accessibility and Historic Preservation Resource Guide

A guide to accessible design in older facilities.
   Historic Windsor, Inc.
   P.O. Box 1777
   Windsor, VT 05089-0021
   802-674-6752 or 1-800-376-6882

Achieving Physical and Communication Accessibility

A practical guide to accommodating people with disabilities in public accommodations and places of employment. Provides factual information on four major types of disabilities—physical, hearing and speech, visual, and cognitive—and suggests cost-effective accessibility modifications.
   Adaptive Environments Center
   374 Congress St., Suite 301
   Boston, MA 02210
   617-695-1225 (V/TDD)
   617-482-8099 (Fax)

Recommendations for Accessibility Standards for Children's Environments

A summary of proposed changes to UFAS for children's dimensions from a study conducted for the U.S. Architectural and Transportation Barriers Compliance Board (ATBCB).
   The Center for Accessible Housing
   North Carolina State University
   219 Oberlin Road
   Raleigh, NC 27695-8613

Workplace Workbook 2.1

An illustrated guide to workplace adaptations for people with various functional limitations. Includes universally designed solutions to meet a range of needs.
   James Mueller
   RESNA
   1101 Connecticut Ave., N.W.,
   Suite 700
   Washington, DC 20036
   202-857-1199 (voice)
   202-857-1140 (TDD)

Access Checklists

ADAAG Checklist

A technical assistance checklist for surveying buildings and facilities for compliance with Titles II and III of ADA.
 U.S. Architectural and Transportation Barriers Compliance Board
 1331 F Street, N.W., Suite 1000
 Washington, DC 20004-1111
 1-800-USA-ABLE (voice)
 202-272-5434 (voice)
 202-272-5449 (TDD)

The Americans with Disabilities Act Facilities Compliance Workbook

An extensive guide and checklist to all aspects of ADA compliance.
 John Wiley & Sons, Inc.
 605 Third Avenue
 New York, NY 10157-0228
 1-800-526-5368

BOMA ADA Compliance Checklist

A detailed building checklist for meeting Title III provisions of ADA.
 Building Owners and Managers Association, Marketing Department
 1201 New York Ave., N.W., Suite 300
 Washington, DC 20005
 202-408-2685

ADA Survey Software (System George™)

Software provides site surveys for ADA compliance. Includes: customized checklist, automatic and preliminary reports, summary and priority forms.
 R. S. Means Company, Inc.
 100 Construction Plaza
 Kingston, MA 02364
 1-800-448-8182

Readily Achievable Checklist

Detailed, easy-to-use survey tool to identify barriers in existing facilities. Each survey question is followed by suggestions for easily-accomplished, "readily achievable" access solutions. Questions are organized into four sections that correspond with the priorities recommended by the Department of Justice, and ADAAG requirements are referenced throughout.
 Adaptive Environments Center
 374 Congress St., Suite 301
 Boston, MA 02210
 617-695-1225 (V/TDD)
 617-482-8099 (Fax)

Title II Action Guide

Guide for compliance with Title II, State and Local Governments, of ADA. Provides clear and detailed steps with worksheets, for conducting self-evaluation of employment, programs, communications, and facility accessibility, and formulating a transition plan for compliance.
 LRP Publications
 747 Dresher Rd., Suite 500
 Horsham, PA 19044
 215-784-0860
 215-784-9639 (Fax)

ADA Regulations and Technical Assistance Manuals

Title I Regulations and Technical Assistance Manual with Supplements

Complete regulations, technical standards, and explanatory supplements on Title I (Employment)
 Equal Employment Opportunities Commission (EEOC)
 ADA Services Office
 1801 L St., N.W.
 Washington, DC 20507
 800-669-3362 (ADA publications)

Title II Regulations and Technical Assistance Manual with Supplements
Title III Regulations (including ADAAG) and Technical Assistance Manual with Supplements

Complete regulations, technical standards, and explanatory supplements on Title II (State and Local Government) and Title III (Public Accommodations and Commercial Facilities). Available from:
 Department of Justice (DOJ)
 Civil Rights Division
 Public Access Section
 P.O. Box 66738
 Washington, DC 20035-9998
 202-514-0301 (voice)
 202-514-0383 (TDD)

ADA Resources

Federal Agencies

The federal agencies listed below are responsible for implementing various facets of the ADA. These resources provide valuable information through their hotlines and printed materials.

Department of Justice (DOJ)
Civil Rights Division
Public Access Section
P.O. Box 66738
Washington, DC 20035-9998
202-514-0301 (voice)
202-514-0383 (TDD)

Responsible for developing and enforcing the ADA state and local government (Title II) and public accommodations (Title III) regulations. They publish the ADA Standards for Technical Design in the Title III Final Rule. Any changes to ADAAG must be approved by DOJ before they are final. DOJ also coordinates technical assistance programs for federal agencies. The telephone numbers given are information lines on the ADA and the regulatory process.

In addition to producing as well as supporting numerous technical assistance materials, the Equal Employment Opportunity Commission and the Department of Justice have jointly produced the *Americans with Disabilities Act* Handbook, which provides background, summary, rule-making history, overview of the regulations, section-by-section analysis of comments and revisions, P.L. 101-336 and annotated regulations of Titles I, II, and III, plus appendices and related federal disability laws. One copy free upon request from EEOC, DOJ, or from Disability and Business Technical Assistance Centers. Multiple copies can be purchased from:
 U.S. Government Printing Office
 Superintendent of Documents
 Mail Stop: SSOP
 Washington, DC 20402-9328
 (202) 783-3238 (voice)
 (202) 512-1426 (TDD)

Equal Employment Opportunity Commission (EEOC)
1801 L Street, N.W.
Washington, DC 20507
1-800-669-EEOC (voice)
1-800-800-3302 (TDD)

Responsible for enforcing ADA employment regulations. Investigates charges of employment discrimination and resolves problems through conciliation or legislation. The 800 number provides referrals to local EEOC offices and information on discrimination laws in English and Spanish. Free publications available.

Department of Transportation
400 Seventh Street, SW, Room 10424
Washington, DC 20590
(202) 366-9305 (voice)
(202) 755-7687 (TDD)

Developed and enforces the regulations to implement the transportation requirements of the ADA. Contact the Department of Transportation for information on the Title II and Title III requirements for public and specified private transportation. Publishes the *Paratransit Handbook*.

U.S. Architectural and Transportation Barriers Compliance Board
1331 F Street, N.W., Suite 1000
Washington, DC 20004-1111
1-800-USA-ABLE (voice)
202-272-5434 (voice)
202-272-5449 (TDD)

An independent federal agency that developed U.F.A.S., the *ADA Accessibility Guidelines* and other architectural accessibility guidelines for the government. Provides technical assistance and information on the architectural requirements of the ADA and other access-related legislation, and architectural, communication and transportation accessibility. List of free publications available.

Disability and Business Technical Assistance Centers (DBTACs) Regional Offices

The National Institute on Disability and Rehabilitation Research (NIDRR) has funded a network of ten regional Disability and Business Technical Assistance Centers (DBTACs). These centers provide information, training, and technical assistance to businesses and agencies covered by the Americans with Disabilities Act and to people with disabilities who have rights under the ADA. You can contact any center by calling the telephone number listed, or call 1-800-949-4ADA (voice/TDD) to be automatically connected to the center in your region.

Employment

Cornell University
School of Industrial and Labor Relations
106 Extension
Ithaca, NY 14853
607-255-7727 (voice/TDD)
Contact: Susanne Bruyere.

International Association of Machinists
Center for Administering, Rehabilitation, and Employment Services (IAMCARES).
Contact: Regional DBTAC.

These and other publications are available through IAMCARES:
- *A Guide to Selected Forms of Accommodations—Rescheduling Work Hours, Restructuring a Job, or Reassigning Employees*
- *A Guide to Selected Forms of Accommodations—Modified and Specialized Equipment*

Following is a list of the ten technical assistance centers. You can contact any center by calling the telephone number listed, or call 1-800-949-4ADA (voice/TDD) to be automatically connected to the center in your region.

New England (DBTAC) (Region I: CT, ME, MA, NH, RI, VT)
University of Southern Maine
Muskie Institute of Public Affairs
145 Newbury Street
Portland, ME 04101
207-874-6535 (voice/TDD)

Northeast (DBTAC) (Region II: NJ, NY, PR)
United Cerebral Palsy Association of New Jersey
354 South Broad Street
Trenton, NJ 08608
609-392-4004 (voice)
609-392-7044 (TDD)

Mid-Atlantic (DBTAC) (Region III: DC, DE, MD, PA, VA, WV)
Independence Center of Northern Virginia
2111 Wilson Boulevard, Suite 400
Arlington, VA 22201
703-525-3268 (voice/TDD)

Southeast (DBTAC) (Region IV: AL, FL, GA, KY, MS, NC, SC, TN)
United Cerebral Palsy Association, Inc.
National Alliance of Business
1776 Peachtree Street, Suite 310N
Atlanta, GA 30309
404-888-0022 (voice)
404-888-9007 (TDD)

Great Lakes (DBTAC) (Region V: IL, IN, MI, MN, OH, WI)
University of Illinois at Chicago
Affiliated Program in Developmental Disabilities
1640 West Roosevelt Road (M/C627)
Chicago, IL 60608
312-413-1407 (voice)
312-413-0453 (TDD)

Southwest (DBTAC) (Region VI: AR, LA, NM, OK, TX)
The Institute for Rehabilitation and Research (TIRR)
2323 South Shepherd Boulevard
Suite 1000
Houston, TX 77019
713-520-0232 (voice)
713-520-5136 (TDD)

Great Plains (DBTAC) (Region VII: IA, KS, NE, MO)
University of Missouri at Columbia
4816 Santana Drive
Columbia, MO 65203
314-882-3600

Rocky Mountain (DBTAC) (Region VIII: CO, MT, ND, SD, UT, WY)
Meeting the Challenge, Inc.
3630 Sinton Road, Suite 103
Colorado Springs, CO 80907-5072
719-444-0252 (voice)
719-444-0268 (TDD)

Pacific (DBTAC) (Region IX: AZ, CA, HI, NV, PB)
Berkeley Planning Associates
440 Grand Avenue, Suite 500
Oakland, CA 94610
510-465-7884 (voice)
510-465-3172 (TDD)

Northwest (DBTAC) (Region X: AK, ID, OR, WA)
Washington State Governor's Committee
P.O. Box 9046
Olympia, WA 98507-9046
206-438-3168 (voice)
206-438-3167 (TDD)

General Access Resources

Some of these agencies offer materials or consulting specifically about the ADA. Others are resources about disability, access, or related issues.

AbleData
National Rehabilitation Information Center
8455 Colesville Road, Suite 935
Silver Spring, MD 20910
(800) 346-2742 (voice/TDD)

Database funded by NIDRR with over 17,000 listings of adaptive equipment for people with all types of disabilities. Assists with finding technological solutions for specific functional limitations. Produces the most detailed listing of ADA resource material produced by public and private resources.

Adaptive Environments Center
374 Congress St., Suite 301
Boston, MA 02210
617-695-1225 (voice/TDD)
617-482-8099 (Fax)

Founded in 1978 as a nonprofit organization. Develops and conducts educational programs, and produces ADA technical assistance materials and publications on accessibility, including award-winning design guidelines. Manages the Universal Design Education Project which awards stipends to design college faculty, and manages the Universal Design Information Network on Internet. Publication list is available.

Barrier Free Environments
P.O. Box 30634
Raleigh, NC 27622
919-782-7823 (voice/TDD)

Provides consulting and design services, produces accessibility guidelines, and presents educational seminars. Produces technical assistance materials on the Fair Housing Amendments Act and the ADA. Publication list is available.

Center for Accessible Housing
School of Design
North Carolina State University
Box 8613
Raleigh, NC 27695-8613
919-737-3082 (voice/TDD)
919-737-3023 (Fax)

Federally supported center geared towards improving the usability, availability and affordability of housing for people with disabilities. Provides information and technical assistance on housing through telephone hotline and wide range of publications. Publication list available.

Disability Rights and Education Defense Fund (DREDF)
2212 Sixth Street
Berkeley, CA 94710
(510) 644-2555 (voice)
(510) 644-2629 (TDD)
1633 Q Street, NW, Suite 220
Washington, DC 20009
(202) 986-0375 (voice/TDD)
Toll-free information line:
   (800) 466-4ADA (voice/TDD)

Legal resource center provides training, technical assistance, and informed analysis of requirements under disability law. Maintains a toll-free information service for ADA Titles II and III, funded by a grant from the U.S. Department of Justice.

Independent Living Centers (ILCs)
Independent Living Centers are a national network of more than two hundred community-based service and advocacy programs run by people with disabilities. ILCs are often a good source of information and assistance on questions and issues related to the ADA. If you are unable to find an Independent Living Center in your phone book, contact your state vocational rehabilitation agency or any of the following for assistance in locating one near you:

- National Council on Independent Living
  2111 Wilson Boulevard, Suite 405
  Arlington, VA 22201
  (703) 525-3406 (voice)
  (703) 525-3407 (TDD)
- Independent Living Research Utilization Center
  2323 South Shepherd, Suite 1000
  Houston, TX 77019
  (713) 520-0232 (voice)
  (713) 520-5136 (TDD)

Job Accommodation Network (JAN)
West Virginia University
809 Allen Hall
Morgantown, WV 26506
1-800-ADA-WORK (voice/TDD)
1-800-526-7234 (voice/TDD)
304-293-7186 (voice/TDD)

A service of the President's Committee on Employment of People with Disabilities, JAN is an international information and consulting resource for employers and job applicants. This group helps solve specific job accommodation problems through their toll-free hotline.

Accessible Products Guides

The Accessibility Book
A Building Code Summary and Products Directory
Julee Quarve-Peterson, Inc.
P.O. Box 28093
Crystal, MN 55428
612-553-1246

Directory of Accessible Building Products
National Association of Home Builders Research Center
400 Prince George's Boulevard
Upper Marlboro, MD 20772-8731
301-249-4000

Enabling Products: A Sourcebook
Institute for Technology Development
Oxford, MS 38655

The Illustrated Directory of Disability Products
3600 W. Timber Ct.
Lawrence, KS 66049-2149

Sweet's Accessible Building Products
Sweet's Group
1221 Avenue of the Americas
New York, NY 10020
212-512-6566

Universal Design Newsletter: Accessibility and the Americans with Disabilities Act
1700 Rockville Pike, Suite 110
Rockville, MD 20852
301-770-7890 (voice/TDD)
301-770-4338 (Fax)

Costing Resources

R. S. Means Company, Inc.
100 Construction Plaza
Box 100
Kingston, MA 02364-0800
617-585-7880

R.S. Means Co. provides annual construction cost data in the form of 23 publications on all construction types and specialties. Means also publishes over 60 reference books including

The New ADA: Compliance and Costs, which overviews the ADA and provides guidance for surveying a facility and planning modifications that will not only meet ADA requirements, but best meet the needs of disabled people.

Cross-Reference: Pricing Guide/ADAAG

Pricing Guide Table of Contents		ADAAG Table of Contents
1. Parking Spaces	4.6	Parking and Passenger Loading Zones
2. Passenger Drop-Off	4.6	Parking and Passenger Loading Zones
3. Pathway, New	4.3	Accessible Route
4. Pathway, Graded	4.3	Accessible Route
5. Pathway, Modify	4.3	Accessible Route
6. Gratings	4.5	Ground and Floor Surfaces
7. Curb Cuts	4.7	Curb Ramps
8. Ramps, New: Straight	4.8	Ramps
9. Ramps, New: Switchback	4.8	Ramps
10. Ramps, New: Dog-Leg	4.8	Ramps
11. Ramps, New: Below Grade	4.8	Ramps
12. Ramps, Modify	4.8	Ramps
13. Ramp Handrails	4.26	Handrails, Grab Bars, and Tub and Shower Seats
14. Vertical Platform Lifts	4.11	Platform Lifts (Wheelchair Lifts)
15. Stairway Inclined Lifts	4.11	Platform Lifts (Wheelchair Lifts)
16. Stairs, New	4.9	Stairs
17. Steps, Modify	4.9	Stairs
18. Stair Handrails	4.26	Handrails, Grab Bars, and Tub and Shower Seats
19. Under-Stair Barriers	4.4	Protruding Objects
20. Door Opening: Drywall	4.13	Doors
21. Door Opening: Masonry	4.13	Doors
22. Door Opening: Storefront	4.13	Doors
23. Install Automatic Door Opener	4.13	Doors
24. Double Leaf Doors	4.13	Doors
25. Door, Modify	4.13	Doors
26. Sliding Doors	4.13	Doors
27. Vestibule, Enlarged	4.14	Entrances
28. Buzzers/Intercoms	4.27	Controls and Operating Mechanisms
29. Partitions, Remove or Modify	4.3	Accessible Route
30. Install Wing Walls	4.3	Accessible Route
31. Flooring Materials	4.5	Ground and Floor Surfaces
32. Elevator, New: Exterior Shaft	4.10	Elevators
33. Elevator, New: Interior Shaft	4.10	Elevators
34. Elevator Cab, Modify	4.10	Elevators
35. Elevator Hall Signals	4.10	Elevators
36. Elevator, Raised and Braille	4.30	Signage
37. Public Telephones	4.31	Telephones

Pricing Guide Table of Contents	ADAAG Table of Contents
38. Public Text Telephones	4.31 Telephones
39. Drinking Fountains	4.15 Drinking Fountains and Water Coolers
40. Controls	4.27 Controls and Operating Mechanisms
41. Outlets	4.27 Controls and Operating Mechanisms
42. Switches	4.27 Controls and Operating Mechanisms
43. Audible/Visual Fire Alarms	4.28 Alarms
44. Signage	4.30 Signage
45. Signage, Electric Display	10.3 Fixed Facilities and Stations
46. Single-User Toilet Rooms	4.22 Toilet Rooms
47. Multi-Stall Toilet Rooms	4.22 Toilet Rooms
48. Accessible Stalls	4.17 Toilet Stalls
49. Toilet, Replace	4.16 Water Closets
	4.18 Urinals
50. Toilet, Modify	4.16 Water Closets
	4.18 Urinals
51. Sink, Replace	4.19 Lavatories and Mirrors
52. Sink, Modify	4.19 Lavatories and Mirrors
53. Grab Bars	4.26 Handrails, Grab Bars, and Tub and Shower Seats
54. Toilet Room Dispensers	4.23 Bathrooms, Bathing Facilities, and Shower Rooms
55. Children's Bathroom Fixtures	No technical standards in ADAAG. Refer to project for recommendations.
56. Roll-In Showers	4.21 Shower Stalls
57. Shower, Modify	4.21 Shower Stalls
58. Tub, Replace	4.21 Shower Stalls
59. Tub, Modify	4.20 Bathtubs
60. Gang Showers	4.21 Shower Stalls
61. Counters	7. Business and Mercantile
62. Aisles	7. Business and Mercantile
63. Dining Area, Modify	4.32 Fixed or Built-In Seating and Tables
	5. Restaurants and Cafeterias
64. Dressing Rooms	4.35 Dressing and Fitting Rooms
65. Theater Seating	4.33 Assembly Areas
66. Assistive Listening Systems	4.33 Assembly Areas
67. Detectable Warnings	4.29 Detectable Warnings
68. Emergency Communications	4.10 Elevators
	4.27 Controls and Operating Mechanisms
69. Kitchens, Modify	9. Accessible Transient Lodging
70. Closets, Modify	4.25 Storage
71. Play Area Pathways	4.3 Accessible Route
	4.5 Ground and Floor Surfaces
72. Beach Access	4.3 Accessible Route
	4.5 Ground and Floor Surfaces
73. Swimming Pool Access	No technical standards in ADAAG. Refer to project for recommendations.
74. Accessible Trails	4.3 Accessible Route
	4.5 Ground and Floor Surfaces
75. Accessible Sleeping Rooms	9. Accessible Transient Lodging

Cross-Reference: ADAAG/Pricing Guide

ADAAG Table of Contents   **Pricing Guide Table of Contents**

4.3   Accessible Route
- 3. Pathway, New
- 4. Pathway, Graded
- 5. Pathway, Modify
- 29. Partitions, Remove or Modify
- 30. Install Wing Walls
- 71. Play Area Pathways
- 72. Beach Access
- 74. Accessible Trails

4.4   Protruding Objects
- 19. Under-Stair Barriers

4.5   Ground and Floor Surfaces
- 6. Gratings
- 31. Flooring Materials
- 71. Play Area Pathways
- 72. Beach Access
- 74. Accessible Trails

4.6   Parking and Passenger Loading Zones
- 1. Parking Spaces
- 2. Passenger Drop-Off

4.7   Curb Ramps
- 7. Curb Cuts

4.8   Ramps
- 8. Ramps, New: Straight
- 9. Ramps, New: Switchback
- 10. Ramps, New: Dog-Leg
- 11. Ramps, New: Below Grade
- 12. Ramps, Modify

4.9   Stairs
- 16. Stairs, New
- 17. Steps, Modify

4.10 Elevators
- 32. Elevator, New: Exterior Shaft
- 33. Elevator, New: Interior Shaft
- 34. Elevator Cab, Modify
- 35. Elevator Hall Signals
- 68. Emergency Communication

4.11 Platform Lifts (Wheelchair Lifts)
- 14. Vertical Platform Lifts
- 15. Stairway Inclined Lifts

4.13 Doors
- 20. Door Opening: Drywall
- 21. Door Opening: Masonry
- 22. Door Opening: Storefront
- 23. Install Automatic Door Opener
- 24. Double Leaf Doors

ADAAG Table of Contents

4.14 Entrances
4.15 Drinking Fountains and Water Coolers
4.16 Water Closets

4.17 Toilet Stalls
4.18 Urinals

4.19 Lavatories and Mirrors

4.20 Bathtubs
4.21 Shower Stalls

4.22 Toilet Rooms

4.23 Bathrooms, Bathing Facilities, & Shower Rooms
4.25 Storage
4.26 Handrails, Grab Bars, and Tub and Shower Seats

4.27 Controls and Operating Mechanisms

4.28 Alarms
4.29 Detectable Warnings
4.30 Signage

4.31 Telephones

4.32 Fixed or Built-in Seating and Tables
4.33 Assembly Areas

4.35 Dressing and Fitting Rooms
5. Restaurants and Cafeterias
7. Business and Mercantile

9. Accessible Transient Lodging

10.3 Fixed Facilities and Stations

Pricing Guide Table of Contents

25. Door, Modify
26. Sliding Doors
27. Vestibule, Enlarged
39. Drinking Fountains
49. Toilet, Replace
50. Toilet, Modify
48. Accessible Stalls
49. Toilet, Replace
50. Toilet, Modify
51. Sink, Replace
52. Sink, Modify
59. Tub, Modify
56. Roll-In Showers
57. Shower, Modify
58. Tub, Replace
60. Gang Showers
46. Single-User Toilet Rooms
47. Multi-Stall Toilet Rooms
54. Toilet Room Dispensers
70. Closets, Modify
13. Ramp Handrails
18. Stair Handrails
53. Grab Bars
28. Buzzers/Intercoms
40. Controls
41. Outlets
42. Switches
43. Audible/Visual Fire Alarms
67. Detectable Warnings
36. Elevator Raised and Braille
44. Signage
37. Public Telephones
38. Public Text Telephones
63. Dining Area, Modify
65. Theater Seating
66. Assistive Listening Systems
64. Dressing Rooms
63. Dining Area, Modify
61. Counters
62. Aisles
69. Kitchens, Modify
75. Accessible Sleeping Rooms
45. Signage, Electric Display
67. Detectable Warnings

Abbreviations

A	Area Square Feet; Ampere	Bk.	Backed	Compr.	Compressor
ABS	Acrylonitrile Butadiene Stryrene; Asbestos Bonded Steel	Bkrs.	Breakers	Conc.	Concrete
		Bldg.	Building	Cont.	Continuous; Continued
A.C.	Alternating Current; Air-Conditioning; Asbestos Cement; Plywood Grade A & C	Blk.	Block	Corr.	Corrugated
		Bm.	Beam	Cos	Cosine
		Boil.	Boilermaker	Cot	Cotangent
		B.P.M.	Blows per Minute	Cov.	Cover
A.C.I.	American Concrete Institute	BR	Bedroom	CPA	Control Point Adjustment
A.C.T.	Acoustical Ceiling Tile	Brg.	Bearing	Cplg.	Coupling
AD	Plywood, Grade A & D	Brhe.	Bricklayer Helper	C.P.M.	Critical Path Method
ADA	Americans with Disabilities Act	Bric.	Bricklayer	CPVC	Chlorinated Polyvinyl Chloride
ADAAG	Americans with Disabilities Act Accessibility Guidelines	Brk.	Brick	C.Pr.	Hundred Pair
		Brng.	Bearing	CRC	Cold Rolled Channel
Addit., Add.	Additional	Brs.	Brass	Creos.	Creosote
Adj.	Adjustable	Brz.	Bronze	Crpt.	Carpet & Linoleum Layer
af	Audio-frequency	Bsn.	Basin	CRT	Cathode-ray Tube
a.f.f.	Above Finished Floor	Btr.	Better	CS	Carbon Steel
A.G.A.	American Gas Association	BTU	British Thermal Unit	Csc	Cosecant
Agg.	Aggregate	BTUH	BTU per Hour	C.S.F.	Hundred Square Feet
A.H.	Ampere Hours	BX	Interlocked Armored Cable	CSI	Construction Specifications Institute
A hr.	Ampere-hour	c	Conductivity	C.T.	Current Transformer
A.H.U.	Air Handling Unit	C	Hundred; Centigrade	CTS	Copper Tube Size
A.I.A.	American Institute of Architects	C/C	Center to Center	Cu	Cubic
AIC	Ampere Interrupting Capacity	Cab.	Cabinet	Cu. Ft.	Cubic Foot
Allow.	Allowance	Cair.	Air Tool Laborer	cw	Continuous Wave
alt.	Altitude	Calc	Calculated	C.W.	Cool White; Cold Water
Alum.	Aluminum	Cap.	Capacity	Cwt.	100 Pounds
a.m.	Ante Meridiem	Carp.	Carpenter	C.W.X.	Cool White Deluxe
Amp.	Ampere, Amplifier	C.B.	Circuit Breaker	C.Y.	Cubic Yard (27 cubic feet)
Anod.	Anodized	C.C.A.	Chromate Copper Arsenate	C.Y./Hr.	Cubic Yard per Hour
Approx.	Approximate	C.C.F.	Hundred Cubic Feet	Cyl.	Cylinder
Apt.	Apartment	cd	Candela	d	Penny (nail size)
Asb.	Asbestos	cd/sf	Candela per Square Foot	D	Deep; Depth; Discharge
A.S.B.C.	American Standard Building Code	CD	Grade of Plywood Face & Back	Dis.;Disch.	Discharge
Asbe.	Asbestos Worker	CDX	Plywood, Grade C & D, exterior glue	Db.	Decibel
A.S.H.R.A.E.	American Society of Heating, Refrig. & AC Engineers	Cefi.	Cement Finisher	Dbl., dbl.	Double
		Cem.	Cement	DC	Direct Current
A.S.M.E.	American Society of Mechanical Engineers	CF	Hundred Feet	Demob.	Demobilization
		C.F.	Cubic Feet	d.f.u.	Drainage Fixture Units
A.S.T.M.	American Society for Testing and Materials	CFM	Cubic Feet per Minute	D.H.	Double Hung
		c.g.	Center of Gravity	DHW	Domestic Hot Water
Attchmt.	Attachment	CHW	Chilled Water; Commercial Hot Water	Diag.	Diagonal
Avg.	Average			Diam., dia.	Diameter
A.W.G.	American Wire Gauge	C.I.	Cast Iron	Distrib.	Distribution
Bbl.	Barrel	C.I.P.	Cast in Place	Dk.	Deck
B. & B.	Grade B and Better; Balled & Burlapped	Circ.	Circuit	D.L.	Dead Load; Diesel
		C.L.	Carload Lot	Do.	Ditto
B. & S.	Bell and Spigot	Clab.	Common Laborer	Dp.	Depth
B. & W.	Black and White	C.L.F.	Hundred Linear Feet	D.P.S.T.	Double Pole, Single Throw
b.c.c.	Body-centered Cubic	CLF	Current Limiting Fuse	Dr.	Driver
B.C.Y.	Bank Cubic Yards	CLP	Cross Linked Polyethylene	Drink.	Drinking
BE	Bevel End	cm	Centimeter	D.S.	Double Strength
B.F.	Board Feet	CMP	Corr. Metal Pipe	D.S.A.	Double Strength A Grade
Bg. cem.	Bag of Cement	C.M.U.	Concrete Masonry Unit	D.S.B.	Double Strength B Grade
BHP	Boiler Horsepower; Brake Horsepower	Col.	Column	Dty.	Duty
		C.O.	Clear Opening	DWV	Drain Waste Vent
B.I.	Black Iron	CO_2	Carbon Dioxide	DX	Deluxe White, Direct Expansion
Bit.; Bitum.	Bituminous	Comb.	Combination	dyn	Dyne

e	Eccentricity	Gal.	Gallon	K.V.A.	Kilovolt Ampere
E	Equipment Only; East	Gal./Min.	Gallon per Minute	K.V.A.R.	Kilovar (Reactance)
Ea.	Each	Galv.	Galvanized	KW	Kilowatt
E.B.	Encased Burial	Gen.	General	KWh	Kilowatt-hour
Econ.	Economy	G.F.I.	Ground Fault Interrupter	L	Labor Only; Length; Long; Medium Wall Copper Tubing
EDP	Electronic Data Processing	Glaz.	Glazier		
EIFS	Exterior Insulation Finish System	GPD	Gallons per Day	Lab.	Labor
E.D.R.	Equiv. Direct Radiation	GPH	Gallons per Hour	lat	Latitude
Eq.	Equation	GPM	Gallons per Minute	Lath.	Lather
Elec.	Electrician; Electrical	GR	Grade	Lav.	Lavatory
Elev.	Elevator; Elevating	Gran.	Granular	lb.; #; lbs.	Pound, pounds
EMT	Electrical Metallic Conduit; Thin Wall Conduit	Grnd.	Ground	L.B.	Load Bearing; L Conduit Body
		Gyp.	Gypsum	L. & E.	Labor & Equipment
Eng.	Engine	H	High; High Strength Bar Joist; Henry	lb./hr.	Pounds per Hour
EPDM	Ethylene Propylene Diene Monomer	H.C.	High Capacity	lb./L.F.	Pounds per Linear Foot
EPS	Expanded Polystyrene	H.D.	Heavy Duty; High Density	lbf/sq.in.	Pound-force per Square Inch
Eqhv.	Equip. Oper., Heavy	H.D.O.	High Density Overlaid	L.C.L.	Less than Carload Lot
Eqlt.	Equip. Oper., Light	Hdr.	Header	Ld.	Load
Eqmd.	Equip. Oper., Medium	Hdwe.	Hardware	LE	Lead Equivalent
Eqmm.	Equip. Oper., Master Mechanic	Help.	Helpers Average	L.F.	Linear Foot
Eqol.	Equip. Oper., Oilers	HEPA	High Efficiency Particulate Air Filter	Lg.	Long; Length; Large
Equip.	Equipment	Hg	Mercury	L & H	Light and Heat
ERW	Electric Resistance Welded	HIC	High Interrupting Capacity	L.H.	Long Span High Strength Bar Joist
Est.	Estimated	H.O.	High Output	L.J.	Long Span Standard Strength Bar Joist
esu	Electrostatic Units	Horiz.	Horizontal	L.L.	Live Load
E.W.	Each Way	HP	Handicapped Person	L.L.D.	Lamp Lumen Depreciation
EWT	Entering Water Temperature	H.P.	Horsepower; High Pressure	lm	Lumen
Excav.	Excavation	H.P.F.	High Power Factor	lm/sf	Lumen per Square Foot
Exp.	Expansion, Exposure	Hr.	Hour	lm/W	Lumen per Watt
Ext.	Exterior	Hrs./Day	Hours per Day	L.O.A.	Length Over All
Extru.	Extrusion	HSC	High Short Circuit	log	Logarithm
f.	Fiber stress	Ht.	Height	L.P.	Liquefied Petroleum; Low Pressure
F	Fahrenheit; Female; Fill	Htg.	Heating	L.P.F.	Low Power Factor
Fab.	Fabricated	Htrs.	Heaters	LR	Long Radius
FBGS	Fiberglass	HVAC	Heating, Ventilation & Air-Conditioning	L.S.	Lump Sum
F.C.	Footcandles	Hvy.	Heavy	Lt.	Light
f.c.c.	Face-centered Cubic	HW	Hot Water	Lt. Ga.	Light Gauge
f'c.	Compressive Stress in Concrete; Extreme Compressive Stress	Hyd.;Hydr.	Hydraulic	L.T.L.	Less than Truckload Lot
		Hz.	Hertz (cycles)	Lt. Wt.	Lightweight
F.E.	Front End	I.	Moment of Inertia	L.V.	Low Voltage
FEP	Fluorinated Ethylene Propylene (Teflon	I.C.	Interrupting Capacity	M	Thousand; Material; Male; Light Wall Copper Tubing
F.F.E.	Furniture, Fixtures, & Equipment	ID	Inside Diameter		
F.G.	Flat Grain	I.D.	Inside Dimension; Identification	m/hr; M.H.	Man-hour
F.H.A.	Federal Housing Administration	I.F.	Inside Frosted	mA	Milliampere
Fig.	Figure	I.M.C.	Intermediate Metal Conduit	Mach.	Machine
Fin.	Finished	In.	Inch	Mag. Str.	Magnetic Starter
Fixt.	Fixture	Incan.	Incandescent	Maint.	Maintenance
Fl. Oz.	Fluid Ounces	Incl.	Included; Including	Marb.	Marble Setter
Flr.	Floor	Int.	Interior	Mat; Mat'l.	Material
F.M.	Frequency Modulation; Factory Mutual	Inst.	Installation	Max.	Maximum
		Insul.	Insulation	MBF	Thousand Board Feet
Fmg.	Framing	I.P.	Iron Pipe	MBH	Thousand BTU's per hr.
Fndtn.	Foundation	I.P.S.	Iron Pipe Size	MC	Metal Clad Cable
Fori.	Foreman, Inside	I.P.T.	Iron Pipe Threaded	M.C.F.	Thousand Cubic Feet
Foro.	Foreman, Outside	I.W.	Indirect Waste	M.C.F.M.	Thousand Cubic Feet per Minute
Fount.	Fountain	J	Joule	M.C.M.	Thousand Circular Mils
FPM, fpm	Feet per Minute	J.I.C.	Joint Industrial Council	M.C.P.	Motor Circuit Protector
FPT	Female Pipe Thread	K	Thousand; Thousand Pounds; Heavy Wall Copper Tubing	MD	Medium Duty
Fr.	Frame			M.D.O.	Medium Density Overlaid
F.R.	Fire Rating	K.A.H.	Thousand Amp. Hours	Med.	Medium
FRK	Foil Reinforced Kraft	KCMIL	Thousand Circular Mils	MF	Thousand Feet
FRP	Fiberglass Reinforced Plastic	KD	Knock Down	M.F.B.M.	Thousand Feet Board Measure
FS	Forged Steel	K.D.A.T.	Kiln Dried After Treatment	Mfg.	Manufacturing
FSC	Cast Body; Cast Switch Box	kg	Kilogram	Mfrs.	Manufacturers
Ft.	Foot; Feet	kG	Kilogauss	mg	Milligram
Ftng.	Fitting	kgf	Kilogram Force	MGD	Million Gallons per Day
Ftg.	Footing	kHz	Kilohertz	MGPH	Thousand Gallons per Hour
Ft. Lb.	Foot Pound	Kip.	1000 Pounds	MH, M.H.	Manhole; Metal Halide; Man-Hour
Furn.	Furniture	KJ	Kiljoule	MHz	Megahertz
FVNR	Full Voltage Non-Reversing	K.L.	Effective Length Factor	Mi.	Mile
FXM	Female by Male	K.L.F.	Kips per Linear Foot	MI	Malleable Iron; Mineral Insulated
Fy.	Minimum Yield Stress of Steel	Km	Kilometer	mm	Millimeter
g	Gram	K.S.F.	Kips per Square Foot	Mill.	Millwright
G	Gauss	K.S.I.	Kips per Square Inch	Min., min.	Minimum, minute
Ga.	Gauge	K.V.	Kilovolt	Misc.	Miscellaneous

Abbr.	Meaning	Abbr.	Meaning	Abbr.	Meaning
ml	Milliliter	Pkg.	Package	S.F.R.	Square Feet of Radiation
M.L.F.	Thousand Linear Feet	Pl.	Plate	S.F. Shlf.	Square Foot of Shelf
Mo.	Month	Plah.	Plasterer Helper	S4S	Surface 4 Sides
Mobil.	Mobilization	Plas.	Plasterer	Shee.	Sheet Metal Worker
Mog.	Mogul Base	Pluh.	Plumbers Helper	Sin.	Sine
MPH	Miles per Hour	Plum.	Plumber	Skwk.	Skilled Worker
MPT	Male Pipe Thread	Ply.	Plywood	SL	Saran Lined
MRT	Mile Round Trip	p.m.	Post Meridiem	S.L.	Slimline
ms	Millisecond	Pord.	Painter, Ordinary	Sldr.	Solder
M.S.F.	Thousand Square Feet	pp	Pages	S.N.	Solid Neutral
Mstz.	Mosaic & Terrazzo Worker	PP; PPL	Polypropylene	S.P.	Static Pressure; Single Pole; Self-Propelled
M.S.Y.	Thousand Square Yards	P.P.M.	Parts per Million		
Mtd.	Mounted	Pr.	Pair	Spri.	Sprinkler Installer
Mthe.	Mosaic & Terrazzo Helper	Prefab.	Prefabricated	Sq.	Square; 100 Square Feet
Mtng.	Mounting	Prefin.	Prefinished	S.P.D.T.	Single Pole, Double Throw
Mult.	Multi; Multiply	Prop.	Propelled	SPF	Spruce Pine Fir
M.V.A.	Million Volt Amperes	PSF; psf	Pounds per Square Foot	S.P.S.T.	Single Pole, Single Throw
M.V.A.R.	Million Volt Amperes Reactance	PSI; psi	Pounds per Square Inch	SPT	Standard Pipe Thread
MV	Megavolt	PSIG	Pounds per Square Inch Gauge	Sq. Hd.	Square Head
MW	Megawatt	PSP	Plastic Sewer Pipe	Sq. In.	Square Inch
MXM	Male by Male	Pspr.	Painter, Spray	S.S.	Single Strength; Stainless Steel
MYD	Thousand Yards	Psst.	Painter, Structural Steel	S.S.B.	Single Strength B Grade
N	Natural; North	P.T.	Potential Transformer	Sswk.	Structural Steel Worker
nA	Nanoampere	P. & T.	Pressure & Temperature	Sswl.	Structural Steel Welder
NA	Not Available; Not Applicable	Ptd.	Painted	St.;Stl.	Steel
N.B.C.	National Building Code	Ptns.	Partitions	S.T.C.	Sound Transmission Coefficient
NC	Normally Closed	Pu	Ultimate Load	Std.	Standard
N.E.M.A.	National Electrical Manufacturers Assoc.	PVC	Polyvinyl Chloride	STP	Standard Temperature & Pressure
NEHB	Bolted Circuit Breaker to 600V.	Pvmt.	Pavement	Stpi.	Steamfitter, Pipefitter
N.L.B.	Non-Load-Bearing	Pwr.	Power	Str.	Strength; Starter; Straight
MN	Non-Metallic Cable	Q	Quantity Heat Flow	Strd.	Stranded
nm	Nanometer	Quan.;Qty.	Quantity	Struct.	Structural
No.	Number	Q.C.	Quick Coupling	Sty.	Story
NO	Normally Open	r	Radius of Gyration	Subj.	Subject
N.O.C.	Not Otherwise Classified	R	Resistance	Subs.	Subcontractors
Nose.	Nosing	R.C.P.	Reinforced Concrete Pipe	Surf.	Surface
N.P.T.	National Pipe Thread	Reconn.	Reconnect	Sw.	Switch
NQOD	Combination Plug-on/Bolt on Circuit Breaker to 240V.	Rect.	Rectangle	Swbd.	Switchboard
		Reg.	Regular	S.Y.	Square Yard
N.R.C.	Noise Reduction Coefficient	Reinf.	Reinforced	Syn.	Synthetic
N.R.S.	Non Rising Stem	Rem.	Remove	S.Y.P.	Southern Yellow Pine
ns	Nanosecond	Repl.	Replace	Sys.	System
nW	Nanowatt	Req'd.	Required	t.	Thickness
OB	Opposing Blade	Res.	Resistant	T	Temperature; Ton
OC; OC	On Center	Resi.	Residential	Tan	Tangent
OD	Outside Diameter	Rgh.	Rough	T.C.	Terra Cotta
O.D.	Outside Dimension	R.H.W.	Rubber, Heat & Water Resistant; Residential Hot Water	T & C	Threaded and Coupled
ODS	Overhead Distribution System			T.D.	Temperature Difference
O & P	Overhead and Profit	rms	Root Mean Square	TDD, (or TTY)	Telecommunications Display Device
Oper.	Operator	Rnd.	Round	T.E.M.	Transmission Electron Microscopy
Opng.	Opening	Rodm.	Rodman	TFE	Tetrafluoroethylene (Teflon)
Orna.	Ornamental	Rofc.	Roofer, Composition	T. & G.	Tongue & Groove; Tar & Gravel
OSB	Oriented Strand Board	Rofp.	Roofer, Precast		
O. S. & Y.	Outside Screw and Yoke	Rohe.	Roofer Helpers (Composition)	Th.; Thk.	Thick
Ovhd.	Overhead	Rots.	Roofer, Tile & Slate	Thn.	Thin
OWG	Oil, Water or Gas	R.O.W.	Right of Way	Thrded	Threaded
Oz.	Ounce	RPM	Revolutions per Minute	Tilf.	Tile Layer, Floor
P.	Pole; Applied Load; Projection	R.R.	Direct Burial Feeder Conduit	Tilh.	Tile Layer, Helper
p.	Page	R.S.	Rapid Start	THW.	Insulated Strand Wire
Pape.	Paperhanger	Rsr	Riser	THWN; THHN	Nylon Jacketed Wire
P.A.P.R.	Powered Air Purifying Respirator	RT	Round Trip	T.L.	Truckload
PAR	Weatherproof Reflector	S.	Suction; Single Entrance; South	Tot.	Total
Part.	Partition	Scaf.	Scaffold	T.S.	Trigger Start
Pc.	Piece	Sch.; Sched.	Schedule	Tr.	Trade
P.C.	Portland Cement; Power Connector	S.C.R.	Modular Brick	Transf.	Transformer
P.C.F.	Pounds per Cubic Foot	S.D.	Sound Deadening	Trhv.	Truck Driver, Heavy
P.C.M.	Phase Contract Microscopy	S.D.R.	Standard Dimension Ratio	Trlr	Trailer
P.E.	Professional Engineer; Porcelain Enamel; Polyethylene; Plain End	S.E.	Surfaced Edge	Trlt.	Truck Driver, Light
		Sel.	Select	TV	Television
		S.E.R., S.E.U.	Service Entrance Cable	T.W.	Thermoplastic Water Resistant Wire
Perf.	Perforated	S.F.	Square Foot	Typ.	Typical
Ph.	Phase	S.F.C.A.	Square Foot Contact Area	UCI	Uniform Construction Index
P.I.	Pressure Injected	S.F.G.	Square Foot of Ground	UF	Underground Feeder
Pile.	Pile Driver	S.F. Hor.	Square Foot Horizontal		

U.H.F.	Ultra High Frequency	Vib.	Vibrating	XFMR	Transformer
U.L.	Underwriters Laboratory	V.L.F.	Vertical Linear Foot	XHD	Extra Heavy Duty
Unfin.	Unfinished	Vol.	Volume	XHHW; XLPE	Cross-Linked Polyethylene Wire Insulation
URD	Underground Residential Distribution	W	Wire; Watt; Wide; West		
V	Volt	w/	With	Y	Wye
V.A.	Volt Amperes	W.C.	Water Column; Water Closet	yd	Yard
V.C.T.	Vinyl Composition Tile	W.F.	Wide Flange	yr	Year
VAV	Variable Air Volume	W.G.	Water Gauge	Δ	Delta
VC	Veneer Core	Wldg.	Welding	%	Percent
Vent.	Ventilation	W. Mile	Wire Mile	~	Approximately
Vert.	Vertical	W.R.	Water Resistant	∅	Phase
V.F.	Vinyl Faced	Wrck.	Wrecker	@	At
V.G.	Vertical Grain	W.S.P.	Water, Steam, Petroleum	#	Pound; Number
V.H.F.	Very High Frequency	WT., Wt.	Weight	<	Less Than
VHO	Very High Output	WWF	Welded Wire Fabric	>	Greater Than

Glossary

Definitions taken directly from the Americans with Disabilities Act Accessibility Guidelines are given in quotes.

Above Finished Floor Refers to the height of an element measured from the top of the existing floor surface.

Access Aisle "An accessible pedestrian space between elements, such as parking spaces, seating, and desks, that provides clearances appropriate for use of the elements."

Accessible As defined by ADA: "Describes a building, facility, or portion thereof that complies with these guidelines (Americans with Disabilities Act Accessibility Guidelines)." In general usage, a facility, building, space, or building element that can be used by people with disabilities.

Accessible Route "A continuous unobstructed path connecting all accessible elements and spaces of a building or facility."

Adaptable A space which can be easily altered to accommodate the needs of individuals with disabilities (such as a sink with cabinets below, where cabinets can be removed to provide knee space).

Addition An enclosed space added to an existing building.

Alteration A physical change to a building or facility that affects the building or facility's usability. (ADA does not include standard maintenance, such as roofing, painting, or wallpapering, as alterations.)

Area of Primary Function See *Primary Function*.

Assistive Listening System A sound transmission system used to increase the audibility of sound for people with hearing impairments through the use of individual receivers or through the T-switch on a hearing aid.

C-Pull A C-shaped door or drawer handle that doesn't require grasping with one's fingers to operate (sometimes called a *D-pull*).

Clear The dimension between two opposite surfaces (such as between two walls or railings).

Clear Opening The dimension between the face of a door open to 90° and the stop on the opposite jamb.

Commercial Facility "A facility whose operations affect commerce and is intended for non-residential use by a private entity." Under Title III, commercial facilities are businesses which are not public accommodations (do not fit into one of the twelve categories of businesses open to the public defined by Department of Justice regulations). Examples are warehouses, factories, and corporate office space.

Cross Slope The slope of a walking surface perpendicular to the direction of travel.

Curb Cut (Curb Ramp) A ramp built up to a curb or cutting into a curb.

Detectable Warning A textured paving material with high visibility that alerts people with visual impairments to the location of a hazardous area.

Disability "... with respect to an individual, a physical or mental impairment that substantially limits one or more of the major life activities of such an individual; a record of such an impairment; or being regarded as having such an impairment."

Equivalent Facilitation Alternative design or technology which provides substantially equivalent or greater access to a facility than required in the technical or scoping sections of the ADA Accessibility Guidelines.

Existing A current condition of a facility or part of a facility.

Facility "... all or any portion of buildings, structures, sites, complexes, equipment, rolling stock or other conveyances, roads, walks, passageways, parking lots, or other real or personal property, including the site where the building, property, structure, or equipment is located."

Level Flat horizontal surface, pitched only for drainage (1:50 maximum).

Maximum Extent Feasible "... applies to the occasional case where the nature of an existing facility makes it virtually impossible to comply fully with applicable accessibility standards through a planned alteration. In these circumstances, the alteration shall provide the maximum physical accessibility feasible. Any altered features of the facility that can be made accessible shall be made accessible. If providing accessibility in conformance with this section to individuals with certain disabilities (e.g., those who use wheelchairs) would not be feasible, the facility shall be made accessible to persons with other types of disabilities (e.g., those who use crutches, those who have impaired vision or hearing, or those who have other impairments."

Modification Physical change to an architectural space or element.

On Center A measurement of the distance between the centers of two repeating members in a structure.

Operable Part A usable part of a public machine (such as the coin slot in a telephone or the push button on a drinking fountain).

Paddle Faucet Faucet handles with extended blades that don't require grasping or twisting of the wrist to operate.

Primary Function "A major activity for which the facility is intended. Areas that contain a primary function include, but are not limited to, the customer services lobby of a bank, the dining area of a cafeteria, the meeting rooms in a conference center, as well as offices and other work areas in which the activities of the public accommodation or other private entity using the facility are carried out. Mechanical rooms, boiler rooms, supply storage rooms, employee lounges or locker rooms, janitorial closets, entrances, corridors, and rest rooms are not areas containing a primary function."

Public Accommodation "A facility operated by a private entity whose operations affect commerce and falls within at least one of the following twelve categories: 1. Places of lodging; 2. Establishments serving food or drink; 3. Places of exhibition or entertainments; 4. Places of public gathering; 5. Sales or rental establishments; 6. Service places; 7. Stations used for public transportation; 8. Places for public display or collection; 9. Places of recreation; 10. Places of education; 11. Social service establishments; 12. Places of exercise and recreation." Essentially, businesses which invite the public in to sell them goods or services.

Public Use "Describes interior or exterior rooms or spaces that are made available to the general public."

Ramp Any walking surface with a slope between 1:20 and 1:12.

Reach Range The widest measure of the reach of an adult seated in a wheelchair (from 9" to 54" above the floor for reaching to the side, and 15" to 48" for reaching in front).

Readily Achievable A barrier removal which is easily accomplished at little expense.

Reasonable Accommodation "Modifications or adjustments to the work environment, or to the manner or circumstances under which the position held or desired is customarily performed, that enable a qualified individual with a disability to perform the essential functions of that position." (This is only part (ii) of the definition; see Section 1630.2 (o) of the Equal Employment Opportunities Commission Title I Regulations for the full definition.)

Renovation See *Alteration*.

Sans Serif A lettering type without serifs (the small lines used to finish off the main stroke of a letter, such as on the top and bottom of this "I"). This line is in sans serif type.

Signage "Displayed verbal, symbolic, tactile, and pictoral information."

Simple Serif A lettering type with small serifs (no firm definition; usually a non-elaborate typeface). This line is in simple serif type. *This line is not.*

Slip-Resistant A surface which provides traction even when wet.

Slope (Running Slope) Angle or pitch of walking surface, usually measured in a ratio of rise to run (or length to height; e.g., 1:12 means that the slope rises one inch vertically for every twelve inches of horizontal length).

Technically Infeasible ". . . an alteration . . . that . . . has little likelihood of being accomplished because existing structural conditions would require removing or altering a load-bearing member which is an essential part of the structural frame; or because other existing physical or site constraints prohibit modification . . ."

Tactile As used in ADAAG, something detectable by touch (such as raised letters).

TDD See *Text Telephone*.

Text Telephone A phone which transmits characters by means of a keyboard, and displays characters typed into it either on a screen or a roll of paper. Commonly referred to as either a TDD (Telecommunications Display Device) or a TTY (Teletypewriter).

TTY See *Text Telephone*.

Typical Refers to identification of a single element that is repeated throughout the facility.

Undue Burden Significant difficulty or expense.

Walk(way) An exterior pedestrian path.

Index

A

Access aisle, 14, 203
Accessible hardware, 76, 85, 88, 99, 102, 106, 224
Accessible route, 11, 16, 20, 31, 34, 58, 59, 60, 76, 85, 88, 90, 102, 106, 114, 123, 127, 138, 160, 162, 164, 184, 186, 190, 192, 196, 198, 201, 203, 206, 208, 209, 221, 226, 238
Acoustical treatment and wood flooring unit costs, 295
ADA Accessibility Guidelines (ADAAG), 3, 5, 6, 7, 11, 21, 31, 35, 43, 47, 58, 59, 63, 66, 70, 76, 77, 85, 88, 116, 136, 154, 159, 169, 181, 214, 218, 227, 235, 238
Additions, 3, 4, 5, 21, 43
Aisle, create accessible, 200-201
    lower wall-mounted information rack, 201
Alterations, 4, 5
Americans with Disabilities Act (ADA), 3-6, 28, 42, 58, 63, 66, 123, 127, 131, 154, 159, 162, 166, 170, 178, 181, 186, 196, 198, 209, 214, 218, 221, 226, 231, 235, 237
Appliances, kitchen, 221, 239
Architectural and Transportation Barriers Compliance Board (ATBCB), 3
Architectural woodwork unit costs, 278
"Area of primary function," 4
Assistive listening systems, install, 210-212
    electrical outlet for, 212
    infrared, 212
    magnetic induction loop, 211
    radio frequency, 212
    signage for, 212
Audible fire alarm, install, 152-153
Audible signals, 131, 134, 238
Audio loop assistive listening system, 211
Auditoriums, 208

B

Balance impairments, 28
Barrier, install under-stair, 74-75
Barrier removal, 3, 144
Beach access, create, 230-232
    concrete pathway, 232
    plastic mats, 231
    wood duckboards, 232
Beach chairs, 231
Boxes and wiring devices unit costs, 318
Braille, 6, 131, 154, 155, 160, 162
    characters, in elevator, 136-137
Budgeting, project, 7-8
Buzzer or intercom, modify, 110-112
    in block wall, 111
    in sheet rock/metal stud wall, 112

C

Call button, 59, 61
Carpet, 119
Case studies, 243-248
Children's accessible bathroom fixtures, 180-182
Children's play areas, 226-227
Closet, modify, 224-225
    install accessible shelf and pole, 225
    install adjustable shelf and pole, 225
    lower coat hooks, 225
Coat hooks, 206
Commercial facilities, 5
Communication
    non-voice, 218, 219
    voice, 218, 219
Concrete, cast-in-place, unit costs, 266
Conductors and grounding unit costs, 316
Control panel, 131
Controls, accessible, 60, 111, 178, 186, 191, 192, 218, 221
    emergency, 131
    install or modify, 146
        lower fire alarm boxes, 147
    volume, 138
Convention centers, 159
Counter, create accessible, 198-199
    install fold-down, 199
    lower section, 199
Counters, kitchen, 221
Curb cuts, 14
    install concrete, 15
    install asphalt, 15
    install or modify, 30-33
        flared sides, asphalt sidewalk, 33
        flared sides, concrete sidewalk, 33
        in asphalt sidewalk, 32
        in brick paver sidewalk on concrete, 32
        in brick paver sidewalk on stone dust, 32
        in concrete sidewalk, 31
        new curb ramp, 33
        patch lip, 33

D

Detectable warning materials, 215
Detectable warnings, install, 214-216
    ceramic tile in existing floor, 216
    ceramic tile in new construction, 216

metal tile on concrete, 215
plastic strip on concrete, 215
rubber strip, 216
Dining area, modify, 202-205
   install ramp, 205
   lower section of bar, 204
   lower tray slide, 203
   replace fixed seating, 204
   widen food service line, 204
Dispensers, toilet room, 170, 172
   install or modify, 178-179
Doors, entrance, 119, 162
Door, install new
   glass storefront, 84-86
      remove glass panels, 86
   in drywall, 76-77
   in masonry, 78-82
      hollow metal in block exterior wall, 82
      hollow metal in brick exterior wall, 81
      hollow metal in masonry veneer wall, 81
      solid core in block exterior wall, 80
      solid core in brick exterior wall, 79
      solid core in masonry veneer exterior wall, 80
Door, modify existing, 98-101
   adjust door-closing speed, 101
   bevel 2 wood thresholds, 101
   install automatic opener, 101
   install push plates, 100
   install push plates and panic bar, 100
   remove threshold, 100
   remove 2 thresholds, 101
   replace hinges with swing-clear hinges, 100
   replace lockset with lever-handled lockset, 100
   reverse swing, wood, 99
Door opener, install new, 88-89
   automatic exterior, 89
   automatic interior, 89
   infrared activated automatic, 89
   power assist interior, 89
Door, sliding, install, 102-104
   hollow core wood pocket in metal stud wall, 103
   hollow core wood pocket in wood stud wall, 104
   solid core wood pocket in metal stud wall, 103
   solid core wood pocket in wood stud wall, 103
Doors, double-leaf, modify existing, 90-97
   install automatic opener, 96
   install magnetic hold-open devices, 97
   remove doors, 97
   widen brick/block opening and replace
      with double hollow metal door, 94
      with double hollow wood door, 93
      with double solid wood door, 93
      with hollow wood door, 92
      with metal door, 92
      with solid wood door, 91
   widen stud wall opening and replace
      with double hollow metal door, 96
      with double hollow wood door, 96
      with double solid wood door, 95
      with hollow wood door, 95
      with metal door, 95
      with solid wood door, 94
Dressing rooms, create accessible, 206-207
Drinking fountains, 116
   install or modify, 144-145
      install high/low, 145
      install low, wall-mounted, 145
      lower existing, 145

E

Earthwork unit costs, 264
Electronic display signage, install, 158-159
Elevator cab, modify existing, 130-133
   add aluminum railing, 133
   add audible signal, 132
   add raised character/Braille signage to panel, 133
   add stainless steel railing, 133
   add visual signal, 132
   add wood railing, 133
   convert freight to passenger, 131
   lower panel, 132
   replace emergency communications system, 133
   replace non-compliant doors, 132
   replace panel, 132
   replace self-leveler, 131
Elevator hall signals, modify existing, 134-135
   add audible signal, 135
   add visual signal, 135
   lower call buttons, 135
   replace call buttons, 135
Elevators, 58, 60
   install new exterior shaft, 122-124
   install new interior shaft, 126-128
   unit costs, 305
Emergency communication device, install, 218-219
Emergency communication system, 131, 136, 219
Employee-only space, 3, 4
Entrance, 106, 154, 200
   public or main, 34, 39, 42, 43, 47
   secondary, 47
Entrances and storefronts unit costs, 286
Excavation, 47
Existing facilities, 4, 5, 6

F

Faucets, 6, 170, 221
Finish carpentry unit costs, 276
Flooring and carpet unit costs, 295
Flooring, slip-resistant, install, 118-120
   absorbent mat, 120
   carpet, 120
   carpet runner with beveled edges, 120
   unglazed quarry tile, 119
FM assistive listening system, 211

G

Gang showers, create accessible, 196-197
Government facilities, 4
Grab bars, 160, 161, 164, 168, 169, 186, 191, 192, 196, 197, 238
   install or modify, 176-177
      in gypsum/metal stud wall, 177
      in gypsum/metal stud wall with ceramic tile, 177
Gratings, 16, 24, 25
   install compliant, 28-29

replace cast iron tree grate, 29
replace catch basin frame and cover, 29
replace trench drain, 29

H

Handrails, 35, 39, 42, 47, 51, 63, 66, 131, 235
  install or modify stair, 70-73
    add extension to freestanding metal, 73
    add extension to freestanding pipe, 72
    add extension to freestanding wood, 73
    add extension to wall-mounted metal, 73
    add extension to wall-mounted pipe, 72
    add extension to wall-mounted wood, 72
    add freestanding metal, 72
    add freestanding pipe, 71
    add wall-mounted metal, 72
    add wall-mounted pipe, 71
    add wall-mounted wood dowel, 71
  metal, 54
  pipe, 54
Hardware unit costs, 286
Hearing impairments, 134, 142, 152, 159, 211, 218, 238
Hinges, swing-clear, 99
Historic buildings, 21, 70, 198
  entry modification, case study, 246-247
Hotels, 159

I

Identifying and pedestrian control devices, 303
In-tub seat, 191, 192, 197
Infrared assistive listening system, 211
International symbol of accessibility, 155, 160, 162, 206

J

Joint sealers unit costs, 281

K

Kitchen, 239

Kitchenette, 239
  modify, 220-222
Knee space, 144, 160, 163, 171, 172, 221

L

Landscaping, 35, 39, 43, 47
Lath, plaster and gypsum board, 291
Lift, platform, 123
Lift, stair, 58
Lighting, 131, 134
  unit costs, 318
Lockers, protective covers and postal specialties unit costs, 303

M

Maneuvering space, 76, 85, 88, 106, 111, 114, 139, 162, 196, 206
  for children, 181
Masonry unit costs, 268
Masonry restoration, cleaning and refractories unit costs, 269
Metal doors and frames unit costs, 282
Metal fabrications, 271
Metal materials, coatings and fastenings unit costs, 269
Mirrors, 160, 161, 162, 170, 171, 172

N

New construction, 3, 4, 5, 21, 43, 150, 218, 238

O

Outlets, 221, 239
  install or modify, 148-149
  raise, 149

P

Painting and wall coverings unit costs, 298
Parking lot, 11
  re-stripe existing, 12
Parking spaces, 154
  install accessible, 10-12
  van-accessible, 11
Partitions, remove or relocate, 114-115
Passenger drop-off, create accessible, 14-15
Pathway

construct graded entrance, 20-23
  asphalt, 22
  asphalt paver, 22
  brick paver over concrete, 23
  brick paver over stone dust, 23
  brick paver over tamper earth, 22
  concrete, 21
construct new, 16-18
  asphalt, 17
  brick paver over concrete base, 18
  brick paver over stone dust base, 18
  brick paver over tamped earth base, 18
  concrete, 17, 26
graded, 35, 43
modify existing, 24-27
  add concrete to, 26
  asphalt, 25
  asphalt block paver, 25
  brick paver, 26
  concrete, 25
  install detectable warning, 27
  patch existing asphalt, 27
  patch existing asphalt paver, 27
  patch existing brick, 27
  patch existing concrete, 26
  relocate objects in path, 27
public, 237
Paving and surfacing unit costs, 265
Pipe and fittings unit costs, 305
Play area pathways, install accessible, 226-229
  asphalt, 228
  rubber mats, interlocking on concrete, 228
  rubber mats, woven on sand, 228
  wood mulch on sand, 228
  woven plastic mats on grass, 229
Play areas, 226-227
Plumbing appliances unit costs, 313
Plumbing fixtures unit costs, 310
Pool transfer device, 235
Public accommodations, 4, 5, 237
Public assembly areas, 208
Public dining areas, 203
Public bathing facilities, 184, 186
Public beaches, 231
Publicly-used space, 3

R

Raceways unit costs, 313

Raised characters. *See* Braille.
Ramps, 21, 60, 123, 235
   construct new below-grade, 46-48
      concrete, 48
   construct new dog-leg, 42-45
      concrete, 45
      painted wood, 44
      pressure-treated wood, 45
   construct new straight, 34-37
      concrete, 36
      painted wood, 36
      pressure-treated wood, 36
   construct new switch-back, 38-41
      concrete, 41
      painted wood, 40
      pressure-treated wood, 41
   covering, 51
   dog-leg, 39
   modify existing, 50-53
      by enlarging pressure-treated wood, 53
      by repaving asphalt, 52
      by repaving concrete, 51
      by resurfacing with sand paint, 52
      by resurfacing with sandpaper strips, 52
      by widening concrete platform, 53
      by widening switch-back concrete, 52
   straight, 34
   straight-run, 35, 39, 42
   switch-back, 39
Ramp railings, install or modify, 54-57
   attach extensions to end posts, 57
   attach pipe to brick wall, 56
   attach wood dowel to brick wall, 56
   install bolt pipe, 55
   mount wood dowel on brass brackets, 55, 56
   set hand-forged wrought iron in concrete, 55
   set pipe in concrete, 55
   weld extensions to existing pipe, 56
"Readily achievable," 4, 6
"Reasonable accommodations," 4
Renovations, 4, 21, 150, 218
Restaurants, 203
   case study, 244-245

Rest rooms, 106, 154, 160, 181, 237
Rough carpentry unit costs, 273

S

Shingles and roofing tiles unit costs, 280
Shower, modify existing, 186-188
   add fold-down seat and grab bars, 188
   remove concrete lip, 187
   replace fixed shower head, 187
   replace floor pan, 187
Shower, roll-in, 176, 190, 238
   install, 184-185
      custom, 185
      prefabricated, 185
Signage, 11, 15, 143, 160, 162, 206, 211
   install, 154-156
      plastic exterior to masonry wall, 155
      metal exterior to masonry wall, 155
      metal interior to gypsum board/stud wall, 156
      metal room to gypsum board/stud wall, 156
Sink, 160, 162, 239
   install new, 170-171
   modify existing, 172-174
      remove apron below counter, 174
      remove base cabinets, 173
      replace knob faucets, 173
Sleeping rooms, create accessible, 238-241
   modify bedroom area of hotel room, 240
   modify bathroom of hotel room, 241
Slip resistant surfaces, 11, 15, 16, 21, 24, 31, 35, 39, 42, 47, 51, 58, 60, 63, 66, 119, 131, 160, 162, 164, 196, 206, 215, 221, 237
Special systems unit costs, 319
Stairs
   dog-leg, 70
   switch-back, 70
Stairs, install new, 62-64
   concrete, 64
   metal pan, 64
   painted wood, 63

   pressure-treated wood, 64
Stairs, modify existing, 66-68
   add non-slip surface to tread, 68
   bevel stair nosing, concrete, 68
   bevel stair nosing, metal pan, 68
   bevel stair nosing, wood, 67
   fill in open stair risers, 67
Stairway inclined lift, install, 60-61
   one turn, 61
   straight run, 61
Stall, create accessible, 164-165
   add new toilet partition, 165
   install grab bar, 165
Storage. *See* Closet, modify.
Subsurface investigation and demolition unit costs, 254
Surveying a facility, 6
Swimming pool access, create, 234-235
Switches, install or modify, 150

T

"Technically infeasible," 4, 5
Telephones, 17, 25, 154, 231, 239
   hearing aid compatibility, 138
   public, install or modify, 138-140
      add counter and outlet for portable TDD, 140
      add signage, 140
      lower, 139
      provide interior text, 139
      provide new accessible, 140
      provide volume control, 139
      remove booth, 139
   public text, install, 142-143
Telephone specialties unit costs, 304
Theater seating, create accessible, 208-209
Tile
   unit costs, 293
   vinyl, 119
Thresholds, 99, 102, 106
Toilet, 160, 163, 164, 176
   install new, 166-167
      lower urinal, 167
      relocate flush valve, 167
      relocate toilet, 167
   modify existing, 168-169
Toilet and bath accessories and scales unit costs, 304
Toilet paper dispenser, 160, 165, 168

Toilet room, 178
　install multiple-stall, 162
　install single-user, 160-161
Trails, create accessible, 236-237
Train platforms, 214
Tub
　modify existing, 192-194
　　add grab bars, 193
　　add seat, 193
　　extend enclosure, 193
　　relocate controls, 194
　　replace fixed shower head, 194
　replace with roll-in shower, 190-191
　　accessible fiberglass shower, 191
　　custom roll-in shower, 191
Tubs and showers, 176, 231, 239

U
Unit costs, 249-319
　how to use, 250-251
Unit pavers, 21, 24
Urinals, 163, 167, 169

V
Vertical platform lift, 60
　install, 58-59
　　exterior enclosed, 59
　　exterior unenclosed, 59
　　interior unenclosed, 59
Vestibule, enlarge, 106-108
　by adding on, 107
　by extending inside building, 108
　by removing doors, 107
Visual alarms, 6, 238, 239
　install audible and, 152-153
Visual display boards, compartments and cubicles unit costs, 301
Visual impairments, 24, 28, 74, 116, 134, 136, 152, 154, 159, 214, 218, 235
Visual paging systems, 159
Visual signals, 134

W
Waterproofing and dampproofing unit costs, 280
Wing walls, install, 116
Wood and plastic doors unit costs, 283